Lecture Notes in Computer Science 644

Edited by G. Goos and J. Hartmanis

Advisory Board: W. Brauer   D. Gries   J. Stoer

A. Apostolico M. Crochemore Z. Galil
U. Manber (Eds.)

Combinatorial
Pattern Matching

Third Annual Symposium
Tucson, Arizona, USA, April 29-May 1, 1992
Proceedings

Springer-Verlag
Berlin Heidelberg New York
London Paris Tokyo
Hong Kong Barcelona
Budapest

A. Apostolico   M. Crochemore   Z. Galil
U. Manber (Eds.)

Combinatorial Pattern Matching

Third Annual Symposium
Tucson, Arizona, USA, April 29-May 1, 1992
Proceedings

Springer-Verlag
Berlin Heidelberg New York
London Paris Tokyo
Hong Kong Barcelona
Budapest

Series Editors

Gerhard Goos
Universität Karlsruhe
Postfach 69 80
Vincenz-Priessnitz-Straße 1
W-7500 Karlsruhe, FRG

Juris Hartmanis
Department of Computer Science
Cornell University
5149 Upson Hall
Ithaca, NY 14853, USA

Volume Editors

Alberto Apostolico
Dept. of Computer Sciences, Purdue University
1398 Comp. Sc. Bldg., West Lafayette, IN 47907-1398, USA

Maxime Crochemore
L. I. T. P., Université Paris VII
F-75251 Paris CX 05, France

Zvi Galil
Columbia University, New York, NY 10027, USA
and
Tel Aviv University
Ramat Aviv, Tel Aviv, Israel

Udi Manber
Computer Science Dept., University of Arizona
Gould-Simpson 721, Tucson, AZ 85721, USA

CR Subject Classification (1991): F.2.2, I.5.4, I.5.0, I.7.3, H.3.3, E.4

ISBN 3-540-56024-6 Springer-Verlag Berlin Heidelberg New York
ISBN 0-387-56024-6 Springer-Verlag New York Berlin Heidelberg

© Springer-Verlag Berlin Heidelberg 1992
Printed in the United States of America

Typesetting: Camera ready by author/editor
45/3140-543210 - Printed on acid-free paper

Foreword

The papers contained in this volume were presented at the third annual symposium on Combinatorial Pattern Matching, held April 29 to May 1, 1992 in Tucson, Arizona. They were selected from 39 abstracts submitted in response to the call for papers.

Combinatorial Pattern Matching addresses issues of searching and matching of strings and more complicated patterns such as trees, regular expressions, extended expressions, etc. The goal is to derive nontrivial combinatorial properties for such structures and then to exploit these properties in order to achieve superior performances for the corresponding computational problems. In recent years, a steady flow of high-quality scientific study of this subject has changed a sparse set of isolated results into a full-fledged area of algorithmics. Still, there is currently no central place for disseminating results in this area. We hope that CPM can grow to serve as the focus point.

This area is expected to grow even further due to the increasing demand for speed and efficiency that comes especially from molecular biology and the Genome project, but also from other diverse areas such as information retrieval (e.g., supporting complicated search queries), pattern recognition (e.g., using strings to represent polygons and string matching to identify them), compilers (e.g., using tree matching), data compression, and program analysis (e.g., program integration efforts). The stated objective of CPM gatherings is to bring together once a year the researchers active in the area for an informal and yet intensive exchange of information about current and future research in the area.

The first two meetings were held at the University of Paris in 1990 and at the University of London in 1991. These two meetings were informal and no proceedings were produced. We hope that these proceedings will contribute to the success and growth of this area.

The conference was supported in part by the National Science Foundation and the University of Arizona.

Program Committee

A. Apostolico
M. Crochemore
Z. Galil
G. Gonnet
D. Gusfield
D. Hirschberg
U. Manber, *chair*
E. W. Myers
F. Tompa
E. Ukkonen

Standing back: (right to left): Gary Benson, Ricardo Baeza-Yates, Andrew Hume, Jim Knight, Christian Burks, Hal Berghel, Andrzej Ehrenfeucht, David Roach, Udi Manber, Mike Waterman, Ramana Idury, Amihood Amir, Gene Lawler, Ethan Port, William Chang, Martin Vingron, Pavel Pevzner, Esko Ukkonen, Frank Olken, Xin Xu, Alberto Apostolico, Thierry Lecroq, and George Havas. **Standing First Row:** Boris Pittel, Mudita Jain, Dominique Revuz, Gene Myers, Alejandro Schaffer, Steve Seiden, Dinesh Mehta, Tariq Choudhary, Tak Cheung Ip, L.J. Cummings, Rob Irving, Sampath Kannan, John Kececioglu, Pekka Kilpelainen, Kaizhong Zhang, Sun Wu, Lucas Hui, Tandy Warnow, and Yuh Dauh Lyuu. **Sitting second row:** Robert Paige, Ladan Rostami, Jong Yong Kim, Martin Farach, Haim Wolfson, Gad Landau, Jeanette Schmid, Glen Herrmannsfeldt, David Sankoff, Ramesh Hariharan, Laura Toniolo, Cari Soderlund, Dan Gusfield, and Wojciech Szpankowski. **Sitting first row:** Maxime Crochemore, Deborah Joseph, Xiaoqui Huang, Mereille Regnier, Dan Hirschberg, Marcella McClure, Gary Lewandowski, Taha Vasi, Brenda Baker, Campbell Fraser, and Philippe Jacquet. **On the floor:** John Oommen, Guy Jacobson, and Kiem Phong Vo.

Table of Contents

Probabilistic Analysis of Generalized Suffix Trees 1
   Wojciech Szpankowski

A Language Approach to String Searching Evaluation 15
   Mireille Régnier

Pattern Matching with Mismatches: A Probabilistic Analysis
and a Randomized Algorithm 27
   Mikhail J. Atallah, Philippe Jacquet and Wojciech Szpankowski

Fast Multiple Keyword Searching 41
   Jang Yang, Xxx and Xxx Shaw Yaunt

Heaviest Increasing/Common Subsequence Problems 52
   Guy Jacobson and Kiem-Phong Vo

Approximate Matching by String Pattern Matching with
.... 66
   James R. Knight and Eugene W. Myers

.... 78
   Daniel ....
   Daniel S. Hirschberg and Lawrence L. Larmore

From Regular .... DFA's Using Compressed ....
   Chia-Hsiang Chang and Robert Paige

....
   Ramana M. Idury and Alejandro A. Schäffer

.... for Genome Comparison Based on 118
   ....

2-D Substructure Matching in Protein 153
   Daniel Fischer, Ruth Nussinov and Haim ....

Fast Serial and Parallel Algorithms for ...
Tree Matching with VLDC's 169
   Xinxiong Zhang, Dennis Shasha and ....

Obstruction of Tree Matching 179
   Frank Hoffmann and Heike Sauer

Table of Contents

Probabilistic Analysis of Generalized Suffix Trees 1
  Wojciech Szpankowski

A Language Approach to String Searching Evaluation 15
  Mireille Régnier

Pattern Matching with Mismatches: A Probabilistic Analysis
and a Randomized Algorithm 27
  Mikhail J. Atallah, Philippe Jacquet and Wojciech Szpankowski

Fast Multiple Keyword Searching 41
  Jong Yong Kim and John Shawe-Taylor

Heaviest Increasing/Common Subsequence Problems 52
  Guy Jacobson and Kiem-Phong Vo

Approximate Regular Expression Pattern Matching with
Concave Gap Penalties 66
  James R. Knight and Eugene W. Myers

Matrix Longest Common Subsequence Problem,
Duality and Hilbert Bases 77
  Pavel A. Pevzner and Michael S. Waterman

From Regular Expressions to DFA's Using Compressed NFA's 88
  Chia-Hsiang Chang and Robert Paige

Identifying Periodic Occurrences of a Template with
Applications to Protein Structure 109
  Vincent A. Fischetti, Gad M. Landau, Jeanette P. Schmidt,
  and Peter H. Sellers

Edit Distance for Genome Comparison Based on Non-Local Operations ...118
  David Sankoff

3-D Substructure Matching in Protein Molecules 133
  Daniel Fischer, Ruth Nussinov and Haim J. Wolfson

Fast Serial and Parallel Algorithms for Approximate
Tree Matching with VLDC's 148
  Kaizhong Zhang, Dennis Shasha and Jason T. L. Wang

Grammatical Tree Matching 159
  Pekka Kilpelainen and Heikki Mannila

Theoretical and Empirical Comparisons of Approximate
String Matching Algorithms 172
   William I. Chang and Jordan Lampe

Fast and Practical Approximate String Matching 182
   Ricardo A. Baeza-Yates and Chris H. Perleberg

DZ: A Text Compression Algorithm for Natural Languages 190
   Dominique Revuz and Marc Zipstein

Multiple Alignment with Guaranteed Error Bounds
and Communication Cost 202
   Pavel A. Pevzner

Two Algorithms for the Longest Common Subsequence of
Three (or More) Strings 211
   Robert W. Irving and Campbell B. Fraser

Color Set Size Problem with Applications to String Matching 227
   Lucas C. K. Hui

Computing Display Conflicts in String and Circular String Visualization ..241
   Dinesh P. Mehta and Sartaj Sahni

Efficient Randomized Dictionary Matching Algorithms 259
   Amihood Amir, Martin Farach and Yossi Matias

Dynamic Dictionary Matching with Failure Functions 273
   Ramana M. Idury and Alejandro A. Schäffer

Probabilistic Analysis of Generalized Suffix Trees
(Extended Abstract)

Wojciech Szpankowski

Department of Computer Science, Purdue University,
W. Lafayette, IN 47907, U.S.A.

Abstract. Suffix trees find several applications in computer science and telecommunications, most notably in algorithms on strings, data compressions and codes. We consider in a probabilistic framework a family of generalized suffix trees – called b-suffix trees – built from the first n suffixes of a random word. In this family of trees, a noncompact suffix trees (i.e., such that every edge is labeled by a single symbol) is represented by $b = 1$, and a compact suffix tree (i.e., without unary nodes) is asymptotically equivalent to $b \to \infty$. Several parameters of b-suffix trees are of interest, namely the typical depth, the depth of insertion, the height, the external path length, and so forth. We establish some results concerning typical, that is, *almost sure* (a.s.), behavior of these parameters. These findings are used to obtain several insights into certain algorithms on words and universal data compression schemes.

1. Introduction

In recent years there has been a resurgence of interest in algorithmic and combinatorial problems on words due to a number of novel applications in computer science, telecommunications, and most notably in molecular biology. In computer science, several algorithms depend on a solution to the following problem: given a word X and a set of $b + 1$ arbitrary suffixes $S_1, \ldots, S_{b+1}$ of X, what is the longest common prefix of these suffixes (cf. [2], [6], [8], [15], [16], [27], [28]). In coding theory (e.g., prefix codes) one asks for the shortest prefix of a suffix S_i which is not a prefix of any other suffixes S_j, $1 \le j \le n$ of a given sequence X. In data compression schemes, the following problem is of prime interest: for a given "data base" subsequence of length n, find the longest prefix of the $n + 1$st suffix S_{n+1} which is not a prefix of any other suffixes S_i $(1 \le i \le n)$ of the underlying sequence X (cf. [21], [29], [17]). And last, but not least, in comparing molecular sequences (e.g., finding homology between DNA sequences) one may search for the longest run of a given motif (pattern) (cf. [11]). These, and several other problems on words, can be efficiently solved and analyzed by a clever manipulation of a data structure known as *suffix tree* [2], [19], [27]).

In general, a suffix tree is a digital tree built from suffixes of a given word X, and therefore it fits into the class of digital search indexes ([14]). A digital tree stores n

* This research was supported in part by NSF Grants CCR-8900305 and INT-8912631, and AFOSR Grant 90-0107, NATO Grant 0057/89, and Grant R01 LM05118 from the National Library of Medicine.

strings $\{S_1, \ldots, S_n\}$ built over a finite alphabet Σ. If the strings $\{S_1, \ldots, S_n\}$ are statistically independent and every edge is labelled by a single symbol from Σ, then the resulting digital tree is called a regular (or independent) trie ([1], [9], [14]). If all unary nodes of a trie are eliminated, then the tree becomes the PATRICIA trie (cf. [9], [14], [23]). Finally, if an external node in a regular trie can store up to b strings (keys), then such a tree is called a b-trie. As mentioned above, a suffix tree is a special trie in which the strings $\{S_1, \ldots, S_n\}$ are suffixes of a given sequence X. Note that in this case the strings are statistically dependent!

As in the case of regular tries, there are several modifications of the standard suffix tree. In a *noncompact suffix tree* – called also spread suffix tree and position tree – each edge is labelled by a letter from the alphabet Σ. If all unary nodes are eliminated in the noncompact version of the suffix tree, then the resulting tree is called a *compact suffix tree* (cf. [2]). Gonnet and Baeza-Yates [9] coined a name PAT for such a suffix tree to resemble the name PATRICIA used for compact tries. Here, we also adopt this name. In addition, we introduce a family of suffix trees – called b-suffix trees – parametrized by an integer $b \geq 1$. A tree in such a family is constructed from the noncompact suffix tree by eliminating all unary nodes b levels above the fringe (bottom level) of the tree (later we slightly modify this definition). These trees have several useful applications in algorithms on words, data compressions, and so forth, but more importantly b-suffix trees form a spectrum of trees with noncompact suffix trees ($b = 1$) at one extreme and compact suffix trees ($b \to \infty$) at the other extreme. This allows to assess some properties of PAT trees in a unified and substantially easier manner (cf. [23]).

In this extended abstract, we offer a characterization of generalized suffix trees in a probabilistic framework. (Most of the proofs are omitted, and can be found in the extended version [26].) Our probabilistic model is a very general one, namely we allow symbols of a string to be dependent. Moreover, instead of concentrating on a specific algorithm we present a list of results concerning several parameters of suffix trees, namely: the typical depth $D_n^{(b)}$, depth of insertion $L_n^{(b)}$, height $H_n^{(b)}$ and the shortest feasible path $s_n^{(b)}$. For example, the typical depth D_n^{PAT} for the PAT tree built from the string $P\$T$ where P and T are the pattern and the text strings respectively, is used by Chang and Lawler [6] in their design of an approximate string matching algorithm. On the other hand, the depth of insertion $L_n^{(1)}$ of a noncompact suffix tree is of prime interest to the complexity of the Lempel-Ziv universal compression scheme (cf. [24]), and $L_n^{(1)}$ is responsible for a *dynamic* behavior of many algorithms on words. Furthermore, the height and the shortest feasible path path indicate how balanced a typical suffix tree is, that is, how much one has to worry about worst-case situations.

Our main results can be summarized as follows. For a b-suffix tree built over an *unbounded word* X, we prove that the normalized height $H_n^{(b)}/\log n$, the normalized shortest feasible path $s_n^{(b)}/\log n$ and the normalized depth $D_n^{(b)}/\log n$ almost surely (a.s.) converge to $1/h_2^{(b)}$, $1/h_1$ and $1/h$ respectively, where for every $1 \leq b \leq \infty$ we have $h_2^{(b)} < h < h_1$. In the above, h is the entropy on the alphabet Σ, while the parameters h_1 and $h_2^{(b)}$ depend of the underlying probabilistic model. If the word has *finite* length, then the above results also hold except for the shortest feasible path which clearly becomes $s_n^{(b)} = 1$ (see Remark 2(iii)). The most interesting behavior reveals the depth of insertion $L_n^{(b)}$ which converges *in probability* (pr.) to $(1/h) \log n$ but not almost surely. We prove that almost surely $L_n^{(b)} / \log n$ oscillates between $1/h_1$ and $1/h_2^{(b)}$. More interestingly, almost sure behavior of the *compact suffix tree* (i.e., PAT tree) can be deduced from the appropriate asymptotics of the b-suffix trees by taking $b \to \infty$. It is worth mentioning that all these results are obtained in a uniform manner by a technique that encompasses the so called string-ruler approach (cf. [13], [20]) and the mixing condition technique. Finally, using our results, we establish the average complexity of some exact and approximate pattern matching algorithms such as Chung-Lawler [6] and others (cf. [6]), etc. In addition, the results for noncompact suffix trees (cf. [25]) were used by us to settle in the negative the conjecture of Wyner and Ziv [29] concerning the length of the repeated pattern in a universal compression scheme (cf. [24]). In this paper, we prove the results already announced in [25] concerning the length of the last block in the Lempel-Ziv parsing algorithm [17].

Asymptotic analyses of suffix trees are very scanty in literature, and most of them deal with *noncompact suffix trees*. To the best of our knowledge, there are no probabilistic results on b-suffix trees and *compact suffix trees*. This can be easily verified by checking Section 7.2 of Gonnet and Baeza-Yates' book [9] which provides an up-to-date compendium of results concerning data structures and algorithms. The average case analysis of noncompact suffix trees was initialized by Apostolico and Szpankowski [3]. For the Bernoulli model (independent sequence of letters from a finite alphabet) the asymptotic behavior of the height was recently obtained by Devroye *et al.* [7], and the limiting distribution of the typical depth in a suffix tree is reported in Jacquet and Szpankowski [13]. Recently, Szpankowski [25] extended these results to a more general probabilistic model for *noncompact* suffix trees, that is, with $b = 1$. Finally, heuristic arguments were used by Blumer *et al.* [5] to show that the average number of internal nodes in a suffix tree is a linear function of n, and a rigorous proof of this can be found in [13]. Some related topics were discussed by Guibas and Odlyzko in [11].

2. Main Results and Their Consequences

In this paper, we consider a family of suffix trees called b-suffix trees. A tree in such a family has no unary nodes in all b levels above the fringe level of the corresponding noncompact suffix tree. Note that noncompact and compact suffix trees lie on two extremes of the spectrum of b-suffix trees; namely, a 1-suffix tree is a noncompact suffix tree, and a b-suffix tree becomes a compact suffix tree when $b \to \infty$. For the purpose of our analysis, however, a modified definition of b-suffix trees is more convenient. Hereafter, by b-suffix tree we mean a suffix tree *built from n first suffixes of an unbounded sequence $X = \{X_k\}_{k=1}^{\infty}$ that can store up to b suffixes in an external node*. We denote such a suffix tree by $\mathcal{S}_n^{(b)}$.

In this paper, we analyze six parameters of b-suffix trees $\mathcal{S}_n^{(b)}$; namely, the mth depth $L_n^{(b)}(m)$, the *height* $H_n^{(b)}$ and the *shortest feasible path* $s_n^{(b)}$, the *typical depth* $D_n^{(b)}$, the *depth of insertion* $L_n^{(b)}$ and the *external path length* $E_n^{(b)}$. The depth of the mth suffix is equal to the number of internal nodes in a path from the root to the external node containing this suffix. Then,

$$H_n^{(b)} = \max_{1 \le m \le n} \{L_n^{(b)}(m)\} , \tag{2.1}$$

that is, the height the longest path in $\mathcal{S}_n^{(b)}$. The shortest feasible path $s_n^{(b)}$ is defined as the length of the shortest path from the root to an *available* (feasible) node, that is, a node that is *not* in the tree $\mathcal{S}^{(b)}$ but whose predecessor node (either an internal or an external one) is in $\mathcal{S}^{(b)}$. In the performance evaluation of algorithms on words and data compression schemes, the typical depth $D_n^{(b)}$, the depth of insertion $L_n^{(b)}$, and the external path length $E_n^{(b)}$ are even more important. The depth of insertion $L_n^{(b)}$ is the depth of the $n+1$-st external node after insertion of the $(n+1)$st suffix S_{n+1} into the suffix tree $\mathcal{S}_n^{(b)}$, that is, $L_n^{(b)} = L_{n+1}^{(b)}(n+1)$. Finally, $D_n^{(b)}$ is defined as the depth of a *randomly* selected suffix, and the external path length $E_n^{(b)}$ is the sum of all depths $L_n^{(b)}(m)$ for $1 \le m \le n$. In other words,

$$E_n^{(b)} = \sum_{m=1}^{n} L_n^{(b)}(m) \quad \text{and} \quad D_n^{(b)} = \frac{E_n^{(b)}}{n} . \tag{2.2}$$

Note that $D_n^{(b)}$ can be interpreted as a successful search length in a suffix tree.

Our purpose is to investigate the behavior of a random b-suffix tree in a general probabilistic framework. The probabilistic model is the same as in [25], but for the completeness we provide here some details. We assume that $\{X_k\}_{k=1}^{\infty}$ is a *stationary ergodic* sequence of symbols generated from a finite alphabet Σ. In such a model, define a partial sequence X_m^n as $X_m^n = (X_m, ..., X_n)$ for $m < n$, and let for every $n \ge 1$ the nth order probability distribution for $\{X_k\}$ be

$$P(X_1^n) = \Pr\{X_k = x_k, 1 \le k \le n, x_k \in \Sigma\} . \tag{2.3}$$

The *entropy* of $\{X_k\}$ is defined in a standard manner as

$$h = \lim_{n \to \infty} \frac{E \log P^{-1}(X_1^n)}{n} , \tag{2.4}$$

and we introduce three additional parameters (cf. [20]), namely

$$h_1 = \lim_{n \to \infty} \frac{\max\{\log P^{-1}(X_1^n) , P(X_1^n) > 0\}}{n} = \lim_{n \to \infty} \frac{\log(1/\min\{P(X_1^n) , P(X_1^n) > 0\}}{n} , \tag{2.5}$$

$$h_2^{(b)} = \lim_{n \to \infty} \frac{\log(E\{P^b(X_1^n)\})^{-1}}{(b+1)n} = \lim_{n \to \infty} \frac{\log\left(\sum_{X_1^n} P^{b+1}(X_1^n)\right)^{-1/(b+1)}}{n} . \tag{2.6a}$$

$$h_3 = \lim_{n \to \infty} \frac{\min\{\log P^{-1}(X_1^n) , P(X_1^n) > 0\}}{n} = \lim_{n \to \infty} \frac{\log(1/\max\{P(X_1^n) , P(X_1^n) > 0\}}{n} , \tag{2.6b}$$

The existence of the limit in (2.5) is guaranteed by Shannon Theorem (cf. [4]), and the existence of h_1, h_3 and $h_2^{(1)}$ was established by Pittel [20] who also noticed that

$0 \leq h_2^{(1)} \leq h \leq h_1$. A generalization to an arbitrary b is easy, and left for the reader. In passing, we note that by the *inequality on means* we have $\lim_{b \to \infty} h_2^{(b)} = h_3$.

Remark 1.

(i) *Bernoulli Model.* In this widely used model (cf. [3], [5], [6], [7], [11], [14], [22], and [25]) symbols from the alphabet Σ are generated independently, that is, $P(X_1^n) = P^n(X_1^1)$. In particular, we assume that the ith symbol from the alphabet Σ is generated according to the probability p_i, where $1 \leq i \leq V$ and $\sum_{i=1}^V p_i = 1$. It is easy to notice that $h = \sum_{i=1}^V p_i \log p_i^{-1}$ ([4]), and from definition (2.3) we find that $h_1 = \log(1/p_{min})$, $h_3 = \log(1/p_{max})$ and $h_2^{(b)} = 1/(b+1) \log(1/P_b)$ where $p_{min} = \min_{1 \leq i \leq V}\{p_i\}$, $p_{max} = \max_{1 \leq i \leq V}\{p_i\}$, and $P_b = \sum_{i=1}^V p_i^{b+1}$. The probability P_b can be interpreted as the probability of a match of $b + 1$ strings in a given position (cf. [22]).

(ii) *Markovian Model.* In this model (cf. [12], [20]) the sequence $\{X_k\}$ forms a stationary Markov chain, that is, the $(k + 1)$st symbol in $\{X_k\}$ depends on the previously selected symbol. We define a transition probability as $p_{i,j} = \Pr\{X_{k+1} = j \in \Sigma | X_k = i \in \Sigma\}$. The transition matrix is denoted by $\mathbf{P} = \{p_{i,j}\}_{i,j=1}^V$. It is well known that the entropy h can be computed as $h = -\sum_{i,j=1}^V \pi_i p_{i,j} \log p_{i,j}$ where π_i is the stationary distribution of the Markov chain. The other quantities are a little harder to evaluate. Szpankowski [22] (see also Pittel [20] for $b = 1$) evaluated the height of regular tries with Markovian dependency, and showed that the parameter $h_2^{(b)}$ is a function of the largest eigenvalue θ_b of the matrix $\mathbf{P}_{[b+1]} = \mathbf{P} \circ \mathbf{P} \ldots \circ \mathbf{P}$ where $\circ$ represents the Schur product of $b+1$ matrices $\mathbf{P}$ (i.e., elementwise product). More precisely, $h_2^{(b)} = 1/(b + 1) \cdot \log \theta_b^{-1}$. With respect to h_1 and h_3 we need to refer to Pittel [20] who cited a nonprobabilistic result of Romanovski who proved that $h_1 = \min_{\mathcal{C}}\{\ell(\mathcal{C})/|\mathcal{C}|\}$ and $h_3 = \max_{\mathcal{C}}\{\ell(\mathcal{C})/|\mathcal{C}|\}$ where the minimum and the maximum are taken over all simple cycles $\mathcal{C} = \{\omega_1, \omega_2, ..., \omega_v, \omega_1\}$ for some $v \leq V$ such that $\omega_i \in \Sigma$, and $\ell(\mathcal{C}) = -\sum_{i=1}^V \log p_{i,i+1}$. $\square$

To complete our description of the probabilistic model, we add some *mixing conditions* (cf. [4]) on the sequence $\{X_k\}_{k=-\infty}^\infty$. Let $\mathcal{F}_m^n$ be a σ-field (i.e., history) generated by $\{X_k\}_{k=m}^n$ for $m \leq n$. It is said that $\{X_k\}$ satisfies the mixing condition if there exist two constants $c_1 \leq c_2$ and an integer d such that for all $-\infty \leq m \leq m + d \leq n$ the following holds

$$c_1 \Pr\{\mathcal{A}\}\Pr\{\mathcal{B}\} \leq \Pr\{\mathcal{AB}\} \leq c_2 \Pr\{\mathcal{A}\}\Pr\{\mathcal{B}\} \tag{2.7a}$$

where $\mathcal{A} \in \mathcal{F}_{-\infty}^m$ and $\mathcal{B} \in \mathcal{F}_{m+d}^n$. In some statements of our results, we need a stronger form of the above mixing condition, namely *strong α-mixing condition* which becomes

$$(1 - \alpha(d))\Pr\{\mathcal{A}\}\Pr\{\mathcal{B}\} \leq \Pr\{\mathcal{AB}\} \leq (1 + \alpha(d))\Pr\{\mathcal{A}\}\Pr\{\mathcal{B}\} \tag{2.7b}$$

where the function $\alpha(d)$ is such that $\alpha(d) \to 0$ as $d \to \infty$.

Finally, for compact suffix trees (i.e., PAT trees) we need one more condition. Let $\omega_i \in \Sigma$ for $1 \leq i \leq n$. Define $P(\omega_1, \ldots, \omega_n) = \Pr\{X_1^n = (\omega_1, \ldots, \omega_n)\}$. Then, for PAT trees we shall require the following

$$P(\omega_1, \ldots, \omega_n) \leq \rho P(\omega_1, \ldots, \omega_{n-1}) \tag{2.7c}$$

for some $0 < \rho < 1$.

Now we ready to present our first main result concerning the typical height and the shortest feasible path, which is further used to prove our next findings. The proof of this theorem is omitted, except for part (ii) regarding PAT trees which is a simple consequence of part (i), and it is proved in Remark 2 (ii) below.

Theorem 1. *Let $\{X_k\}$ be a stationary ergodic sequence satisfying the strong α-mixing condition (2.7b) together with $h_1 < \infty$ and $h_2 > 0$.*

(i) *b*-**Suffix Trees.** *Fix b. Then*

$$\lim_{n \to \infty} \frac{s_n^{(b)}}{\log n} = \frac{1}{h_1} \quad \text{(a.s.)} \tag{2.8}$$

provided

$$\alpha(d) = O(d^\beta \rho^d) \tag{2.9}$$

for some constants $0 < \rho < 1$ and $\beta > 0$. For the height $H_n^{(b)}$ we have

$$\lim_{n \to \infty} \frac{H_n^{(b)}}{\log n} = \frac{1}{h_2^{(b)}} \quad \text{(a.s.)} \tag{2.10}$$

provided the coefficient $\alpha(d)$ in (2.7b) fulfills the following

$$\sum_{d=0}^{\infty} \alpha^2(d) < \infty . \tag{2.11}$$

(ii) **Compact Suffix Tree.** *Almost sure behavior of a compact suffix tree follows from the (a.s.) behavior of b-suffix trees by taking in (2.8) and (2.9) the limit as $b \to \infty$, that is,*

$$\lim_{n \to \infty} \frac{s_n^{PAT}}{\log n} = \frac{1}{h_1} \quad \text{(a.s.)} \qquad \lim_{n \to \infty} \frac{H_n^{PAT}}{\log n} = \frac{1}{h_3} , \tag{2.12}$$

provided (2.7c) holds together with condition (2.9) for s_n^{PAT} and condition (2.11) for H_n^{PAT} respectively. ∎

Theorem 1 can be proved along the same lines as we did in [25] for $b = 1$. Details can be found in [26]. In Section 3 we present a combinatorial lemma that is a new ingredient of the proof for general b.

Our next main results deal with the typical depth $D_n^{(b)}$ and the depth of insertion $L_n^{(b)}$. The proof of Theorem 2 is sketched in Section 3 except part (i) which is discussed in Remark 2 (ii).

Theorem 2. *Let $\{X_k\}$ be a stationary ergodic and mixing sequence in the strong sense of (2.7b), and let (2.9) hold too. Assume also that $1 \le b < \infty$.*

(i) **Convergence in Probability.** *For $h < \infty$ we have*

$$\lim_{n \to \infty} \frac{L_n^{(b)}}{\log n} = \lim_{n \to \infty} \frac{D_n^{(b)}}{\log n} = \frac{1}{h} \quad \text{(pr.)} . \tag{2.14}$$

The same holds for a compact suffix tree provided (2.7c) is true.

(ii) **Almost Sure Convergence of the Typical Depth D_n.** *Let above hypotheses hold. Then,*

$$\lim_{n\to\infty} \frac{D_n^{(b)}}{\log n} = \lim_{n\to\infty} \frac{E_n^{(b)}}{n\log n} = \frac{1}{h} \qquad (a.s.) \,. \qquad (2.15)$$

The above is true also for the compact suffix tree provided (2.7c) is satisfied.

(iii) **Almost Sure Behavior of the Depth of Insertion L_n.** *As in (ii) we assume strong mixing condition (2.7b) together with (2.9) and $h_1 < \infty$ as well as $h_2 > 0$. Then, we have the following result concerning the depth of insertion for $b < \infty$*

$$\lim_{n\to\infty} \inf \frac{L_n^{(b)}}{n \, \log n} = \frac{1}{h_1} \qquad (a.s) \qquad \lim_{n\to\infty} \sup \frac{L_n^{(b)}}{n \, \log n} = \frac{1}{h_2^{(b)}}. \qquad (2.16)$$

For the compact suffix tree (2.16) holds with $h_2^{(b)}$ replaced by h_3, that is, we formally obtain almost sure behavior for the compact suffix tree by taking $b \to \infty$ and assuming (2.7c). ∎

Remark 2

(i) *How to prove part (iii) of Theorem 2 ?* One can view the behavior of $D_n^{(b)}$ and $L_n^{(b)}$ as a surprise. Both quantities characterize the depth of a tree, but $L_n^{(b)}$ is responsible for a dynamic behavior of the suffix tree while $D_n^{(b)}$ is responsible for the "average" one. It is also easy to notice that the main reason for $L_n^{(b)}$'s oscillation is a "tiny" unbalance in the height and the shortest feasible path discovered in Theorem 1 (the typical depth $D_n^{(b)}$ behaves nicely since these oscillations are smoothed by the sum in (2.2)). This was already observed by Pittel [20] for independent tries, and by Szpankowski [25] for noncompact suffix trees. In passing, we note that the only b-suffix tree that has (a.s.) limit for the depth of insertion $L_n^{(b)}$ is PAT tree with the *symmetric* alphabet (i.e., $p_i = 1/V$ for $il \le i \le V$). Indeed, by Theorem 2 (iii) in this case $\lim_{n\to\infty} L_n^{PAT}/\log n = \log V$ (a.s.).

(ii) *Compact Suffix Tree as a Limit of b-Suffix Tree.* We conjecture that there exists a sequence a_n such that for several parameters $P_n^{(b)}$ of b-suffix trees such as the height, the depth, etc., the following holds

$$\lim_{n\to\infty} P_n^{PAT}/a_n = \lim_{b\to\infty} \lim_{n\to\infty} P_n^{(b)}/a_n \,, \qquad (2.17a)$$

where P_n^{PAT} is the corresponding parameter for the PAT tree. We can easily give a formal proof of this interesting fact for every parameter discussed in this section. We consider first all parameters except the height. It is easy to see that

$$\lim_{n\to\infty} \frac{P_n^{PAT}}{\log n} \le \lim_{n\to\infty} \frac{P_n^{(1)}}{\log n} \,. \qquad (2.17b)$$

This immediately establishes the upper bound part of (2.17a) for the above parameters (excluding the height). For the height H_n^{PAT}, following Pittel [20] we note that the event $\{H_n^{PAT} > k + b\}$ implies that there exists a set of b suffixes such that all of them share the same first k symbols. In other words, the event $\{H_n^{PAT} > k + b\}$ implies $\{H_n^{(b)} > k\}$. Therefore,

$$\lim_{n \to \infty} \frac{H_n^{PAT}}{\log n} \le \lim_{b \to \infty} \lim_{n \to \infty} \frac{H_n^{(b)}}{\log n} = \frac{1}{h_3} \qquad (2.17c)$$

This completes the upper bound in (2.17a).

For the lower bound we use condition (2.7c). We need a separate discussion for every parameter. Following Pittel [20], for the height and the shortest feasible path we argue as follows. We try to find a path (called a feasible path) in a suffix tree such that the length of it is (a.s.) asymptotically equal to $\log n/h_3$ and $\log n/h_1$ respectively. But, this is immediate from (2.6a) and (2.6c), and Pittel [20] Lemma 2. For the depth we consider a path for which the initial segment of length $O(\log n)$ is such that all nodes are branching (i.e., no unary nodes occur in it). Naturally, such a path after compression will not change, and the depth in the compact suffix tree is at least as large as the length of this path. It can be proved that (a.s.) such a path is smaller or equal to $\log n/h$ which completes the lower bound arguments in the proof for the depth.

Despite our formal proof, it is important to understand intuitively why a compact suffix tree can be considered as a limit of b-suffix trees. There are at least three reasons supporting this claim: (1) b-suffix trees do not possess unary node in any place that is b levels above the fringe of the noncompact suffix tree; (2) unary nodes tend to occur more likely at the bottom of a suffix tree, and it is very unlikely in a typical suffix tree to have a unary node close to the root (e.g., in the Bernoulli model the probability that the root is unary node is equal to $\sum_{i=1}^{V} p_i^n$); (3) on a typical path the compression is of size $O(1)$ (e.g., comparing the depth of regular tries and PATRICIA we know that $ED_n^P - ED_n^T = O(1)$ [22], [23], but for the height we have $EH_n^P - EH_n^T = \log n$ [20], and therefore, we can expect troubles only with the height, and this is confirmed by our analysis).

(iii) *Finite Strings.* In several computer science applications (cf. [2]) the string $\{X_k\}_{k=1}^n$ has finite length n, and it is terminated by a special symbol that does not belong to the alphabet Σ, e.g., $X\$$ with $\$ \notin \Sigma$. Most of our results, however, can be directly applied to such strings. Let s_n', H_n' and D_n' denote the shortest feasible path, the height and the depth in a suffix tree (b-suffix tree or compact suffix tree) built over such a finite word, respectively. Then, it is easy to see that $s_n' = 1$, but the other two parameters have exactly the same asymptotics as for the infinite string case, that is, $H_n'/\log n \sim 1/h_2^{(b)}$ (a.s.) and $D_n'/\log n \sim 1/h$ (a.s.) under the hypotheses of Theorems 1 and 2. Indeed, our analysis reveals that only $O(\log n)$ last suffixes contribute to the difference between the finite and infinite string analysis. But, building a suffix tree from the first $n' = n - O(\log n)$ suffixes will lead to the same asymptotics as for an infinite string, except for $s_n^{(b)}$. Details are left to the interested reader. □

Theorems 1 and 2 find several applications in combinatorial problems on words, data compressions and molecular biology. As an illustration, we solve here two problems on words using Theorem 2. One of them deals with the average time-complexity of the exact string matching algorithm proposed recently by Chang and Lawler [6], while the other sheds some lights on the typical behavior of a phrase in the Lempel-Ziv parsing scheme [17]. The solution to the last problem was already announced in [25], but here we provide a proof. We also point out that several other practical algorithms on words (i.e., with good average complexity) are of immediate consequence of our findings. The reader is referred to Apostolico and Szpankowski [3] for more details.

PROBLEM 1. *String Matching Algorithms*

Recently, Chang and Lawler [6] demonstrated how to use PAT trees to design *practical* and still efficient algorithms for approximate string matching algorithms. They formulated several conclusions based on a heuristic analysis of PAT trees under the *symmetric* Bernoulli model. Our Theorems 1 and 2 immediately generalize results of [6] to a more general probabilistic model, and additionally provide stronger results. For example, consider the exact string matching algorithm (cf. Section 2.3 in [6]) in which we search for all occurrences of the pattern string P of length m in the text string T of length n. The heart of Chang-Lawler's analysis is an observation that there exists such $d_{m,n}$ that a substring of the text T of length $d_{m,n}$ is not a substring of the pattern P. This can be verified by building first a compact suffix tree for P and then insert suffixes of T (i.e., constructing a compact suffix tree for $P\$T$ where $\$$ is a special character). But then, one may observe that $d_{m,n}$ is equivalent to the typical depth D_n^{PAT}, and therefore $d_{m,n} \sim (1/h)\log(m+n)$ (a.s.). This further implies that the complexity C_n of the algorithm becomes $C_n \sim (n/hm)\log(m+n)$ (a.s.), which is a much stronger version of Chang-Lawler's result for a more general probabilistic model. Several other approximate string matching algorithms can be analyzed in a similar manner.

PROBLEM 2. *Block Length in the Lempel-Ziv Parsing Algorithm*

The heart of the Lempel-Ziv compression scheme is a method of parsing a string $\{X_k\}_{k=1}^n$ into blocks of different words. The precise scheme of parsing the first n symbols of a sequence $\{X_k\}_{k=1}^\infty$ is complicated and can be found in [17]. For example, for $\{X_k\} = 110101001111\cdots$ the parsing looks like $(1)(10)(10100)(111)(1\cdots)$. In Figure 1 we show how to perform the parsing using a sequence of noncompact suffix trees (cf. [10]). Note that the length of a block is a subsequence of depth of insertions $L_{n_k}^{(1)}$. More precisely, if ℓ_n is the length of the nth block in the Lempel-Ziv parsing algorithm, then Figure 1 suggests the following relationship $\ell_n = L_{\sum_{k=0}^{n-1}\ell_k}^{(1)}$. For example, in Fig. 1 we have $\ell_0 = L_0^{(1)} = 1$, $\ell_1 = L_1^{(1)} = 2$, $l_2 = L_{\ell_0+\ell_1}^{(1)} = L_3^{(1)} = 5$, and $\ell_3 = L_{1+2+5} = 3$, and so forth. To obtain (a.s.) behavior of the block length ℓ_n, we note that

$$\lim_{n\to\infty}\frac{\ell_n}{\log n} = \lim_{n\to\infty}\frac{L_{\sum_{k=0}^{n-1}\ell_k}^{(1)}}{\log\left(\sum_{k=0}^{n-1}\ell_k\right)}\cdot\frac{\log\left(\sum_{k=1}^{n-1}\ell_k\right)}{\log n}. \qquad (2.18)$$

We first estimate the second term in (2.18). One immediately obtains

$$1 \le \frac{\log\left(\sum_{k=0}^{n-1}\ell_k\right)}{\log n} \le \frac{\log\left(\sum_{k=0}^{n-1}L_k^{(1)}\right)}{\log n} \le \frac{\log\left(\sum_{m=0}^{n}L_n^{(1)}(m)\right)}{\log n} \to 1 \qquad (a.s)\,,$$

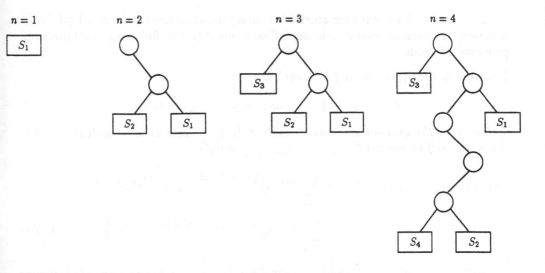

Fig. 1. First four suffix trees used to parse the sequence $X = 110101001111 \cdots$

where the RHS of the above is a direct consequence of Theorem 2(ii). But then, (2.18) leads to the following result.

Corollary 3. *Let* $\{X_k\}_{k=1}^{\infty}$ *be a mixing stationary sequence. Then*

$$\frac{1}{h_1} \leq \liminf_{n \to \infty} \frac{\ell_n}{n \, \log n} \leq \limsup_{n \to \infty} \frac{\ell_n}{n \, \log n} \leq \frac{1}{h_2^{(1)}} \qquad (a.s.) . \qquad (2.19)$$

provided (2.9) holds. ∎

3. Sketch of Proofs

Due to the lack of space, we only briefly discuss Theorem 1 (i.e., the main differences between the case $b = 1$ and general b), and give some details for the proof of Theorem 2(i).

Throughout the proof of Theorem 1 we use a novel technique that encompasses the mixing condition and another technique called the *string-ruler approach* that was already applied by Pittel [20] and extended by Jacquet and Szpankowski [13]. The idea of the string-ruler approach is to measure a correlation between words by another *nonrandom* word w belonging to a set of words $\mathcal{W}$.

To provide some details, we need some new notation. Let $\mathcal{W}_k$ be the set of all strings w_k of length k, that is, $\mathcal{W}_k = \{w \in \Sigma^k : |w| = k\}$, where $|w|$ is the length of w. We write w_k^ℓ to mean a concatenation of ℓ strings w_k from $\mathcal{W}_k$, and if $X_m^{m+k} = w_k$, then we denote $P(w_k) = P(X_m^{m+k})$. For a function $f(w_k)$ of w_k, we write $\sum_{\mathcal{W}_k} f(w_k) = \sum_{w_k \in \mathcal{W}_k} f(w_k)$ where the sum is over all strings w_k.

The proof of Theorem 1 for general b is along the same lines as for $b = 1$ (cf. [25]), however, it requires several new ingredients, notably the following combinatorial property of words.

Lemma 4. *Let $d_1, \ldots, d_b$ and k be such that*

$$d_0 = 0 \leq d_1 \leq \cdots \leq d_i \leq k \leq d_{i+1} \leq \cdots \leq d_b .$$

Define d as the greatest common divisor of $\{d_i\}_{i=1}^b$, that is, $d = \gcd(d_1, \ldots, d_b)$. Then, the self-alignment $C_{1,1+d_1,\ldots,1+d_1+\cdots+d_b}$ satisfies

$$\Pr\{C_{1,1+d_1,\ldots,1+d_1+\cdots+d_b} \geq k\} = \sum_{w_d} P\left(w_d^{\lfloor \frac{k}{d} \rfloor + \frac{d_1+\cdots+d_i}{d}} \overline{w}_d (w_d^{\lfloor \frac{k}{d} \rfloor} \overline{w}_d)^{b-i}\right)$$

$$= \sum_{w_d} P\left(w_d^{(b+1-i)\lfloor \frac{k}{d} \rfloor + \frac{d_1+\cdots+d_i}{d}} \overline{w}_d^{b+1-i}\right) , \tag{3.0a}$$

where $\overline{w}_d$ is a prefix of w_d', and $\lfloor x \rfloor$ is the floor function. Two cases are of particular interest, namely: (i) if $k \leq d_1 \leq \cdots \leq d_b$, then

$$\Pr\{C_{1,1+d_1,\ldots,1+d_1+\cdots+d_b} \geq k\} = \sum_{w_k} P(w_k^{b+1}) ; \tag{3.0}$$

(ii) if $d_1 \leq \cdots \leq d_b \leq k$, then

$$\Pr\{C_{1,1+d_1,\ldots,1+d_1+\cdots+d_b} \geq k\} = \sum_{w_d} P\left(w_d^{\lfloor \frac{k}{d} \rfloor + \frac{d_1+\cdots+d_i}{d}} \overline{w}_d\right) . \tag{3.0c}$$

Proof. It is illustrative to start with $b = 1$. In such a case, it is well known [18] that for any pair of suffixes S_1 and S_{1+d} there exists a word w_d such that the common prefix Z_k of length k of S_1 and S_{1+d} can be represented as $Z_k = w_d^{\lfloor \frac{k}{d} \rfloor} \overline{w}_d$. Then, (3.0c) is a simple consequence of this. The above rule is easy to extend to b suffixes. Let Z_k be the common prefix of length k of the following b suffixes $\{S_1, S_{1+d_1}, \ldots, S_{1+d_1+\cdots+d_b}\}$. To avoid heavy notation, we consider three cases separately. If $k \leq d_1 \leq \cdots \leq d_b$, then all suffixes are separated by more than k symbols, so certainly there exists a word w_k such that $Z_k = w_k^{b+1}$, which further implies (3.0b). Let us now consider the case $d_1 \leq \cdots \leq d_b \leq k$, that is, there are mutual overlaps between any two consecutive suffixes. Then, there must exist a word w_d of length $d = \gcd(d_1, \ldots, d_b)$ such that $Z_k = w_d^{\lfloor \frac{k}{d} \rfloor + \frac{d_1+\cdots+d_i}{d}} \overline{w}_d$, which leads to (3.0c). Finally, the general solution (3.2a) is a combination of the two above cases. ∎

We provide some more details regarding the proof of Theorem 2, particularly part (i). The proof of this theorem is based on "counting" and it is typical for the information theory community. It uses the following property of the entropy called the *Asymptotic Equipartition Property* (AEP) [4]. It reads as follows: *For a stationary and ergodic sequence $\{X_k\}_{k=1}^n$, the state space Σ^n can be partitioned into two sets, namely "good states" set G_n and "bad states" set B_n such that for $X_1^n \in G_n$ and for sufficiently large n we have $P(X_1^n) \geq 1 - \varepsilon$ for any $\varepsilon > 0$, and $P(B_n) \leq \varepsilon$. Moreover,* for a typical X_1^n (i.e., $X_1^n \in G_n$) we have $e^{-n(h+\varepsilon)} \leq P(X_1^n) \leq e^{-n(h+\varepsilon)}$ where h is the entropy.

We concentrate on L_n, but the proof for D_n follows the same steps (cf. [20]). Define an event A_n such that $A_n = \{X_1^\infty : | L_n/\log n - 1/h | \geq \varepsilon/h\}$. For Theorem 2 (i) it suffices to prove that $\Pr\{A_n\} \to 0$ as $n \to \infty$. Also, for some $\varepsilon_1 > 0$ and $n_0 \geq n$ we define another event (i.e., set of "good states")

$$G_{n_0} = \{X_1^\infty : | n^{-1}\log P^{-1}(X_1^n) - h | < \varepsilon_1 h , \quad n > n_0\} . \tag{3.1}$$

By the total probability formula we have

$$P(A_n) \leq \Pr\{A_n G_{n_0} \text{ and } L_n < \delta\log n\} + \Pr\{L_n \geq \delta\log n\} + P(B_{n_0}) \tag{3.2}$$

where $\delta > 1/h_2$ and

$$B_{n_0} = \sup_{n \geq n_0} \{X_1^\infty : |n^{-1}\log P^{-1}(X_1^n) - h| \geq \varepsilon_1 h , \quad n > n_0\} .$$

By AEP we know that $\lim_{n_0 \to \infty} P(B_{n_0}) = 0$. In addition, from by Theorem 1(ii) $\Pr\{L_n \geq \delta\log n\} \leq c/n^{\delta - 1/h_2}$ for $\delta > 1/h_2$, hence the second probability in the above also tends to zero.

In the view of the above, we can now deal only with the first term in (3.2) which we denote for simplicity by $P_1(A_n G_n)$. This probability can be estimated as follows

$$P_1(A_n G_n) \leq \sum_{r \in C_n} \Pr\{L_n = r ; |\log(P^{-1}(X_1^r))/r - h| < \varepsilon_1 h , r \geq n_0\} = \sum_{r \in C_n} P_n^{(r)} , \tag{3.3}$$

where

$$C_n = \{r : | r/\log n - 1/h | \geq \varepsilon/h \text{ and } r \leq \delta\log n\} .$$

Note that in (3.3) we restrict the summation only to "good states" represented by G_n. Therefore, for a word $w_r \in G_n$ we have with high probability

$$c_1 \exp\{-(1 - \varepsilon_1)hr\} \leq P(w_r) \leq c_2 \exp\{-(1 + \varepsilon_1)hr\} .$$

The next step is to estimate the probability $\Pr\{L_n = r\}$. This is rather intricate, and details are omitted. We can show that for $r \geq (1 + \varepsilon)\frac{\log n}{n}$ the rth term in the sum of (3.3) becomes $P_n^{(r)} \leq c/n^\varepsilon$. On the other hand, when $r \leq (1 - \varepsilon)\frac{\log n}{n}$ we need the strong mixing condition to prove that $P_n^{(r)} \leq c\exp(n^{b\varepsilon/2}/\log^b n)$ provided (2.9) holds.

Putting everything together, we note that the cardinality of the set C_n in (3.3) is bounded from the above by $\delta\log n$, hence

$$P(A_n) \leq c\log n \left(\exp(-n^{b\varepsilon/2}/\log^b n) + n^{-\varepsilon}\right) + P(B_{n_0})$$

which suffices for the proof of Theorem 2 (i) since $P(B_{n_0}) \to 0$ as $n_0 \to \infty$ by AEP.

References

1. A.V. Aho, J.E. Hopcroft and J.D. Ullman, *The Design and Analysis of Computer Algorithms*, Addison-Wesley (1974).
2. A. Apostolico, The Myriad Virtues of Suffix Trees, *Combinatorial Algorithms on Words*, 85-96, Springer-Verlag, ASI F12 (1985).
3. A. Apostolico and W. Szpankowski, Self-alignments in Words and Their Applications, *J. of Algorithms*, 13 (1992), in press.
4. P. Billingsley, *Ergodic Theory and Information*, John Wiley & Sons, New York 1965.
5. A. Blumer, A. Ehrenfeucht and D. Haussler, Average Size of Suffix Trees and DAWGS, *Discrete Applied Mathematics*, 24, 37-45 (1989).
6. W. Chang, E. Lawler, Approximate String Matching in Sublinear Expected Time, *Proc. of 1990 FOCS*, 116-124 (1990).
7. L. Devroye, W. Szpankowski and B. Rais, A note of the height of suffix trees, *SIAM J. Computing* 21, 48-54 (1992).
8. Z. Galil, K. Park, An Improved Algorithm for Approximate String Matching, *SIAM J. Computing*, 19, 989-999 (1990)
9. G.H. Gonnet and R. Baeza-Yates, *Handbook of Algorithms and Data Structures*, Addison-Wesley, Workingham (1991).
10. P. Grassberger, Estimating the Information Content of Symbol Sequences and Efficient Codes, *IEEE Trans. Information Theory*, 35, 669-675 (1991).
11. L. Guibas and A. W. Odlyzko, String Overlaps, Pattern Matching, and Nontransitive Games, *Journal of Combinatorial Theory*, Series A, 30, 183-208 (1981).
12. P. Jacquet and W. Szpankowski, Analysis of Digital Tries with Markovian Dependency, *IEEE Trans. Information Theory*, 37, 1470-1475 (1991).
13. P. Jacquet and W. Szpankowski, Autocorrelation on Words and Its Applications. Analysis of Suffix Tree by String-Ruler Approach, INRIA TR-1106 (1989); also submitted to a journal.
14. D. Knuth, *The Art of Computer Programming. Sorting and Searching*, Addison-Wesley (1973).
15. G.M. Landau and U. Vishkin, Fast String Matching with k Differences, *J. Comp. Sys. Sci.*, 37, 63-78 (1988)
16. G.M. Landau and U. Vishkin Fast Parallel and Serial Approximate String Matching, *J. Algorithms*, 10, 157-169 (1989).
17. A. Lempel and J. Ziv, On the Complexity of Finite Sequences, *IEEE Information Theory* 22, 1, 75-81 (1976).
18. M. Lothaire, *Combinatorics on Words*, Addison-Wesley (1982)
19. E.M. McCreight, A Space Economical Suffix Tree Construction Algorithm, *JACM*, 23, 262-272 (1976).
20. B. Pittel, Asymptotic growth of a class of random trees, *The Annals of Probability*, 18, 414 - 427 (1985).
21. M. Rodeh, V. Pratt and S. Even, Linear Algorithm for Data Compression via String Matching, *Journal of the ACM*, 28, 16-24 (1981).
22. W. Szpankowski, On the Height of Digital Trees and Related Problems, *Algorithmica*, 6, 256-277 (1991).
23. W. Szpankowski, Patricia tries again revisited, *Journal of the ACM*, 37, 691-711 (1991).
24. W. Szpankowski, A Typical Behavior of Some Data Compression Schemes, *Proc. of Data Compression Conference*, pp. 247-256, Snowbirds (1991).
25. W. Szpankowski, (Un)Expected Behavior of Typical Suffix Trees, *Proc. Third Annual ACM-SIAM Symposium on Discrete Algorithms*, pp. 422-431, Orlando 1992.

26. W. Szpankowski, Suffix Trees Revisited: (Un)Expected Asymptotic Behaviors, Purdue University, CSD-TR-91-063 (1991).
27. P. Weiner, Linear Pattern Matching Algorithms, *Proc. of the 14-th Annual Symposium on Switching and Automata Theory*, 111 (1973).
28. U. Vishkin, Deterministic Sampling – A New Technique for fast Pattern Matching, *SIAM J. Computing*, 20, 22-40 (1991).
29. A. Wyner and J. Ziv, Some Asymptotic Properties of the Entropy of a Stationary Ergodic Data Source with Applications to Data Compression, *IEEE Trans. Information Theory*, 35, 1250-1258 (1989).

A LANGUAGE APPROACH TO STRING SEARCHING EVALUATION

Mireille Régnier

INRIA, 78153 Le Chesnay, France

Abstract. We propose a general framework to derive average performance of string searching algorithms that preprocess the pattern. It relies mainly on languages and combinatorics on words, joined to some probabilistic tools. The approach is quite powerful: although we concentrate here on Morris-Pratt and Boyer-Moore-Horspool, it applies to a large class of algorithms. A fairly general character distribution is assumed, namely a Markovian one, suitable for applications such as natural languages or biological databases searching. The average searching time, expressed as the number of text-pattern comparisons, is proven to be asymptotically Kn and the linearity constant is given.

1 Introduction

This paper is devoted to the evaluation of average performance of string searching algorithms. Our approach is algebraic. As a matter of fact, the main problem for the analysis is the huge number of states of the automaton defining each algorithm: q^m where m is the size of the searched pattern and q the cardinality of the alphabet. We propose a general framework based on languages and combinatorics of words, that drastically reduces the combinatorial explosion of the problem. We propose a *bootstrapping* approach. We concentrate first on the main contributions, defined by a partition of the states into a few number of states, and refine later. Practically, the convergence to the actual cost is very fast. Moreover, one can consider simultaneously all possible lengths m. More precisely, performance show a simple dependancy on this parameter and are simply obtained, for finite m, from the asymptotic result. This scheme applies for a large class of algorithms, namely the ones based on a preprocessing of the searched pattern, and fairly general distributions. More precisely, we assume 1-order Markovian dependencies between the characters. This is suitable for applications to natural languages or molecular biology. We consider here two examples: Morris-Pratt and Boyer-Moore-Horspool. We prove that the average searching time, expressed as the number of text-pattern comparisons, is linear in the size n of the text and compute the linearity constant.

So far, quite a few results were available. Attempts in [Sch88, BY89a, BY89b] used Markov chains and hence were limited by the exponential number of states. For a pattern length m greater than $3, 4$, only upper or lower bound could be derived. First asymptotics came out recently. The naive algorithm (respectively Morris-Pratt[KMP77] and Boyer-Moore-Horspool [BM77, Hor80]) were analyzed, for *uniform* character distributions for text and pattern, in [Bar85] (respectively [Rég89]

* This work was partially supported by the ESPRIT II Basic Research Actions Program of the EC under contract No. 3075 (project ALCOM).

and [BYGR90, BYR92]). These analyses are extended for biased stationary distributions in [Rég91] that allow very precise asymptotic developments of the linearity constant. Other results on Morris-Pratt can also be found in [Han91]. The combinatorial explosion stucks them to small patterns or asymptotic order and prevents from a generalization to Markovian dependencies.

2 Algebraic tools

2.1 Probabilistic models and definitions

The aim of this section is to provide tools for the average analysis of string searching algorithms. The text and the pattern are words in A^*. Hence, defining our text or pattern distribution is equivalent to the definition of two *probability measures* $\mathcal{Q}$ or $\mathcal{P}$ on A^* associated to the text and the pattern. In the following, $w[k]$ denotes the k-th character of a word w in A^*, counted from left to right.

Definition 1. We assume that the text and the pattern define two Markov processes on A^*, with transition matrices $Q = ||q_{i,j}||$ and $P = ||p_{i,j}||$. Let $\mathcal{Q}$ and $\mathcal{P}$ be the probability measures so defined on A^*. We assume both processes admit a limiting process, and note $(q_i)_{i=1...q}$ and $(p_i)_{i=1...q}$ the stationary probabilities.

Our results will be expressed as functions of the quantities defined below.

Definition 2. We note:

* the generalized stationary diagonal matrices: $S_{\alpha,\beta} = ||p_1^\alpha q_1^\beta, \ldots, p_q^\alpha q_q^\beta||$,
* the matching transition matrix: $F = ||p_{i,j} q_{i,j}||$,

Then, w being a subword in the text (or the pattern), we have:

$$E_{\mathcal{Q}}(w[k+1] = a_j / w[k] = a_i) = q_{i,j} .$$

Moreover, if w is "far enough" to the right in the text: $E_{\mathcal{Q}}(w[1] = a_i) = q_i$. If the size $|m|$ of the pattern is "big enough" and w a subword starting "far enough" to the right: $E_{\mathcal{P}}(w[1] = a_i) = p_i$. The special cases of stationary distributions and uniform distributions were considered in [Rég91] and [Rég89, BYGR90, BYR92]. For a sake of simplicity and clarity, we assume in the following that all transition probabilities $p_{i,j}, q_{i,j}$ are non-zero, deferring to an extended paper the generalization. Remark that the fundamental matrix of F, $(I - F)^{-1}$ always exists for "non pathological" distributions.

We now introduce some notation to deal with probability measures on subsets of A^*, i.e. languages.

Definition 3. Let $\mathcal{L}$ be a language on alphabet A, with probability measure $\mathcal{P}$ and $\mathcal{L}_{i,j} = \mathcal{L} \cap a_i.A^* \cap A^*.a_j$. Let $L_{\mathcal{P}}^{(1)}$ be the diagonal matrix:

$$L_{\mathcal{P}}^{(1)} = ||p_i.1_{\mathcal{L} \cap a_i.A^* \neq 0}||_{i=1..q} \tag{1}$$

Let $L_{\mathcal{P}}^{(2)}$ be a matrix satisfying the equation:

$$||E_{\mathcal{P}}(\mathcal{L}_{i,j})|| = L_{\mathcal{P}}^{(1)} \circ L_{\mathcal{P}}^{(2)} . \tag{2}$$

Then $(L_{\mathcal{P}}^{(1)}, L_{\mathcal{P}}^{(2)})$ is a matricial expectation couple for $\mathcal{L}$.

Remark that when $\mathcal{P}$ changes to $\mathcal{Q}$, $L_{\mathcal{Q}}^{(i)}$ is derived from $L_{\mathcal{P}}^{(i)}$ by the set of substitutions: $p_{i,j} \to q_{i,j}, p_i \to q_i$. Similarly, one notes $L_{\mathcal{PQ}}^{(i)}$ the matrix derived from $L_{\mathcal{Q}}^{(i)}$ by the substitutions:

$$q_{i,j} \to p_{i,j} q_{i,j}, q_i \to p_i q_i .$$

To express our results, we also define a set of *operators*:

Definition 4. Given $M = ||m_{i,j}||$ a matrix and a a character from the alphabet A, we define the operators:

* $Line_a = ||m_{i,j}.1_{a_i=a}||$ and $Col_a = ||m_{i,j}.1_{a_j=a}||$.
* $\overline{Line}_a(M) = M - Line_a(M)$ and $\overline{Col}_a(M) = M - Col_a(M)$
* $S(M) = \sum_{i,j} m_{i,j}$

Finally, assume $m_{i,j} = \phi_{i,j}(x_1, \ldots, x_k)$. We note $M_m = ||\phi_{i,j}(x_1^m, \ldots, x_k^m)||$.

Remark: $Line_a(M \circ N) = Line_a(M) \circ N$ and $Col_a(N \circ M) = N \circ Col_a(M); N \circ \overline{Line}_a(M) = \overline{Col}_a(N) \circ M$, and $N \circ \overline{Line}_a(M) = \overline{Col}_a(N) \circ M$. Practically, M_m will be used with variables x_k ranging on $(q_{i,j}), (p_{i,j}), (q_i)$ and (p_i).

2.2   Languages and automata

Through this paper, we consider algorithms that are based on a *preprocessing of the pattern* and not of the text, as for instance the one in [KR87]. Such algorithms are currently seen as a set of automata $\{A_p\}$ [HU79, Tho68]: automaton A_p recognizes in a text t, the input, all the occurrences of pattern p. Each text-pattern comparison leads to a state change, that also determines the next comparison to be performed. Hence, the complexity is measured by the number of transitions. To deal with the average complexity, our approach computes first the expectation of the number of transitions for a given p and a random text, and afterwards averages when p ranges. Note that the reverse would give equivalent results, but is not adapted to these algorithms that preprocess the pattern. As the length of the pattern is a sensible parameter, we are led to the following definitions:

Definition 5. Let p be a pattern of length $|p| = m$. Let $C_{n,p}(Alg)$ be the average number of text-pattern comparisons performed by algorithm Alg on a random text of size n. Let $Cost_p(Alg)$ be $\frac{C_{n,p}(Alg)}{n}$. We note $Cost(Alg)$ the expectation of $Cost_p(Alg)$ when p ranges on the set of patterns of length m.

One approach to the first step is to write down for every p the associated automaton [Bar85, BY89a, Han91] and compute its steady state, hence $Cost_p(Alg)$. As extensively discussed in [Rég89, Rég91], the combinatorial explosion of the number of states quickly makes this model untractable. We advocate in this paper that string searching performance evaluation reduces to word enumeration and should rely on combinatorics on words. We propose here to consider the *language* $\mathcal{L}_p$ associated to each automaton A_p [Eil74]. For example, assume $p = 0103$ is searched by Morris-Pratt algorithm. The associated automaton A_p has 4 states, with a backward (fail) transition from 4 to 2. It recognizes language $(01)^*0x, x \neq 1$. Success (fail) occurs if $x = 3$ $(x \neq 3)$. Multiple comparisons on a character occur only for backward transition from 4 to 2. Hence, we define the *cost function* ϕ for a word $w = (01)^m 0x$ to be $m - 1$ (resp. m) for a success (resp. a fail). Now, $E_Q(w) = (\frac{1}{q^2})^m \cdot \frac{1}{q^2}$. Hence, as every character is read at least once:

$$C_{0103}(MP) = 1 + \sum_{m \geq 1} (\frac{1}{q^2})^m \cdot \frac{1}{q^2}[(m-1) + (q-1)m] = 1 + \frac{1}{q(q^2-1)} .$$

More generally, the sum

$$\sum_{w \in \mathcal{L}_p} E_Q(w)\phi(w) \tag{3}$$

where $E_Q(w)$ is the probability for w to occur in the text represents the number of multiple comparisons. Finally, the sum

$$\sum_{p \in A^m} E_P(p) \sum_{w \in \mathcal{L}_p} E_Q(w)\phi(w) \tag{4}$$

yields the average cost $Cost(Alg)$. So far, the problem is not simplified, as language $\mathcal{L}_p$ may be large and the computational complexity of (3) and (4) still exponential. Nevertheless, we show below how this complexity can be reduced in a systematic manner. We define an *approximate automaton* A'_p and a cost function ϕ' on the recognized language $\mathcal{L}'_p$ providing a good approximation of (3) and (4).

Our main constraint on languages $\mathcal{L}'_p$ is the easiness of computation of (3) and (4). They will be defined from basic languages via word constructors such as concatenation, exponentiation,...Hence, our first step is to translate language constructors onto computation rules on the probability space. This will apply to the evaluation of (3).

Theorem 6. *Let $\mathcal{L}$ be a language on alphabet A defined as:*

(i) the union of two disjoint languages U and V,
(ii) the concatenation of two languages U and V, in a unique manner,
(iii) the repetition of words of U, l times.

Then, given $\mathcal{P}$, a matricial expectation couple for $\mathcal{L}$ can be computed from the expectation couples $(U^{(1)}, U^{(2)})$ and $(V^{(1)}, V^{(2)})$ by the following translation rules:

(i) If $U^{(1)} = V^{(1)}$, then:

$$(L^{(1)}, L^{(2)}) = (U^{(1)}, U^{(2)} + V^{(2)})$$

(ii) If $V = B.V'$, with $B \subset A$, then:

$$(L^{(1)}, L^{(2)}) = (U^{(1)}, U^{(2)} \circ \sum_{a \in B} Col_a(P) \circ V^{(2)}) ,$$

(iii) If $U = \{a\}.U'.\{b\}$, then:

$$(L^{(1)}, L^{(2)}) = (U^{(1)}, q_{b,a}^{l-1} U_l^{(2)}) .$$

Moreover, $U^{(2)}$ may be rewritten as $V^{(2)} \circ Col_b(P)$.

Hint for the proof: We first consider (ii) and assume $B = \{a\}$. Here:

$$w \in \mathcal{L}_{i,j} \Leftrightarrow w = uav', u \in \mathcal{L}_{i,k}, av' \in \mathcal{L}_{a,j} .$$

As u and v' are separated by one character, events $\{u\}$ and $\{v'\}$ are independent and we can rewrite:

$$P(av' \in \mathcal{L}_{a,j}/u \in \mathcal{L}_{i,k}) = q_{k,a} \cdot \frac{E_{\mathcal{Q}}(av' \in \mathcal{L}_{a,j})}{q_a} = q_{k,a} V_{a,j}^{(2)}$$

Hence, $\sum_{k,a} P(u \in \mathcal{L}_{i,k}, av' \in \mathcal{L}_{a,j}) = \sum_{k,a} U^{(1)} \circ Col_k(U^{(2)}) . q_{k,a} V_{a,j}^{(2)}$ also is $\sum_{k,a} U^{(1)} \circ Col_k(U^{(2)}) \circ Col_a(P) \circ V_{a,j}^{(2)} = U^{(1)} \circ U^{(2)} \circ \sum_a Col_a(P) \circ V^{(2)}$. Second, we consider (iii). First,

$$v = (aub)^l \in \mathcal{L}_{i,j} \Leftrightarrow a = a_i \text{ and } b = a_j .$$

Then, rewriting $aub = aa_1 \ldots a_{l-1} b$, we get:

$$E_{\mathcal{Q}}(v) = q_a (q_{a,a_1} \ldots q_{a_{l-1},b})^l q_{b,a}^{l-1}$$

Hence, $\sum_v E_{\mathcal{Q}}(v) = U^{(1)} \circ V_l^{(2)} . q_{b,a}^{l-1}$. $\qquad\qquad\qquad\qquad\qquad\qquad \square$

Let us consider here the special and important case of A, A^j and A^*. From the definition, the expectation couple for A, under measure $\mathcal{P}$ is $(S_{1,0}; Id)$. Applying first case of (ii) yields for A^2: $(S_{1,0}; P)$ and for A^j: $(S_{1,0}, P)$. Finally, applying (i) yields for $A^* = \epsilon + A + \ldots + A^j + \ldots$ the couple $(S_{1,0}; (I - P)^{-1})$.

Let us deal now with the evaluation of (4), where p ranges over all possible patterns of length m. Practically, all algorithms test a matching condition between p or a subword and t and some algorithmic decision is taken when a mismatch occurs. Hence, the computation of $E_{\mathcal{Q}}(w) E_{\mathcal{P}}(w)$ reduces to two cases only:

* $p = w$
* $p = p'a, w = p'b, b \neq a$.

Theorem 7. *Let $\mathcal{L}$ be a subset of A^* with transition couple $(L^{(1)}; L^{(2)})$. Assume probability $\mathcal{Q} \times \mathcal{P}$ on $A^* \times A^*$. Then:*

$$E_{\mathcal{Q} \times \mathcal{P}}(Diag(\mathcal{L} \times \mathcal{L})) = \mathcal{S}(U_{pq}^{(1)} \circ U_{pq}^{(2)}) \quad , \tag{5}$$

$$E_{\mathcal{Q}\mathcal{P}}(A \times A - Diag(A)) = \mathcal{S}(S_{0,1} - S_{1,1}) . \tag{6}$$

Finally, combining results above,

$$E_{\mathcal{Q}\mathcal{P}}(\mathcal{L} \times \mathcal{L}(A \times A - Diag(A)) = \mathcal{S}(U_{pq}^{(1)} \circ U_{pq}^{(2)} \circ (I - F)) . \tag{7}$$

Proof. To prove (5), let $w = a_1 a_2 \ldots a_j \in \mathcal{L}$. Then:

$$E_{\mathcal{Q}}(w) = q_{a_1} q_{a_1,a_2} \ldots q_{a_{j-1},a_j} ,$$

$$E_{\mathcal{Q}\mathcal{P}}(w, w) = (q_{a_1} p_{a_1}) . (q_{a_1,a_2} p_{a_1,a_2}) \ldots (q_{a_{j-1},a_j} p_{a_{j-1},a_j}) .$$

Hence, $E_{\mathcal{Q}\mathcal{P}}(Diag(\mathcal{L} \times \mathcal{L})) = \sum_w E_{\mathcal{Q}\mathcal{P}}(w, w)$ is derived from $\sum_w E_{\mathcal{Q}}(w) = \mathcal{S}(U^{(1)} \circ U^{(2)})$ by the set of substitutions: $q_{i,j} \to q_{i,j} p_{i,j}, q_i \to q_i p_i$ which yield precisely:

$\mathcal{S}(U_{pq}^{(1)} \circ U_{pq}^{(2)})$. Now, $E_{\mathcal{QP}}(A \times A - Diag(A)) = \sum_a \sum_b E_{\mathcal{Q}}(a) E_{\mathcal{P}}(b) - \sum_a E_{\mathcal{QP}}(a,a)$, i.e. $1 - \mathcal{S}(S_{1,1}) = \mathcal{S}(S_{0,1} - S_{1,1})$. Finally,

$$E_{\mathcal{QP}}(\mathcal{L} \times \mathcal{L}.(A \times A - Diag(A)) = \sum_{w \in \mathcal{L}, a \in A} E_{\mathcal{QP}}(w,w) - E_{\mathcal{QP}}(wa, wa) .$$

Applying twice (5), we get: $\mathcal{S}(U_{pq}^{(1)} \circ U_{pq}^{(2)} \circ (I - F))$.   □

3   Knuth-Morris-Pratt algorithm

We deal here with our first example: performance of Morris-Pratt algorithm and its variants, notably the Knuth-Morris-Pratt one [KMP77]. 3.1 defines a language $\mathcal{L}_p$ built by constructors of Theorem 6 that allows a good approximation of (3). This formalizes the former approach of [Rég89] and allows the definition of $Cost_p(MP)$ or $Cost_p(KMP)$. We assume Markovian distributions for text and pattern. I.e. p is allowed to range over all possible patterns.

3.1   An associated language

Our approach is based on the *memoryless property of Knuth-Morris-Pratt*. I.e. given a position in the text, the number of comparisons performed at that position only depends on a few left neighbours, and at most $|p|$. Hence, we will characterize the subsets of preceding words inducing k comparisons. This incremental definition of language $\mathcal{L}$ allows a fast convergence of the expression in (3). We first introduce the basic notion of quasi-mismatch.

Definition 8. Given p, a quasi-mismatch is a word $p'a$ such that:

$$p' \in A^{*+}, a \in A \text{ and } p' \preceq p, p'a \npreceq p$$

Let $\mathcal{L}_p^{(1)}$ be the subset of quasi-mismatches.

   Note first that any extra-comparison implies a quasi-mismatch. As a matter of fact, every character is read at least once. Extra-comparisons only occur when a comparison $a?p[j], j \neq 1$ ends with a mismatch. Denote $p' = p[1] \ldots p[j-1]$. Such an event implies that p' occurred as a matching sequence immediately to the left of a and that $p'a \npreceq p$. Second, several extra-comparisons on a character are taken into account by the appearance of several quasi-mismatches with various substrings p'. For instance, let $p = 012012013$ and $t = *012013 * **$. Then, aligning p with the first 0 would imply 2 extra-comparisons on 3 for which 2 quasi-mismatches are counted, associated to $p'_1 = 01201$ and $p'_2 = 01$. Hence, with cost function $\phi(w) = 1$, $\sum_{\mathcal{L}_p^{(1)}} E_{\mathcal{Q}}(w)$ provides a first approximation to (3). This *upper bound* is strict, as some quasi-mismatches may not be associated to multiple comparisons. As a matter of fact, two kinds of *false quasi-mismatches* may occur. First, a character that belongs to a matching sequence may be associated to a quasi-mismatch. For example, with the same p, let $t = **012012013 **$. Two quasi-mismatches are counted for 3, which belongs to a matching sequence. ¿From the Defect Theorem [Lot83], this first case is associated to language $\mathcal{L}_p^{(2)}$, defined below.

Definition 9. P being the set of primitive words, let

$$\mathcal{M}^{(2)} = \{z.w^l; w \in P, z \in A^{*+}, z \subseteq w, l \in \mathcal{N}\} ,$$

$$\mathcal{L}^{(2)} = \mathcal{M}^{(2)}.A, \; \mathcal{L}_p^{(2)} = \mathcal{L}^{(2)} \cap \{p - \text{prefixes}\} .$$

Let us consider the second case of false quasi-mismatches. For a mismatching character, it may happen for some of the associated quasi-mismatches (if more than one) that the associated comparison is "skipped". For example, let $s = a.baaba = aba.aba$ and $p = sc, c \neq a, c \neq b$, and $t = **sa**$. Two quasi-mismatches are counted, for only one extra-comparison. This second case is related to the multiple decomposition property defined by $\mathcal{L}_p^{(3)}$.

Definition 10. Given $l \in \mathcal{N}$, note $\mathcal{M}_l^{(2)}$ the restriction of $\mathcal{M}^{(2)}$ to decompositions $z.w^l$. Note:

$$\mathcal{M}^{(3)} = \mathcal{M}_1^{(2)} \cap (\cup_{l \geq 2} \mathcal{M}_l^{(2)}) ,$$

$$\mathcal{L}_p^{(3)} = \mathcal{M}^{(3)} \cap \{p - \text{prefixes}\}.A - \{p - \text{prefixes}\} .$$

We define $\phi_2(z.w^l a) = l$ if $a \neq w[1]$, else 0. Also, for $w = u(vu)^l.a = z(tz)a \in \mathcal{L}_p^{(3)}$, we note:

$$\phi_3(w) = \begin{cases} 0 & \text{if } a \npreceq vu, a \npreceq tz \\ l & \text{if } a \npreceq vu, a \preceq tz \\ l_1 & \text{if } a \preceq vu, a \npreceq tz, |u(vu)^{l_1-1}| < |z| < |u(vu)^{l_1}|. \end{cases}$$

It is beyond the scope of this extended abstract to detail these relations, derived from word combinatorics. All this formalizes in language terms the reasoning of [Rég89, Rég91] and steadily leads to:

Theorem 11. *With the notation above:*

$$Cost_p(MP) \equiv 1 + \sum_{w \in \mathcal{L}_p^{(1)}} E_Q(w) - \sum_{w \in \mathcal{L}_p^{(2)}} E_Q(w)\phi_2(w) - \sum_{w \in \mathcal{L}_p^{(3)}} E_Q(w).\phi_3(w) .$$

We are ready to give the performance of Morris-Pratt algorithm, from which performance of Knuth-Morris-Pratt variant follow. Deriving these expectations is the aim of next Section quite powerful, as the two first terms are easily computable and provide a tight approximation.

3.2 Knuth-Morris-Pratt performance

We group our results in Theorem 12, where an asymptotic development of the linearity constant is given. They only depend of the initial and stationary distributions, of the fundamental matrix of F and of length m. The order of approximation is a simple function of the distribution.

Notation: We note:

$$\alpha = q.max(\sum p_{c,a}q_{c,a}, \sum p_c q_c)$$

Theorem 12. *Assume Markovian distributions for the text and the pattern. The average number of comparisons performed by Morris-Pratt algorithm is linear and the linearity constant satisfies:*

$$c = 1 + \sum_a q_a p_a - \mathcal{S}(S_{1,1} \circ Diag(F) \circ (I - Diag(F))^{-2} \circ F) + O(\alpha^5) \ .$$

This reduces in the stationary case to:

$$c \sim 1 + (\sum_a p_a q_a)(1 - \sum_a (p_a q_a)^2) \ .$$

Proof. We first derive the dominating term. As detailed above, the expectation couple for A^j is $(S_{0,1}; Q^{j-1})$. Summing yields $\sum Q^{j-1} = (I - Q^{m-1}) \circ (I - Q)^{-1}$. Now, to take into account the condition $p'a \in t, p'a \not\preceq p$, we apply rule (7) and get:

$$E_{QP}(Diag(\cup_j A^j).(A \times A - Diag(A)) = \mathcal{S}(S_{1,1} \circ (I - F^{m-1})) = \sum_{a \in A} p_a q_a - \mathcal{S}(S_{1,1} \circ F^{m-1}) \ .$$

where the second term is $O(\alpha^m)$.

We gave a closed formula for $\sum_{w \in \mathcal{L}_p^{(2)}} E_Q(w)$ and uniform distributions in [Rég91], based on the generating function of primitive words. For biased distributions, less simplifications occur. Here, 1-order dependencies increase the complexity, as one must consider small words independently of the general case. Namely, let $\mathcal{L} = \{c^{l+1}, c \in A, l \geq 1\}$. Repeatedly applying rule (ii) yields the expectation couple $(Col_c(S_{0,1}); Col_c(Q)^l)$ for a given c. Concatenation with $a \neq c$ yields a multiplication by $Q - Col_c(Q)$ or $Q - Diag(Q)$. Applying now (7), we get:

$$\sum_{l \geq 1} l[S_{1,1} \circ Diag(F)^l \circ (F - Diag(F))] = S_{1,1} \circ Diag(F) \circ (I - Diag(F))^{-2} \circ (F - Diag(F)) \ .$$

We have neglected primitive words of size greater than 1. The smallest neglected sequence is $b.cb.a, a \neq c$ whose expectation is upper bounded by:

$$\mathcal{S}(S_{1,1} \circ [Diag(F^2) - Diag(F)^2] \circ F) = O(\alpha^5) \ .$$

Finally, we upper bound $\sum_{w \in \mathcal{L}^{(3)}} \phi_3(w) E_{QP}(w)$. Noting that the smallest sequences in $\mathcal{L}^{(3)}$ are $cbccbcc$ and $cbccbcb$, we get an approximation order α^{11}. $\square$

Further terms in the development can be obtained in a similar manner. This will be interesting for very biased distributions. We defer to an extended paper the discussion on the right choices of subsets of $\mathcal{L}^{(2)}$ to get the fastest convergence of the asymptotic development of the linearity constant as well as an efficient (notably non exponential!) computation of it. Finally, Knuth-Morris-Pratt performance are easily derived from above, by a slight modification of $\mathcal{L}_p^{(2)}$, as detailed for stationary distributions in [Rég91].

4 Boyer-Moore-Horspool algorithm

This section is devoted to the Horspool variant [Hor80] of Boyer-Moore [BM77]. The main difficulty comes from the disappearance of the memoryless property of Knuth-Morris-Pratt. The originality of our approach is to point out a stationary process. We proceed in two stages. First, we define a **head** [BYGR90] as a character read immediately after a shift and derive the expected number of heads in Theorem 13 and Theorem 4.2. Second, we consider the number of right to left comparisons starting from a given head. Results are joined in Theorem 14 and Theorem

4.1 Fundamental events

For Markovian distributions we have the extension of a theorem proved for uniform distributions in [BYGR90] and extended for stationary distributions in [Rég91].

Lemma 13. *For a given pattern p of length m, let H_l be the probability that the l-th character be a head. Then, H_l converges to a stationary probability H_p^∞ defined as the waiting time:*

$$H_p^\infty = \frac{1}{E_p[shift]} = \frac{1 + q_{p[m-1]} - q_{p[m-1],p[m-1]}}{\sum_{a \in A} P_a s_p(a)} .$$

with: $P_a = q_a + q_{p[m-1],a}(q_{p[m-1]} - q_a)$. Here, $s_p(a)$ is the shift performed when a is found as a head in the text and $E_p[shift]$ denotes the average shift when the aligned character ranges over the q values in the alphabet.

Proof. We consider the event $\{t[n] = a \text{ is a head}\}$ and $H_{n,a}$ its probability. Assume $b \in A$ occurred in the closest head to the left, say $n - j$. The complementary event occurs either when $1 \le j < s_p(b)$ or when $j = s_p(b)$ with the additional condition $t[n] \ne a$, of probability q_a or $q_{p[m-1],a}$. This yields a set of equations ,for $\{H_{n,a}\}$, defining stationary probabilities $\{H_{\infty,a}\}$ that satisfy in turn:

$$1 - H_{\infty,a} = \sum_{b \ne p[m-1]} H_{\infty,b}(s_p(b) - q_a) + H_{\infty,p[m-1]}(1 - q_{p[m-1],a}) \ , a \in A .$$

The problem reduces to find the inverse of $(1, \ldots, 1)$ in the $q \times q$ matricial equation so defined. One can easily check that $\left(\frac{P_a}{\sum_{a \in A} P_a s_p(a)}\right)_{a \in A}$ is a solution. Remark that we get in passing the probability of $s_p(a)$. □

We turn now to the number of right to left comparisons.

Theorem 14. *Given p, let us consider the following subsets of A^{m+1}:*

$$\mathcal{L}_p^{(i)} = \{w; p[m - i + 1] \sqsubseteq w\}, 1 \le i \le m ,$$
$$\mathcal{M}_p^{(i)} = \{w; p[m - i] \ldots p[m] \sqsubseteq w\}, 1 \le i \le m - 1 .$$

We note: $C_p^{(i)} = \sum_{w \in \mathcal{L}^{(i)}} E_Q(w)$, $D_p^{(i)} = \sum_{w \in \mathcal{M}^{(i)}} E_Q(w)$. Then:

$$Cost_p(BMH) = H_p^\infty[1 + C_p^{(1)} + 3D_p^{(1)} - C_p^{(1)}D_p^{(1)} - C_p^{(2)}D_p^{(1)} - C_p^{(3)}D_p^{(1)}1_{p[m-2] \ne p[m-1]}$$

$$+D_p^{(2)}1_{p[m-2]\neq p[m-1]}]+O(\sum_{i=2}^{m-1}\sum_{w\in M_p^{(i)}}E_Q(w))\ .$$

Proof. Given p and l a position in the text, the expected number of right to left comparisons starting in l is:

$$H_l.(1+\sum_{j\geq1}P(\#Comp>j))\ .$$

Denote w the word of size m ending at l. First, one notices that:

$$\cup_{j\geq3}\{\#Comp>j\}\subseteq\cup_{j\geq3}\{p[m-j+1]\ldots p[m]\subseteq w\}=\cup_{j\geq3}M_p^{(j)}\ .$$

Second, one writes:

$$\{\#Comp>1\ and\ l=head\}=\cup_{a\in A}\{p[m]=t[l],p[m-s_p(a)]=a\ and\ l-s_p(a)=head\}\ .$$

Let us denote:

$$\mathcal{L}_{p,a}=\{a\}.A^{|s_p(a)-1|}.\{p[m]\}\ .$$

Now, events to the right of $l-s_p(a)$ are independent of $\{l-s_p(a)=head\}$ that only depends on the past. Moreover, in the stationary state, $H_{l-s_p(a)}\equiv H_p^\infty$. Hence, we have:

$$E[\#right\ to\ left\ comparisons>1/l=head]=\sum_{w\in\mathcal{L}_{p,a}}E_Q(w)\ .$$

This is rewritten:

$$\sum_{w\in\mathcal{L}_p^{(1)}}E_Q(w)+\sum_{w\in\mathcal{L}_p^{(4)}}E_Q(w)-\sum_{w\in\mathcal{L}_p^{(1)}}E_Q(w)\sum_{w\in\mathcal{L}_p^{(2)}}E_Q(w)\ ,$$

as, for a 1-order Markov dependency, events $\{t[l-s_p(a)]=a\}$ and $\{p[m]=t[l]\}$ are independent except for $s_p(a)=1$. Notice that in the stationary case, the last two terms cancel [Rég91]. We proceed similarly for $E(\#Comp>2)$, defining :

$$\mathcal{L}_{p,a}=\{a\}.A^{|s_p(a)-2|}.\{p[m-1]p[m]\}\ ,a\neq p[m-1],\ and\ \mathcal{L}_{p,p[m-1]}=\mathcal{L}_p^{(4)}\ .$$

$\square$

4.2 Boyer-Moore-Horspool Performance

Theorem 15. *When p ranges over all patterns, one has:*

$$E_{\{allpatterns\}}[E_p(shift)]=\sum_a Col_a(S_{1,1})\circ P\circ(I-\overline{Line}_a(P))^{-2}=\frac{1}{\mathcal{H}}\ ,$$

which reduces to $\sum_{a\in A}\frac{q_a}{p_a}$ in the stationary case. Let H^∞ be $E_{\{all\ patterns\}}(H^\infty)$. Then:

$$\mathcal{H}\leq H^\infty\leq\frac{2}{q(q+1)min_a q_a}\ .$$

Finally, for the number of right to left comparisons, we get:

$$E_{\{allpatterns\}}(S_p^\infty)=1+\mathcal{S}(S_{1,1})+2\mathcal{S}(S_{1,1}\circ F)+O(\alpha^4)$$

Conjecture: The lower bound is tight, and $H_p^\infty - > \mathcal{H}$ when $q - > \infty$.

Proof. To derive the expectation, we consider for a given a the event $\{s_p(a) = k\} = \{a\}.(A - \{a\})^{k-1}.A \cap \{p - suffixes\}$. Hence, $P(s_p(a) = k) = Col_a(S_{1,1}) \circ Col_a(P)^{k-1} \circ P = Col_a(S_{1,1}) \circ P \circ \overline{Line}_a(P)^{k-1}$. Formal summation yields the result. Note that for stationary distributions, $P \circ \overline{Line}_a(P) = (1 - p_a)P$. The upper bound is a consequence of the straightforward lower bound for $E_p[shift]$: $1.q_{a_1} + 2.q_{a_2} + \ldots + q.q_{a_q} \geq \frac{q(q+1)}{2} \min_{a \in A} q_a$ [BYR92]. General inequality $E(1/s) \geq 1/E(s)$ yields the lower bound. Finally, we apply second order translation scheme to translate the result in Theorem 14. $\qquad\square$

Theorem 16. *When p ranges over all patterns, the average cost is:*

$$Cost(BMH) \sim E_{\{allpatterns\}}(H_p^\infty) \times (1 + ß1)$$

Proof. The main difficulty is the dependency between the events defining H_p^∞, i.e. the character positions in p, and the ones defining the number of right to left comparisons. Notably, we rewrite the event $\{\#Comp > 1 \text{ and } l \text{ is a head}\}$ as: $\cup_{a \neq p[m-1]}\{last\ head = a\ and\ p[m]\}$. The associated probability is $q_{p[m]}H_p^\infty + H_{\infty,p[m-1]}[q_{p[m-1],p[m]} - q_{p[m]}]$. First term contributes by $\sigma_1 E_{\{allpatterns\}}(H_p^\infty)$. We also have $E(q_{p[m-1]}(q_{p[m-1],p[m]} - q_{p[m]})) = O(\alpha^2)$. Finally, we isolate events involving $p[m-1]$ from others:

$$E\left(\frac{1}{\sum_{a \in A} P_a s_p(a)}\right) = E\left(\frac{1}{\sum_{a \neq p[m-1]} P_a s_p(a)}\right) + O\left(\frac{q_{p[m-1]}}{(\sum_{a \neq p[m-1]} P_a s_p(a))^2}\right) .$$

$\qquad\square$

5   Conclusion

We have presented an average analysis of the main string searching algorithms, assuming a 1-order Markov dependency between characters. We provide an answer to conjectures over the expected behavior of Knuth-Morris-Pratt and Boyer-Moore algorithms. The expected number of comparisons was proved to be asymptotically cn, and linearity constants were derived. In the second case, one has $c < 1$-around $1/q$ in the uniform case-; hence, we proved the conjectured *sublinearity*. An approach via word enumeration was proposed that proved to be powerful, as it "sticks" to the intrinseque nature of the algorithms. A challenging problem is now to determine different range domains for m, q and the *data distributions* so as the best algorithm may be chosen in any case, and notably the pathological distributions. The scheme should also apply to other algorithms that preprocess the pattern, such as the one in [CGG90] or multidimensional search [BYR90]. It is also worth extending that work to string searching with k mismatches.

References

[Bar85]   G. Barth. An analytical comparison of two string matching algorithms. *IPL*, 30:249–256, 1985.

[BM77]   R. Boyer and S. Moore. A fast string searching algorithm. *CACM*, 20:762–772, 1977.

[BY89a]   R. Baeza-Yates. Efficient text searching. PhD Thesis CS-89-17, Univ. Waterloo, Canada, 1989.

[BY89b]   R.A. Baeza-Yates. String Searching Algorithms Revisited. In *WADS'89*, volume 382 of *Lecture Notes in Computer Science*, pages 75–96. Springer-Verlag, 1989. Proc. WADS'89, Ottawa.

[BYGR90]   R. Baeza-Yates, G. Gonnet, and M. Régnier. Analysis of Boyer-Moore-type string searching algorithms. In *SODA'90*, pages 328–343. SIAM, 1990. Proc. Siam-ACM Symp. on Discrete Algorithms, San Francisco, USA.

[BYR90]   R. Baeza-Yates and M. Régnier. Fast algorithms for two dimensional and multiple pattern matching. In *SWAT'90*, volume 447 of *Lecture Notes in Computer Science*, pages 332–347. Springer-Verlag, 1990. Proc. Swedish Workshop on Algorithm Theory, Bergen, Norway.

[BYR92]   R. Baeza-Yates and M. Régnier. Average running time of Boyer-Moore-Horspool algorithm. *Theoretical Computer Science*, pages 19–31, 1992. special issue.

[CGG90]   L. Colussi, Z. Galil, and R. Giancarlo. On the exact Complexity of string matching. In *FOCS'90*, pages 135–143. IEEE, 1990. Proc. 31-st Annual IEEE Symposium on the Foundations of Computer Science.

[Eil74]   Samuel Eilenberg. *Automata, Languages, and Machines*, Volume A. Academic Press, 1974.

[Han91]   Ch. Hancart. Algorithme de Morris et Pratt et ses raffinements: une analyse en moyenne. Research report 91.56, Université de Paris VII, October, 1991.

[Hor80]   R. N. Horspool. Practical fast searching in strings. *Software-Practice and Experience*, 10:501–506, 1980.

[HU79]   J. E. Hopcroft and J.D. Ullman. *Introduction to Automata Theory*. Addison Wesley, Reading, Mass, 1979.

[KMP77]   D.E. Knuth, J. Morris, and V. Pratt. Fast pattern matching in strings. *SIAM J. on Computing*, 6:323–350, 1977.

[KR87]   R. Karp and M. Rabin. Efficient randomized pattern-matching algorithms. *IBM J. Res. Development*, 31:249–260, 1987.

[Lot83]   Lothaire. *Combinatorics on Words*. Addison-Wesley, Reading, Mass., 1983.

[Rég89]   M. Régnier. Knuth-Morris-Pratt algorithm: an analysis. In *MFCS'89*, volume 379 of *Lecture Notes in Computer Science*, pages 431–444. Springer-Verlag, 1989. Proc. Mathematical Foundations for Computer Science 89, Porubka, Poland.

[Rég91]   M. Régnier. Performance of String Searching Algorithms under Various Probabilistic Models, 1991. submitted. also as INRIA Research Report 1565.

[Sch88]   R. Schaback. On the Expected Sublinearity of the Boyer-Moore Algorithm. *SIAM J. on Computing*, 17:548–558, 1988.

[Tho68]   K. Thompson. Regular expression search algorithm. *CACM*, 11:419–422, 1968.

Pattern Matching With Mismatches:
A Probabilistic Analysis and a Randomized Algorithm
(Extended Abstract)

Mikhail J. Atallah[1], Philippe Jacquet[2], Wojciech Szpankowski[3]

[1] Dept. of Computer Science, Purdue University, W. Lafayette, IN 47907, USA
[2] INRIA, Rocquencourt, 78153 Le Chesnay Cedex, France
[3] Dept. of Computer Science, Purdue University, W. Lafayette, IN 47907, USA

Abstract. Given a text of length n and a pattern of length m over some (possibly unbounded) alphabet, we consider the problem of finding all positions in the text at which the pattern "almost occurs". Here by "almost occurs" we mean that at least some fixed fraction ρ of the characters of the pattern (for example, $\geq 60\%$ of them) are equal to their corresponding characters in the text. We design a randomized algorithm that has $O(n \log m)$ worst-case time complexity and computes with high probability all of the almost-occurrences of the pattern in the text. This algorithm assumes that the fraction ρ is given as part of its input, and it works well even for relatively small values of ρ. It makes no assumptions about the probabilistic characteristics of the input. Our second contribution deals with the issue of which values of ρ correspond to the intuitive notion of similarity between pattern and text, and this leads us to the development of a probabilistic analysis for the case where both input strings are random (in the usual, i.e., Bernoulli, model).

1 Introduction

Pattern matching is one of the most fundamental problems in computer science, and it has various applications in other areas of science, notably molecular biology and speech recognition (cf. [9], [27]). The version of this problem we investigate here is the following one. Consider two strings, a text string $\mathbf{a} = a_1 a_2 ... a_n$ and a pattern string $\mathbf{b} = b_1 b_2 ... b_m$, such that symbols a_i and b_j belong to a V-ary alphabet $\Sigma = \{1, 2, ..., V\}$, and V might be unbounded. Let C_i be the number of positions at which the string $a_i a_{i+1} ... a_{i+m-1}$ agrees with the pattern $\mathbf{b}$ (an index j that is out of

* The first author's research was supported by the Office of Naval Research under Grants N0014-84-K-0502 and N0014-86-K-0689, and in part by AFOSR Grant 90-0107, and the NSF under Grant DCR-8451393, and in part by Grant R01 LM05118 from the National Library of Medicine. The second author was supported by NATO Collaborative Grant 0057/89. The third author's research was supported by AFOSR Grant 90-0107 and NATO Collaborative Grant 0057/89, and, in part by the NSF Grant CCR-8900305, and by Grant R01 LM05118 from the National Library of Medicine.

range is understood to stand for $1+(j \bmod n)$). That is, $C_i = \sum_{j=1}^{m} equal(a_{i+j-1}, b_j)$ where $equal(x, y)$ is 1 if $x = y$, and is 0 otherwise. We are interested in computing all i at which $C_i \geq \rho m$, for some constant ρ $(0 < \rho \leq 1)$. We have two results.

One result is a randomized algorithm that has $O(n \log m)$ worst-case time complexity and computes with high probability all of the almost-occurrences of the pattern in the text. The constant ρ is supplied to the algorithm as part of its input, and the algorithm makes no assumptions about the probabilistic characteristics of the two input strings. We can prove that the error decays exponentially fast, and we also exhibit experimental data showing that the algorithm performs very well in practice.

Another result in this paper is an investigation of the maximum number (we call it $M_{m,n}$) of matches produced by comparison of two *random* strings; that is, $M_{m,n} = \max_{1 \leq i \leq n} \{C_i\}$ with a and b random. Recall that by "random string" we mean the usual Bernoulli model: *symbols from the alphabet Σ are generated independently, and symbol i from the alphabet Σ occurs with probability p_i at each position of the text or the pattern.* This analysis of $M_{m,n}$ could be useful in interpreting the output of the above-mentioned algorithm, or in selecting its input parameter ρ. In order for the algorithm's output to correspond to our intuitive notion of similarity between a pattern and a string, ρm should be significantly *larger* than $M_{m,n}$ (because if even random inputs give rise to a match of $M_{m,n}$, it is reasonable not to attach particular significance to this amount of "similarity" between two strings). This investigation of $M_{m,n}$ is analytically the more difficult of our two results. It is summarized in Theorem 2 (the more challenging part of which is the lower bound). In a nutshell, Theorem 2 shows that, for $\log n = o(m)$, $M_{m,n} = mP + \Theta(\sqrt{m \log n})$, where P is the probability of a match between any two symbols of the text and pattern strings. The proof of Theorem 2 is of independent interest, since it introduces a fairly general approach that can be used to analyze pattern matching in strings.

It is well known that the $O(n \log m)$ time performance for the problem is trivial to achieve in the case of an alphabet of size $O(1)$ (by using convolution), but for possibly large alphabets the best known deterministic algorithm runs in $O(n\sqrt{m}\text{polylog}(m))$ time [1]. Large alphabets arise in many practical situations, for example, when the text represents a time series of physical measurements, of prices, ..., etc. A number of powerful techniques were developed (see [20], [21], [22], [23], [13], [7]) for the more general problem where insertions and deletions are also allowed (i.e., not only mismatches), but since the number of mismatches under our definition of almost-occurrence is proportional to m, these techniques cannot be invoked here since all but that of [7] would result in a quadratic time bound (these methods were geared towards the situation where the number of mismatches k is $O(1)$, since they contain an $O(km)$ term in their time complexity). The elegant probabilistic method of [7] requires that both input strings be random, that the alphabet size be $O(1)$, and that $k < m/\log m + O(1)$, and hence cannot be used in our framework (recall that our algorithm makes no assumptions about the size of the alphabet or the probabilistic characteristics of the input strings, and does not require that k be a constant).

The paper is organized as follows. We next present our probabilistic result, and then we describe our randomized algorithm. Theorem 2 of the next section contains our main probabilistic result concerning the behavior of $M_{m,n}$. All proofs are delayed until Section 3 (we actually only sketch the proofs – the details are given in the full version).

2 Main Results

This section presents our main probabilistic and algorithmic results. The former is derived under the Bernoulli model, but the latter makes no such assumption.

2.1 Probabilistic Results

This section presents our main results derived under the Bernoulli model discussed above. Note that, in such a model, $P = \sum_{i=1}^{V} p_i^2$ represents the probability of a match between a *given* position of the text string **a** and a given one in the pattern string **b**. All our asymptotic results are valid in the so called *right domain asymptotic*, that is, for $\log n = o(m)$. In fact, the most interesting range of values for m and n is such that $m = \Theta(n^\alpha)$ for some $0 \le \alpha \le 1$. However, to cover the whole right domain asymptotics, we assume that α may also slowly converge to zero, that is, $1 \le \alpha^{-1} \le o(m/\log m)$.

Finally, for simplicity of the presentation, it helps to imagine that **a** is written on a cycle of size $n \ge 2m$. Then, we call $R^i(\mathbf{b})$ the version of **b** written on the same cycle, and cyclically shifted by i positions relative to **a**.

Note that C_i is the number of matches between **a** and **b** when **b** is shifted by i positions. In our cyclic representation, C_i can alternatively be thought of as the number of places on the cycle in which **a** and $R^i(\mathbf{b})$ agree. It is easy to see that the distribution of C_i is binomial, hence for any i

$$\Pr\{C_i = \ell\} = \binom{m}{\ell} P^\ell (1-P)^{m-\ell} . \tag{1}$$

Naturally, the average number of matches EC_1 is equal to mP. Furthermore, C_i tends *almost surely* to its mean mP (by the *Strong Law of Large Numbers* [10]).

Now we can present our first result regarding $M_{m,n} = \max_{1 \le i \le n}\{C_i\}$ which can be interpreted as the amount of similarity between **a** and **b**.

Theorem 1. *If m and n are in the right domain asymptotic, that is , $\log n = o(m)$, then*

$$\lim_{m \to \infty} \frac{M_{m,n}}{m} = P \qquad \text{(a.s.)} \tag{2}$$

Proof. A lower bound on $M_{m,n}$ follows from the fact that, by its definition, $M_{m,n}$ must be greater than C_1 which tends – by the *Strong Law of Large Numbers* – almost surely to mP. For the upper bound, we use the following inequality

$$\Pr\{M_{m,n} > r\} = \Pr\{C_1 > r \ \ or \ \ C_2 > r \ \ or \ \ ... \, C_n > r\} \leq n\Pr\{C_1 > r\} \, . \qquad (3)$$

It suffices to show that for $r \sim mP$ the above probability becomes $o(1)$, that is, $n\Pr\{C_1 > (1 + \varepsilon)mP\} = o(1)$. We note that C_1 can be represented as a sum of m independent Bernoulli distributed random variables X_i, where X_i is equal to one when there is a match at the ith position, and zero otherwise. From the *Central Limit Theorem* we know that $(C_1 - EC_1)/(\sqrt{mP(1 - P)}) \to \mathcal{N}(0, 1)$, where $\mathcal{N}(0, 1)$ is the standard normal distribution. Setting $r = mP + (1 + \varepsilon)\sqrt{m2P(1 - P)\log n}$ in (3), and using the above normal approximation, we finally obtain

$$\Pr\{M_{m,n} > mP + (1 + \varepsilon)\sqrt{2mP(1 - P)\log n}\} \leq \frac{1}{n^{\varepsilon}} \, . \qquad (4)$$

This proves the convergence in probability. The stronger almost sure convergence will be proved in the final version of the paper. ∎

Theorem 1 does not provide much useful information, and an estimate of $M_{m,n}$ based on it would be a very poor one. From the proof of Theorem 1 we learn, however, that $M_{m,n} - EC_1 = O(\sqrt{m\log n})$, hence the next term in the asymptotics of $M_{m,n}$ plays a very significant role, and definitely cannot be omitted in any reasonable computation. The next theorem – our main result – provides an extension of Theorem 1, and shows how much the maximum $M_{m,n}$ differs from the average EC_1. A sketch of its proof will be given in Section 3.

Theorem 2. *Under the hypothesis of Theorem 1 we also have, for every $\varepsilon > 0$,*

$$\lim_{n \to \infty} \Pr\{\underline{\delta}(1 - \varepsilon) \leq \frac{M_{m,n} - mP}{\sqrt{2m\log n}} < \bar{\delta}(1 + \varepsilon)\} = 1 \, , \qquad (5)$$

where

$$\underline{\delta} = \max\{\sqrt{(1 - \alpha)(P - T)}, \ \sqrt{\min\{(P - T), \ (1 - \alpha)\delta_1\}}\} \qquad (6)$$

$$\bar{\delta} = \min\{\sqrt{P - P^2}, \sum_{j=1}^{V} p_j \sqrt{1 - p_j}\} \, . \qquad (7)$$

In the above,

$$\delta_1 = \frac{(P - T)(P - 3P^2 + 2T)}{6(T - P^2)} \, ,$$

where $T = \sum_{i=1}^{V} p_i^3$. ∎

Remark 1. The lower bound (6) and the upper bound (7) are quite different. Neither of them is better than the other for the entire range of the probabilities $0 < p_i < 1$. With respect to the lower bound, we can show that there exist values of the probabilities $\{p_i\}_{i=1}^{V}$ such that either $\underline{\delta}^2 = (1-\alpha)(P-T)$ or $\underline{\delta}^2 = \min\{(P-T), (1-\alpha)\delta_1\}$. The latter case clearly can occur. Surprisingly enough, the former case may happen too. Indeed, let one of the probability p_i dominates all the others, e.g., $p_1 = 1-\varepsilon$. Then, we have either $\underline{\delta}^2 = (1-\alpha)\varepsilon + O(\varepsilon^2)$ or $\underline{\delta}^2 = \min\{2/3 \cdot (1-\alpha)\varepsilon, \ \varepsilon\} + O(\varepsilon^2)$. For α close to one the former bound is tighter. With respect to the upper bound, we have a similar situation. In most cases, $\bar{\delta}^2 = P - P^2$. However, sometimes $\bar{\delta} = \sum_{i=1}^{V} p_i \sqrt{1-p_i}$. Indeed, this occurs for example for a binary alphabet with $p_1 < 0.1$. $\square$

Note that, when all the $p_i \to 1/V$ (the symmetric case) we have $\delta_1 \to \infty$. Thus, for all $\alpha < 1$ we have $\underline{\delta} = \bar{\delta} = \sqrt{1/V - 1/V^2}$. Thus, we obtain the following corollary.

Corollary 3. *If the alphabet is symmetric and $\alpha < 1$, then for every $\varepsilon > 0$ the following holds*

$$\lim_{n \to \infty} \Pr\{1 - \varepsilon < \frac{M_{m,n} - mP}{\sqrt{2m(1/V - 1/V^2)\log n}} < 1 + \varepsilon\} = 1 \,, \tag{8}$$

that is, $M_{m,n} - mP \sim \sqrt{2m(1/V - 1/V^2)\log n}$ (pr.). ∎

In the last subsection we will extend the condition of the corollary to the limiting case $\alpha = 1$.

In order to verify our bounds for $M_{m,n}$ we have performed some simulation experiments. They confirm our theoretical result of Corollary 3. Namely, for the symmetric alphabet, $(M_{m,n} - mP)/\sqrt{m\log n}$ converges to a constant. But – surprisingly – this seems not to be true for an asymmetric alphabet. In fact, based on our simulations we conjecture that $(M_{m,n} - mP)/\sqrt{m\log n}$ converges *in probability* to a random variable, say Z, which is *not* a degenerate one, except for the symmetric alphabet. A study of the limiting distribution of Z seems to be very difficult. Our Theorem 2 provides some information regarding the behavior of the tail of the distribution of Z.

2.2 The Randomized Algorithm

We now present a randomized algorithm that in $O(n \log m)$ worst-case time finds, with high probability, all "almost-occurrences" of the pattern in the text. Recall that by the "almost-occurrence" we mean that the number of matches is at least ρm, that is, the fraction of matches in the pattern is at least ρ.

Although the algorithm works for any constant ρ, we now brief discuss what a suitable choice for ρ might be. Clearly, the main goal of pattern matching with mismatches is to identify a correlation (i.e., homology or similarity) between the pattern and the text (cf. [27], [9]). One can argue that in order for two strings to be correlated, and hence homologous, the similarity between them should be much higher than a correlation between two *randomly selected* strings. Based on our previous probabilistic analysis, we conclude that a "significant" correlation occurs if $\rho m \gg M_{m,n} \sim mP + \Theta(\sqrt{m \log n})$. This condition is typically satisfied even for moderate values of ρ because P is typically small (for example, in the symmetric case we have $P = 1/V$).

Now we are ready to present our randomized algorithm. The algorithm is conceptually very simple, is easy to implement, and performs very well in practice (see Table 1 below). Its main idea is to replace the problem with another problem whose alphabet size is $O(1)$, solve the latter in $O(n \log m)$ time, and from its answer deduce the answer to the original problem. The following is an outline of the algorithm.

1. Generate a random permutation π of $\{1, 2, ..., V\}$.
2. Go through the pattern and the text, replacing every symbol $j \in \Sigma$ encountered by the symbol $\lceil \pi(j)/L \rceil$ where L is an integer *constant* (how to choose L will be discussed in the full version of this paper — the probability of a correct answer increases with L, but so does the time complexity). Note that this second step produces a modified pattern **p** and text **q**, both over an alphabet of constant size L (i.e., $L = O(1)$ whereas the original alphabet size V could be arbitrarily large).[1]
3. Compute the $\widetilde{C}$ array whose ith entry (call it $\widetilde{C}_i$) is $\widetilde{C}_i = \sum_{\ell=1}^{m} equal(p_{i+\ell-1}, q_\ell)$ where $equal(x, y)$ is 1 if $x = y$, 0 otherwise. It is well known that computing such an array when the alphabet is of size $L = O(1)$ can be done within $O(Ln \log m) = O(n \log m)$ time and linear space; this was in fact one of the crucial ingredients in Abrahamson's scheme [1].
4. We go through the $\widetilde{C}$ array and output every i for which

$$\widehat{C}_i = \frac{\widetilde{C}_i - mQ}{1 - Q} \geq \rho m , \tag{9}$$

where Q is the probability of an "illegal match", that is, a match between symbols in the new alphabet of size L that does not have a corresponding "legal" match in the original alphabet of size V. A straightforward analysis reveals that $Q \approx 1/L$. A simple probabilistic analysis shows that the probability of error P_{er} defined as $P_{er} = \Pr\{\widehat{C}_i > \rho m \mid C_i = (\rho - \varepsilon)m\}$ is $O(e^{-\varepsilon^2 m})$, that is, it decays exponentially fast.

We have compiled extensive experimental data on the quality of the answer returned by the algorithm: the data displayed in Table 1 is typical of what we obtained. In that table, the "%Exact" entry is the exact percentage of the pattern that agrees with a typical position of the text at which it almost occurs (i.e., C_i/m),

[1] In the full paper, we also give an alternative scheme, with a different way of assigning new alphabet symbols to old alphabet symbols. It achieves the same complexity bounds as the above one, but with a slightly different analysis.

whereas the "%Estim" entry is the estimate of that percentage as given by our algorithm (i.e., $\widehat{C}_i/m$). Note that the algorithm performs extremely well in practice, even for moderate values of ρ and L; that it performs well for high values of these parameters is in agreement with intuition, but it is somewhat surprising that it does so good for moderate values of these parameters. In fact, even for fairly small values

Table 1. Simulation results for uniform alphabet with $n = 2000$, $m = 400$ and $V = 100$.

L	$\rho = 0.6$		$\rho = 0.7$		$\rho = 0.8$		$\rho = 0.9$	
	%Exact	%Estim.	%Exact	%Estim.	%Exact	%Estim.	%Exact	%Estim.
5	69.75	70.25	75.25	75.75	83.75	84.00	90.50	90.00
10	69.75	69.25	75.25	75.50	83.75	83.50	90.50	90.50
20	69.75	69.50	75.25	75.25	83.75	83.25	90.50	90.50
30	69.75	70.00	75.25	75.00	83.75	83.25	90.50	90.00
40	69.75	70.00	75.25	75.25	83.50	83.50	90.50	90.50
50	69.75	69.75	75.25	75.25	83.75	83.50	90.50	90.25

of ρ and L, the algorithm still finds the locations of the text where the pattern almost occurs, although in such cases the estimate of the actual percentage of agreement is not as good. If for some reason one needs both a small ρ and a small L, *and* a very accurate count of the percentage of agreement at each place of almost-occurrence, then one could view the output of the algorithm as an indication of *where* to apply a more accurate counting procedure (which could even be a brute force one, if there are relatively few such "interesting" positions in the text).

3 Analysis

In this section, we prove our main result, namely Theorem 2. In the course of deriving it, we establish some interesting combinatorial properties of pattern matching that have some similarities with the work of Guibas and Odlyzko [14, 15] (see also [17]).

Before we plunge into technicalities, we present an overview of our approach. An upper bound on $M_{m,n}$ is obtained from (3), or a modification of it, and the real challenge is in establishing a tight lower bound. The idea was to apply the *second moment method*, in the form of Chung and Erdös [8] which states that for events $\{C_i > r\}$, the following holds

$$\Pr\{M_{m,n} > r\} = \Pr\{\bigcup_{i=1}^{n}(C_i > r)\} \geq \frac{(\sum_i \Pr\{C_i > r\})^2}{\sum_i \Pr\{C_i > r\} + \sum_{(i \neq j)} \Pr\{C_i > r \ \& \ C_j > r\}} \ .$$

$$(10)$$

Thus, one needs to estimate the joint distribution $\Pr\{C_i > r \ \& \ C_j > r\}$. Ideally, we would expect that the right-hand side of (10) goes to one when $r = mP + (1 - \varepsilon)\sqrt{m2(P - P^2)\log n}$ for any $\varepsilon > 0$, which would match the upper bound.

Provided the above works, one needs an estimate of $\Pr\{C_i > r \ \& \ C_j > r\}$. This is a difficult task since C_1 and C_j are *strictly* positively correlated random variables for *all* values of $1 < j \leq n$. In fact, there are two types of dependency. For $j > m$, the patterns aligned at positions 1 and j do not overlap, but nevertheless C_1 and C_j are correlated since the same pattern is aligned. Although this correlation is not difficult to estimate, it is strong enough to "spoil" the second moment method. In addition, an even stronger correlation between C_1 and C_j occurs for $j < m$ since in addition overlapping takes place.

To resurrect the second moment method, we introduce a *conditional* second moment method. The idea is to fix a string σ of length m built over the same alphabet as the pattern $\mathbf{b}$, and to estimate C_1 and C_j under the condition $\mathbf{b} = \sigma$. Then, C_1 and C_j are *conditionally independent* for all $j > m$, and the dependency occurs only for $j \leq m$, that is, when C_1 and C_j overlap. This idea we shall develop below.

We introduce some notation. For a given σ, we define

$$G(r, \sigma) = \Pr\{C_1 > r | \mathbf{b} = \sigma\}, \tag{11}$$

$$F(r, \sigma) = \sum_{i=2}^{m} \Pr\{C_1 > r \ \& \ C_i > r \ | \mathbf{b} = \sigma\} \tag{12}$$

$$p(\sigma) = \Pr\{\mathbf{b} = \sigma\}. \tag{13}$$

Considering our Bernoulli model, $p(\sigma) = \prod_{j=1}^{j=V} p_j^{\Omega_j}$ where the Ω_j denotes the number of occurrence of the symbol j in the string σ. When σ varies, we may consider the conditional probabilities $G(r, \sigma)$ and $F(r, \sigma)$ as random variables with respect to σ. We adopt this point of view.

We shall start with some preliminary results regarding the tail of a binomially distributed random variable, that are used in the proof of our main results. If X is a binomially distributed random variable with parameters m and p, then we write $B(m, p, r) = \Pr\{X > r\}$ for the tail of X. We need the following result concerning the tail of C_1.

Lemma 4. *When m and τ both tend to infinity such that $\tau = O(\log m)$, then*

$$B(m, P, mP + \sqrt{m\tau}) = \Pr\{C_1 \geq mP + \sqrt{m\tau}\} \sim \frac{1}{\sqrt{2\pi(P - P^2)\tau}} \exp\left(\frac{-\tau}{2(P - P^2)}\right). \tag{14}$$

Proof: According to (1), C_1 is binomially distributed, that is, $\Pr\{C_1 = r\} = \binom{m}{r} P^r (1 - P)^{m-r}$. Introducing the generating function $C(u) = \sum_{r=1}^{m} \Pr\{C_1 = r\} u^r$ for u complex, we easily get the formula $C(u) = (1 + P(u - 1))^m$. Then, by Cauchy's celebrated formula [16]

$$\Pr\{C_1 \geq r\} = \frac{1}{2i\pi} \oint (1 + P(u-1))^m \frac{1}{u^r(u-1)} du , \tag{15}$$

where the integration is along a path around the unit disk for u complex. The problem is how to evaluate this integral for large m. In this case the best suited method seems to be the saddle point method [16, 11]. This method applies to integrals of the following form

$$I(m) = \int_{\mathcal{P}} \phi(x) e^{-mh(x)} dx , \tag{16}$$

where $\mathcal{P}$ is a closed curve, and $\phi(x)$ and $h(x)$ are analytical functions inside $\mathcal{P}$. We evaluate $I(m)$ for large m. The value of the integral does not depend on the shape of the curve $\mathcal{P}$. The idea is to run the path of integration through the saddle point, which is defined to be a place where the derivative of $h(x)$ is zero. To apply this idea to our integral (15) we represent it in the form of (16) and find the minimum of the exponent.

Using the above approach we obtain

$$\Pr\{C_1 \geq mP + \sqrt{m}x\} =$$

$$\exp\left(\frac{x^2}{2(P - P^2 + x/\sqrt{m})}\right) \frac{1}{2\pi} \int_{-\log m}^{\log m} \frac{\exp(-t^2/2)}{\frac{x}{\sqrt{P-P^2}} + it} dt \left(1 + O(1/\sqrt{m})\right) ,$$

which after some algebra, proves our result. ∎

To estimate the conditional probability $G(r, \sigma)$, we need some additional asymptotics for the binomial distribution, which are summarized in the lemma below.

Lemma 5. *Let* $r_m = mp + \sqrt{m\tau_m}$ *where* $\tau_m = O(\log m)$. *Then for all* $\rho > 1$ *there exists a sequence* $\gamma_m \to +\infty$ *such that the following hold*

$$\liminf_{m\to\infty} B(m - \gamma_m \sqrt{m}, p, r_m) \exp(\frac{\tau_m \rho}{2(p - p^2)}) \geq 1 \tag{17}$$

and

$$\limsup_{m\to\infty} B(m + \gamma_m \sqrt{m}, p, r_m) \exp(\frac{\tau_m}{\rho 2(p - p^2)}) \leq 1 . \tag{18}$$

Proof: These are computational consequences of Lemma 4. Details are deferred to the the final version of the paper. ∎

Our main preliminary result is contained in the next theorem. It provides (a.s.) bounds for the probability $G(r, \sigma)$.

Theorem 6. *Let* $r_m = mP + \sqrt{m\tau_m}$ *with* $\tau_m = O(\log m)$, *and* P *and* T *as defined in Theorem 2. For all* $\rho > 1$, *the following estimates hold:*

$$\lim_{m\to\infty} \Pr\{\sigma : G(r_m, \sigma) \geq \exp\left(-\frac{\tau_m \rho}{2(P-T)}\right)\} = 1, \tag{19}$$

$$\lim_{m\to\infty} \Pr\{\sigma : G(r_m, \sigma) \leq \exp\left(-\frac{\tau_m}{\rho 2 S^2}\right)\} = 1, \tag{20}$$

where $S = \sum_{j=1}^{V} p_j \sqrt{1 - p_j}$.

Proof: We only prove (19). For convenience of presentation we drop the subscript m from r_m and τ_m. Let $r_j = mp_j^2 + \alpha_j \sqrt{m\tau}$ with $\sum_{j=1}^{V} \alpha_j = 1$. By (17) of Lemma 5, for any $\rho > 1$ there exists a sequence $\gamma_m \to \infty$ such that the following holds for m large enough

$$\prod_{j=1}^{V} B(mp_j - \gamma_m \sqrt{m}, p_j, r_j) \geq \prod_{j=1}^{V} \exp(-\frac{\alpha_j^2 \tau \rho}{2 p_j^2 (1 - p_j)}).$$

Taking $\alpha_j = p_j^2 (1 - p_j)/(\sum p_i^2 (1 - p_i))$, after some algebra involving an optimization of the right-hand side (RHS) of the above, we obtain $(-\tau\rho/2(P-T))$ in the exponent of the RHS of the above. But, $G(r, \sigma) \geq \prod_{j=1}^{V} B(mp_j - \gamma_m \sqrt{m}, p_j, r_j)$ where σ is such that for all j: $\Omega_j \geq mp_j - \gamma_m \sqrt{m}$. Thus,

$$\Pr\{\sigma : G(r, \sigma) \geq \exp(-\frac{\tau\rho}{2(P-T)})\} \geq \Pr\{\forall j : \Omega_j \geq mp_j - \gamma_m \sqrt{m}\}.$$

By the Central Limit Theorem, every random variable Ω_j tends to the Gaussian distribution with mean mp_j and with the standard deviation $\sqrt{m(p_j - p_j^2)}$. So, it is clear that

$$\lim_{m\to\infty} \Pr\{\forall j : \Omega_j \geq mp_j - \gamma_m \sqrt{m}\} = 1$$

as $\gamma_m \to \infty$. ∎

Remark 2. If we denote by $\mathcal{G}_m$ the set of $\sigma \in \Sigma^m$ such that $G(r, \sigma) \geq \exp\left(-\frac{\tau\rho}{2(P-T)}\right)$, then (19) asserts that $\Pr\{\mathcal{G}_m\} \to 1$. As above, if we denote by $\mathcal{G}'_m$ a set of $\sigma \in \Sigma^m$ such that $G(r, \sigma) \leq \exp(-\frac{\tau}{\rho 2 S^2})$, then (20) implies that $\Pr\{\mathcal{G}'_m\} \to 1$.

Now, we are ready to establish our bounds. We begin with the upper bound. Part (i) of the theorem below was already proved in Theorem 1, while part (ii) follows directly from Theorem 6, and is omitted.

Theorem 7. *Let* $\varepsilon > 0$ *be an arbitrary non-negative real number. Then*

(i) *If* $\tau = 2(1 + \varepsilon)(P - P^2) \log n$, *then the following holds* $\lim_{n,m\to\infty} \Pr\{M_{m,n} < r\} = 1$ *with* $r = mP + \sqrt{m\tau}$.

(ii) *If* $\tau = 2(1 + \varepsilon)S^2 \log n$, *with* $S = \sum_{i=1}^{V} p_i \sqrt{1 - p_i}$, *then* $\lim_{n,m\to\infty} \Pr\{M_{m,n} < r\} = 1$ *with* $r = mP + \sqrt{m\tau}$. ∎

The lower bounds are much more intricate to prove. We start with a simple one that does *not* take into account overlapping. Then, we extend this bound to include the overlapping. This extension is of interest to us since it leads to the exact constant in the symmetric case.

In the first lower bound we ignore overlapping by considering $M_{m,n}^* \leq M_{m,n}$ where $M_{m,n}^* = \max_{1 \leq i \leq \lfloor n/m \rfloor} \{C_{1+im}\}$, that is, the maximum is taken over all nonoverlapping positions of a. Note also that C_1 and C_{1+m} are *conditionally* independent (i.e., under $\mathbf{b} = \sigma$). Our first lower bound is contained in the theorem below.

Theorem 8. *Let* $0 < \varepsilon < 1$. *If* $\tau = 2(1-\varepsilon)(P-T)(1-\alpha)\log n$ *and* $r = mP + \sqrt{m\tau}$, *then* $\lim_{n,m \to \infty} \Pr\{M_{m,n} < r\} \to 0$.

Proof. Due to conditional independence we can write

$$\Pr\{M_{m,n}^* < r\} = \sum_{\sigma \in \Sigma^m} p(\sigma)(1 - G(r,\sigma))^{n/m}$$

$$\leq \max_{\sigma \in \mathcal{G}_m} \{(1 - G(r,\sigma))^{n/m}\} + \Pr\{\sigma \notin \mathcal{G}_m\} \,,$$

where $\mathcal{G}_m$ is defined in Remark 2, that is, for $\rho = (1-\varepsilon)^{-1/2}$ we have $G(r,\sigma) > \exp(-\tau\rho/(2(P-T)))$ for all $\sigma \in \mathcal{G}_m$. Note also that $\Pr\{\sigma \notin \mathcal{G}_m\} \to 0$. We concentrate now on the first term of the above. Using $\log(1-x) \leq -x$, we obtain

$$(1 - G(r,\sigma))^{n/m} \leq e^{-(n/m)G(r,\sigma)} \,.$$

Now, it suffices to show that $(n/m)G(r,\sigma) \to \infty$ for $\sigma \in \mathcal{G}_m$. By Theorem 6, we have for $\sigma \in \mathcal{G}_m$

$$\frac{n}{m}G(r,\sigma) \geq \frac{n}{m}\exp\left(-\frac{\tau\rho}{2(P-T)}\right) \sim (n/m)^{1-\sqrt{1-\varepsilon}} \to \infty \,,$$

where the convergence is a consequence of our restrictions $\log n = o(m)$. This completes the proof of the "easier" lower bound in Theorem 2. ∎

The second lower bound is more elaborate. It requires an estimate of $\Pr\{C_1 > r \ \& \ C_i > r\}$ also for $i < m$. Let $F_{m,n} = \sum_{i=2}^m \Pr\{C_1 > r \ \& \ C_i > r\}$. Note that $F_{m,n} \leq F_{m,n}^* = \sum_{i=2}^m \Pr\{C_1 + C_i > 2r\}$. We need the following result which is proved in the full version of the paper.

Theorem 9. *For* $\tau_m = O(\log m) \to \infty$ *we have*

$$F_{m,n}^*(mP + \sqrt{m\tau_m}) \sim \frac{m(P - 3P^2 + 2T)^{5/2}}{2(T - P^2)\sqrt{\pi\tau_m^3}}\exp(-\frac{\tau_m}{P - 3P^2 + 2T}) \,. \tag{21}$$

for $m \to \infty$. ∎

Assuming that Theorem 9 is available, we proceed as follows. Using the *second moment method* we have

$$\Pr\{M_{m,n} > r\} \geq \sum_{\sigma \in \Sigma^m} p(\sigma)S(r,\sigma) , \tag{22}$$

where following (10)

$$S(r,\sigma) = \frac{(nG(r,\sigma))^2}{nG(r,\sigma) + nF(r,\sigma) + (n^2 - n(2m+1))(G(r,\sigma))^2} .$$

In the above, the probability $F(r,\sigma)$ is defined in (13) (to recall, $F(r,\sigma) = \sum_{i=2}^{m} \Pr\{C_1 > r \ \& \ C_i > r \ |\mathbf{b} = \sigma\}$). We rewrite $S(r,\sigma)$ as

$$S(r,\sigma) = \left(\frac{1}{nG(r,\sigma)} + \frac{F(r,\sigma)}{n(G(r,\sigma))^2} + 1 - \frac{2m+1}{n}\right)^{-1} . \tag{23}$$

We also have the following identity

$$\sum_{\sigma \in \Sigma^m} p(\sigma)F(r,\sigma) = F_{m,n}(r) ,$$

where the probability $F_{m,n}(r) \leq F_{m,n}^*(r)$ is estimated in Theorem 9. The almost sure behavior of $F(r,\sigma)$ – as a random function of σ – is discussed in the next lemma.

Lemma 10. *If $\tau \to \infty$, then for any constant β we have*

$$\lim_{m \to \infty} \Pr\{\sigma : F(r,\sigma) \leq \beta m \exp\left(-\frac{\tau}{P - 3P^2 + 2T}\right)\} = 1 . \tag{24}$$

Proof: It is a simple consequence of Markov's inequality: Since $\sum_{\sigma \in \Sigma^m} p(\sigma)F(r,\sigma) = F_m(r)$, it is clear that $\Pr\{F(r,\sigma) > \gamma_m F_m(r)\} \leq 1/\gamma_m$. ∎

Remark 3. We denote by $\mathcal{G}_m''$ the set of $\sigma \in \Sigma^m$ such that $F(r,\sigma) \leq \beta m \exp(-\frac{\tau}{P-3P^2+2T})$ holds. The lemma shows that $\lim_{m \to \infty} \Pr\{\sigma \in \mathcal{G}_m''\} = 1$. □

Now we are ready to prove the main result of this subsection that provides the most elaborate lower bound. Let δ_1 be defined as in Theorem 2, that is,

$$\delta_1 = \frac{(P-T)(P - 3P^2 + 2T)}{3(T - P^2)} .$$

Let also $\delta_2 = \min\{(1-\alpha)\delta_1, 2(P-T)\}$.

Theorem 11. *Let $0 < \varepsilon < 1$. If $\tau = (1-\varepsilon)\delta_2 \log n$, then $\lim_{m,n \to \infty} \Pr\{M_{m,n} > r\} = 1$ for $r = mP + \sqrt{m\tau}$.*

Proof: Note that it suffices to prove the theorem for any positive and arbitrarily small ε. By (22) and (23) we need to show that

$$nG(r,\sigma) \to \infty \tag{25}$$

$$\frac{F(r,\sigma)}{nG^2(r,\sigma)} \to 0 . \tag{26}$$

The first identity is easy to prove. For $\sigma \in \mathcal{G}_m$ by Theorem 6 with $\rho \leq (1-\varepsilon)^{-1/2}$ we have

$$nG(r,\sigma) \geq n^{1-1/2\sqrt{1-\varepsilon}\delta_2/(P-T)} \to \infty$$

since $\delta_2/(2(P-T)) \leq 1$.

Now, we deal with (26). Note that $\delta_1 > 0$ since $P^2 < T$. For $\sigma \in \mathcal{G}_m \cap \mathcal{G}_m''$ where $\mathcal{G}_m$ and $\mathcal{G}_m''$ are defined as in (respectively) Theorem 6 and Lemma 10 (with $\beta = 1$), we have

$$\frac{F(r,\sigma)}{nG^2(r,\sigma)} \leq \frac{m}{n} \exp\left(\frac{\tau\rho}{P-T} - \frac{\tau}{P-3P^2+2P}\right)$$

$$\leq \frac{1}{n^{1-\alpha-(1-\varepsilon)\delta_2/\delta_1}} \exp\left((\rho-1)\frac{\tau}{P-T}\right)$$

We know that $1 - \alpha - \delta_2/\delta_1(1-\varepsilon) \geq \varepsilon\delta_2/\delta_1$. Choosing $\rho - 1 = O(\varepsilon^2)$ in the above, we finally obtain

$$\frac{F(r,\sigma)}{nG^2(r,\sigma)} \leq \frac{1}{n^{1-\alpha-\delta_2/\delta_1(1-\varepsilon)+O(\varepsilon^2)}} \to 0 ,$$

since ε can be arbitrary small.

Putting everything together, we have just proved that $S(r,\sigma) \to 1$ for all $\sigma \in \mathcal{G}_m \cap \mathcal{G}_m''$. But, by (22), Theorem 6 and Lemma 10

$$\Pr\{M_{m,n} > r\} \geq \sum_{\sigma \in \Sigma^m} p(\sigma)S(r,\sigma)$$

$$\geq \Pr\{\sigma \in \mathcal{G}_m \cap \mathcal{G}_m''\} \min_{\sigma \in \mathcal{G}_m \cap \mathcal{G}_m''}\{S(r,\sigma)\} \to 1 ,$$

which completes the proof. ∎

References

1. K. Abrahamson, Generalized String Matching, *SIAM J. Comput.*, 16, 1039-1051, 1987.
2. Abramowitz, M. and Stegun, I., *Handbook of Mathematical Functions*, Dover, New York (1964).
3. A.V. Aho, J.E. Hopcroft and J.D. Ullman, *The Design and Analysis of Computer Algorithms*, Addison-Wesley, Reading, Mass., 1974.
4. Aldous, D., *Probability Approximations via the Poisson Clumping Heuristic*, Springer Verlag, New York 1989.
5. Arratia, R., Gordon, L., and Waterman, M., An Extreme Value Theory for Sequence Matching, *Annals of Statistics*, 14, 971-993, 1986.

6. Arratia, R., Gordon, L., and Waterman, M., The Erdös-Rényi Law in Distribution, for Coin Tossing and Sequence Matching, *Annals of Statistics*, 18, 539-570, 1990.
7. Chang, W.I. and Lawler, E.L., Approximate String Matching in Sublinear Expected Time, *Proc. 31st Ann. IEEE Symp. on Foundations of Comp. Sci.*, 116-124, 1990.
8. Chung, K.L. and Erdös, P., On the Application of the Borel-Cantelli Lemma, *Trans. of the American Math. Soc.*, 72, 179-186, 1952.
9. DeLisi, C., The Human Genome Project, *American Scientist*, 76, 488-493, 1988.
10. Feller, W., *An Introduction to Probability Theory and its Applications*, Vol. II, John Wiley & Sons, New York (1971).
11. Flajolet, P., Analysis of Algorithms, in *Trends in Theoretical Computer Science* (ed. E. Börger), Computer Science Press, 1988.
12. Galambos, J., *The Asymptotic Theory of Extreme Order Statistics*, John Wiley & Sons, New York (1978).
13. Galil, Z. and Park, K., An Improved Algorithm for Approximate String Matching, *SIAM J. Comp.*, 19, 989-999, 1990.
14. L. Guibas and A. Odlyzko, Periods in Strings *Journal of Combinatorial Theory*, Series A, 30, 19-43 (1981).
15. L. Guibas and A. W. Odlyzko, String Overlaps, Pattern Matching, and Nontransitive Games, *Journal of Combinatorial Theory*, Series A, 30, 183-208 (1981).
16. Henrici, P., *Applied and Computational Complex Analysis*, vol. I., John Wiley& Sons, New York 1974.
17. Jacquet, P. and Szpankowski, W., Autocorrelation on Words and Its Applications. Analysis of Suffix Trees by String-Ruler Approach, INRIA Technical report No. 1106, October 1989; submitted to a journal.
18. Karlin, S. and Ost, F., Counts of Long Aligned Matches Among Random Letter Sequences, *Adv. Appl. Probab.*, 19, 293-351, 1987.
19. Knuth, D.E., J. Morris and V. Pratt, Fast Pattern Matching in Strings, *SIAM J. Computing*, 6, 323-350, 1977.
20. Landau, G.M. and Vishkin, U., Efficient String Matching with k Mismatches, *Theor. Comp. Sci.*, 43, 239-249, 1986.
21. Landau, G.M. and Vishkin, U., Fast String Matching with k Differences, *J. Comp. Sys. Sci.*, 37, 63-78, 1988.
22. Landau, G.M. and Vishkin, U., Fast Parallel and Serial Approximate String Matching, *J. Algorithms*, 10, 157-169, 1989.
23. E.W. Myers, An O(ND) Difference Algorithm and Its Variations, *Algorithmica*, 1, 252-266, 1986.
24. Noble, B. and Daniel, J., *Applied Linear Algebra*, Prentice-Hall, New Jersey 1988
25. Seneta, E., *Non-Negative Matrices and Markov Chains*, Springer-Verlag, New York 1981.
26. Szpankowski, W., On the Height of Digital Trees and Related Problems, *Algorithmica*, 6, 256-277, 1991.
27. M. Zuker, Computer Prediction of RNA Structure, *Methods in Enzymology*, 180, 262-288, 1989.

Fast Multiple Keyword Searching

Jong Yong Kim and John Shawe-Taylor

Department of Computer Science
Royal Holloway and Bedford New College
University of London
Egham, Surrey TW20 0EX

Abstract. A new multiple keyword searching algorithm is presented as a generalization of a fast substring matching algorithm based on an n-gram technique. The expected searching time complexity is shown to be $O((N/m + ml)\log lm)$ under reasonable assumptions about the keywords together with the assumption that the text is drawn from a stationary ergodic source, where N is the text size, l number of keywords and m the smallest keyword size.

1 Introduction

The multiple keyword searching problem was introduced as a generalization of the substring matching problem. Instead of searching for a single pattern $\mathbf{p}$ in a text $\mathbf{t}$, a finite set of patterns $\mathbf{p}_1, \mathbf{p}_2, \ldots, \mathbf{p}_l$ is given. The output is the set of positions in the text where any one of the patterns occurs. As such it implements the *or* operator of regular expressions and lies somewhere between standard substring matching and the full implementation of regular expression searching described for example by Wu and Manber [14]. The main application areas are bibliographic search and text based information retrieval systems. Aho and Corasick [2] solved this problem using a deterministic finite automaton and the ideas of Knuth, Morris and Pratt [10]. To improve the complexity, Commentz-Walter [6] introduced the Boyer and Moore [5] technique to their algorithm, while Baeza-Yates [3] has used the Boyer, Moore and Horspool [7] technique.

Recently string matching algorithms based on n-gram techniques have been developed and complexity analyses have been performed under certain general statistical assumptions concerning the text strings [8, 9, 12]. The statistical model of language used is more general than previous models, in particular it does not require the symbols to be generated independently at random. Such an assumption is clearly inappropriate for natural language. For the substring searching algorithm [9], it was shown that the expected search time for a pattern of size m in a text of size N is $O((N/m + m)\log m)$. We present a generalization of this algorithm to the multiple keyword searching problem and also analyse its expected complexity.

The paper is organised as follows. In Section 2 we describe the details of the algorithm and data structure. Section 3 gives the background theory of General Sources, which we use in the complexity analysis in Section 4. In Section 5 experiments are reported which confirm the results of the complexity analysis, while the final section discusses conclusions and directions for further investigation.

2 Data Structure and Algorithm

The algorithm follows a similar strategy to the Knuth-Morris-Pratt or Boyer-Moore algorithms in that the first stage sees the construction of a data structure from the pattern which is then used to make the search stage more efficient. The data structure constructed is called *a reverse n-gram tree* and is a trie structure storing the majority of the keyword n-grams together with jump information. An n-gram is simply a contiguous substring of n characters, while the jump information is the jump that can safely be performed if the given n-gram is found aligned with the end of the keywords' current position. We now present the algorithm in more detail.

Algorithm 1. *The Multiple Keyword Searching Algorithm*

Stage 1: Construction of the reverse n-gram tree

Consider the multiple keywords placed artificially into an $l \times m$ two dimensional array called the pattern **p** discarding leftmost characters when aligned to the right, as shown in the Figure 1, where l is the number of keywords and m the size of the shortest keyword.

Fig. 1. Keywords aligned to the right

	$p_{1,1}$	$p_{1,2}$	——	$p_{1,m}$
	$p_{2,1}$	$p_{2,2}$	——	$p_{2,m}$
	$p_{3,1}$	$p_{3,2}$	——	$p_{3,m}$
	$p_{l,1}$	$p_{l,2}$	——	$p_{l,m}$

The *reverse n-gram tree* is a trie containing all the n-grams occurring in the pattern indicating the allowable shift for the rightmost occurrence of each. The *uni*-gram, *di*-gram, ..., $(n-1)$-gram before column n and their shifts are also included in the tree.

The n-grams are inserted from left to right starting with the *uni*-grams from each pattern and finishing with the rightmost n-grams for each pattern in turn. Standard trie insertion is used taking the characters of the n-gram in reverse order and with the following rule governing the computation of the shift values.

Rule

level 1 $\quad = m$
level j $\quad = \ell + 1$
$\qquad$ where $j =$ current level,
$\qquad$ and ℓ is the shift value at the parent node,

level *leaf*   = if $i \le (m - n)$ then $(m - i + 1)$
                    else $-row$
                    where $i =$ the leftmost column of n-gram,
                    $1 \le i \le (m - n + 1)$.

Fig. 2. *tri*-grams of a pattern and the reverse *tri*-gram tree

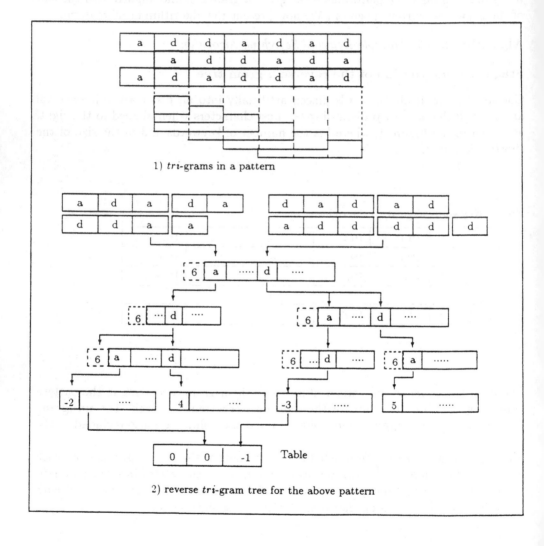

1) *tri*-grams in a pattern

2) reverse *tri*-gram tree for the above pattern

Any internal nodes that already exist should not be overwritten since they could have been leaf nodes for shorter n-grams at the left end of the patterns. Hence shift values are only inserted as nodes are created except at leaf nodes for the inserted n-gram, where the shift value is overwritten. This ensures that the shift is for the rightmost occurrence of the n-gram in the pattern, since the n-grams are inserted from left to right. The n-grams that occur in the rightmost position in the keywords have a shift value of $-row$, where row is the number of the keyword. If the node already exists, the previous value of the shift is written into the array $table$ at position row. In this way a linked list is created of rows which contain the given n-gram. The entries in the array $table$ are initialised to 0 so that a 0 indicates the end of the list for a particular n-gram. This list of rows is used for checking the keywords in turn. An example of the reverse tri-gram tree for three keywords and $n = 3$ is given in Figure 2.

In summary, the n-gram tree built by this method satisfies the following properties:

1. As the traversal path of an n-gram in the tree corresponds to the reverse order of the characters of the n-gram, a right to left text scanning order is possible.
2. The tree returns the shift allowable for the rightmost occurrence of an n-gram if it occurs in the pattern, or that of a suffix of the n-gram if it occurs as a prefix of any pattern rows.

Stage 2: Searching the text

Initially we begin to search for multiple keywords by aligning the pattern with the first m characters of the text. In general to test for matching at a particular position, we use the text n-gram which is aligned with the last n characters of the pattern to obtain a shift value by traversing the trie from the root using the characters in reverse order. If the next character is not present at a node in the trie, we return the shift value at that node. Figure 3 shows four kinds of possible shift values, which we describe in turn.

Case 1. *Neither the n-gram (nor its suffices) occurs in the pattern (pattern prefix).* There is no possibility that the n-gram becomes a substring of any keyword and we can safely move the keyword array completely beyond its present position. The n-gram path leading to a leaf is disconnected at some internal node which will return the shift value m plus the backtracking distance. This is the amount we must increment the index of the last character tested to obtain the rightmost index of the new position of the array.

Case 2. *An ℓ-gram suffix of the n-gram is a prefix of a pattern row.* The array can be shifted $m - \ell$ to the right to align the suffix with the prefix occurring from the first column of the pattern. The n-gram path stops at a leaf or is disconnected at some internal node where the shift value will be returned. The value will be $m - \ell$ plus the backtracking distance.

Case 3. *The n-gram occurs in the middle of the pattern.* Let i be the leftmost position of the (rightmost occurrence of the) n-gram in the pattern. To align the text n-gram with this occurrence, we must shift the pattern $m - i - n + 1$ positions to the right. Hence the index of the end of the new array position is obtained with the returned shift value of $m - i + 1$.

Case 4. *The n-gram occurs at the right end of the pattern.* This means that the
n-gram occurs as the suffix of at least one keyword. The leaf level of the tree
returns the negative shift value giving the index to the linked list of rows where
the same n-gram occurs as a suffix.

Fig. 3. New positions for each case

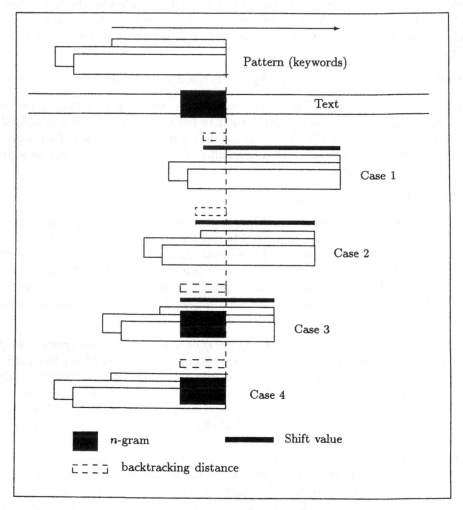

For the purposes of the complexity analysis the algorithm will be applied to the
text by dividing it into sections. In practise this will mean that full shift values

are not used in all cases. Hence in the practical experiments the algorithm was used as presented here. Though our interest is in expected complexity, it is natural to include the Commentz-Walter jump vector and make the maximum of the two jumps in order to guarantee that the search stage of the algorithm is $O(N)$ in a worst case.

3 Background Theory of Sources

In this section we review results from the theory of General Sources. The notation and results may be found in Welsh [13].

A *source* S is an object which emits symbols from a finite alphabet A according to some random mechanism. We will let $\mathbf{X} = (X_1, X_2, \ldots)$ denote the random elements of A emitted by S, where X_i denotes the i-th symbol emitted. We will denote the initial sequence of n characters by $\mathbf{X}_n$.

For a finite alphabet A, we denote by $A^{(n)}$ the set of all sequences of characters from A of length n. For a source S and a sequence $s \in A^{(n)}$ the probability that $\mathbf{X}_n = s$ is denoted by $P_S(\mathbf{X}_n = s)$. Where the source is clear from the context the subscript is often omitted. The *n-stage entropy* $H(X_1, \ldots, X_n)$ is defined to be the sum

$$H(X_1, \ldots, X_n) = \sum_{s \in A^{(n)}} -P(\mathbf{X}_n = s) \log P(\mathbf{X}_n = s).$$

We say a source has *entropy* H if

$$\lim_{n \to \infty} \frac{H(X_1, \ldots, X_n)}{n}$$

exists and equals H. A source is *stationary* if for any positive integers n and h, any $s \in A^{(n)}$ and any non negative indices $i_1, \ldots, i_n$,

$$P(X_{i_1} = s_1, \ldots, X_{i_n} = s_n) = P(X_{i_1+h} = s_1, \ldots, X_{i_n+h} = s_n).$$

This states that the behaviour of the language is time independent, that is that the probabilities of certain combinations of letters are shift independent. As a model of natural language such as English, it is plausible. We have the following result for stationary sources.

Theorem 2. *[13] Any stationary source has entropy.*

For a source S and a sequence s, the *frequency* $f_S^N(s, \mathbf{X})$ is the random variable giving the number of occurrences of s in the first N characters produced by the source. Again if the source is clear from the context this subscript can also be suppressed. A source S is *ergodic* if it is stationary and if, for any finite sequence $s \in A^{(k)}$, we have

$$\lim_{N \to \infty} N^{-1} f^N(s, \mathbf{X}) = P(\mathbf{X}_k = s).$$

This is requiring that $f^N(s, \mathbf{X})/N$ converges with probability 1 to the required constant. Note also that as a result of the stationary property the probability on the right hand side is independent of its position. The following theorem is a powerful result describing the probability distribution of n-grams in stationary ergodic sources.

Theorem 3. *Let S be an ergodic source with entropy $H > 0$. Then, for any $\epsilon > 0$, there exists a positive integer $N_{(\epsilon)}$ such that, if $n > N_{(\epsilon)}$, the set $A^{(n)}$ of possible n-sequences of the source alphabet decomposes into two sets Π and T satisfying*

$$P(\mathbf{X}_n \in \Pi) < \epsilon,$$

$$2^{-n(H+\epsilon)} < P(\mathbf{X}_n = \mathbf{s}) < 2^{-n(H-\epsilon)}$$

for any n-sequence $\mathbf{s} \in T$.

For a proof we refer to Billingsley [4].

4 Expected Complexity

In this section we will give the basic complexity results for the multiple keyword algorithm. As mentioned above we artificially divide the text into sections of length $m - n + 1$, apart from the first and last section. The first section is chosen with length $m - 1$ and the last section contains whatever number of characters remain. We now apply the algorithm to each section in turn positioning the pattern so that its last character is aligned with the first character of the section. When a jump moves the end of the pattern beyond the end of the section, we align the pattern at the beginning of the next section and continue. In this way the last shift made in each section is potentially smaller than it could be. Throughout the complexity analysis we will assume this adaption of the algorithm is used.

Proposition 4. *Assume that for a stationary ergodic source S, there is a constant α, with $0 < \alpha < 1$, such that for sufficiently large n, every n-gram $\mathbf{g}$ contained in the keywords satisfies,*

$$P(\mathbf{X}_n = \mathbf{g}) \leq \alpha^n.$$

In this case the expected running time for Algorithm 1 is

$$O((N/m + lm) \log lm)$$

for l keywords with minimum size m when a text of size N is generated by S.

Proof: We choose the n-gram size to be $n = -2 \log lm / \log \alpha$ and assume that ml is sufficiently large for n to be in the range where the bound on the probability of individual n-grams holds. Smaller values of lm may be covered by suitable adaptation of the constants involved in the upper bound. The first stage of the algorithm is to construct the trie of n-grams occurring in the pattern array with corresponding shift values. This can clearly be done in time $O(mln) = O(ml \log lm)$.

The key to the complexity bound on the main part of the algorithm is to estimate the probability that the n-gram $\mathbf{g}$ which is tested for the i-th block of the algorithm occurs in the pattern array. If this is not the case we can safely conclude that the pattern does not match anywhere within the block and can move to the next block of the text. In order to gain a lower bound on this probability, we seek initially an upper bound on the probability p^* that a randomly chosen n-gram $\mathbf{g}$ does occur in the pattern array. Since the pattern contains less than lm n-grams and each n-gram has probability less than

$$\alpha^n = \alpha^{-2\log lm/\log \alpha} = 1/(lm)^2,$$

we can bound the probability p^* by $1/(lm)$. In order to be able to use this figure in our analysis of the algorithm, we must consider applying the algorithm $m - n + 1$ times to any particular input. In each case the blocking is shifted by one, so that the blocking boundaries occur once between all pairs of characters. This also means that each n-gram in the input will be used once as a test n-gram for a block. As the ergodic property implies that relative frequencies converge to probabilities, the expected number of occurrences of the pattern array n-grams in the text is equal to their probability times the text string length. This means that summing over the $m - n + 1$ applications of the algorithm the expected total work will be bounded by

$$N(cm(1/(lm)) + 2\log lm),$$

where we have assumed a worst case of $cm = O(m)$ time for the application of the algorithm to the block in the case where the test n-gram does occur in the pattern. Using the stationarity of the source, we can average the work equally over the different applications of the algorithm. This implies that the expected amount of work for one application of the search phase of the algorithm is

$$N(c/l + 2\log lm)/(m - n + 1) = O((N/m)\log lm).$$

The result follows. ∎

Corollary 5. *Consider a stationary ergodic source S with non-zero entropy H. Provided that for sufficiently large lm the keywords are chosen so that their $((4/H)\log lm)$-grams fall into the set T of the AEP property for S for $\epsilon = H/2$, the expected running time of the substring search algorithm is*

$$O((N/m + lm)\log lm).$$

Proof: Consider $\epsilon = H/2$ in the AEP property and consider m such that $n = (4/H)\log lm > N(\epsilon)$ given in the theorem. For n-grams in the set T of the theorem we can bound their probability by

$$2^{-n(H-\epsilon)} = \alpha^n,$$

for $\alpha = 2^{-H/2}$. By assumption all n-grams in the search pattern array satisfy this property. ∎

Note that the results so far are expected running time for any patterns satisfying the given requirements. We can also average over different possible search patterns to obtain more general results.

Corollary 6. *Consider a stationary ergodic source S with non-zero entropy H. Provided that for sufficiently large lm the keywords are chosen so that the probability that any of their $((4/H)\log lm)$-grams falls into the set Π of the AEP property for S for $\epsilon = H/2$ is less than $(\log lm)/(lm^2)$, then the expected running time of the Algorithm 1 averaged over all sets of keywords is*

$$O((N/m + lm)\log lm).$$

Proof: The probability that there is an $(n = (4/H)\log lm)$-gram from the set Π of the AEP property in at least one position of the pattern keywords is bounded by $(\log lm)/m$. Assuming that in this case the algorithm has a worst case performance, that is takes $O(N + lm\log lm)$ time, we obtain on averaging with the occasions when no n-gram from Π is contained in the pattern (as covered in Corollary 5) an expected running time of

$$(N + lm\log lm)/(m/\log lm) + (N/m + lm)\log lm = O((N/m + lm)\log lm),$$

as required. ∎

5 Experimental Results

We performed experiments in the practical range of patterns using 1Mb of DNA, and 1 Mb of natural language randomly selected from the Bible. The DNA alphabet size is four: (A,C,G,T). Only the patterns which have the same size of column and row are sought because these seem to represent the average performance of the algorithm well. The pattern $l \times m$ is built from l contiguous parts of the text which are selected randomly but not overlapping each other. Figure 4 gives the actual running time (Search.time) measured by the Unix *times* function. The running time is averaged in seconds taken in seeking 200 random patterns in each of the data on a Sun4 workstation during low system workload. The column (Qty) is the quantity $(search.time \times m)/\log(l \times m)$. The theoretical predictions given in the preceding section indicate that this will be a constant over different pattern sizes. The derived quantity is not constant but is remarkably consistent with the predictions.

The size n is increased at the point where the running time degrades significantly compared with the previous result, considering the above theoretical predictions.

6 Conclusions

We have presented an algorithm for exact pattern matching of multiple keywords with expected running time fast under the assumption that the text is produced by stationary ergodic source together with some restrictions on the n-grams occurring in the keywords. These restrictions govern the probability of the n-grams in

Fig. 4. Search Time of DNA and Natural Language in Seconds on Sun4

Language				DNA			
$l \times m$	Size n	Search.time	Qty	$l \times m$	Size n	Search.time	Qty
6×6	4	62.92	105	14×14	6	56.72	150
7×7	4	58.75	105	16×16	6	52.17	151
8×8	4	55.67	107	18×18	6	49.57	154
9×9	4	52.80	108	20×20	7	45.90	153
10×10	4	50.23	109	22×22	7	44.02	157
11×11	5	48.48	111	24×24	7	41.85	158
12×12	5	47.05	113	26×26	8	39.72	158
13×13	5	45.03	114	28×28	8	37.47	157
14×14	6	43.02	114	30×30	8	35.90	158
15×15	6	41.82	115	32×32	8	34.20	158

the stationary ergodic source. In most cases of interest all n-grams of a source will satisfy the property. This can be seen by considering the Markov approximation to a source, given by generating the next character according to the previous k characters for some constant k. Provided the stationary distribution is taken as the initial distribution the Markov process will deliver a stationary ergodic sequence for which all n-grams satisfy the requirements of Proposition 4 and hence the expected complexity result holds in all cases. Such sources provide an arbitrarily good statistical approximation to the statistics of natural language.

Searching for boolean combinations of keywords has played an important role in text information retrieval systems. Multiple keyword searching has been implemented as part of *grep* in the Unix system and *agrep* [14] which covers regular expressions, wildcards and even typographical errors. These packages are flexible for use in wide variety of applications but not comparable to our algorithm in their expected computational complexity.

The n-gram technique has been successfully applied to a number of string searching problems including approximate string matching, the substring matching problem, two dimensional array matching and here the multiple keyword searching problem. The technique has a strong theoretical background in the theory of general sources, which allows prediction of the expected complexity under natural assumptions about the distribution of inputs. These results appear to predict more accurately the behaviour of the algorithms than those afforded by standard worst case complexity analysis.

References

1. A.V. Aho: Algorithms for finding patterns in strings, Handbook of theoretical computer science. Vol A edited by J.van Leewen (1990) Elsevier 257-300.
2. A.V. Aho and Margaret J. Corasick: Efficient String Matching: An Aid to Bibliographic Search. Comm. ACM **18** (1975) 330–340.
3. R.A. Baeza-Yates: Fast algorithms for two dimensional and multiple pattern matching. Springer-Verlag LNCS 447 SWAT90 332–347.

4. P. Billingsley: Ergodic Theory and Information. John Wiley and Sons 1965.
5. R.S. Boyer and J.S. Moore: A Fast String Searching Algorithm. Com. ACM **20** (1977) 762-772.
6. B. Commentz-Walter: A string matching algorithm fast on the average, Proc 6th international Colloquium on Automata, Languages and programming, Springer-Verlag (1979) 118-132.
7. R.N. Horspool: Practical fast searhing in strings. Software practice and experience **10** (1980) 501–506.
8. J.Y. Kim and J. Shawe-Taylor: An Approximate String Matching Algorithm. Theoretical Computer Science **92** (1992) 107–117.
9. J.Y. Kim and J. Shawe-Taylor: Fast Expected String Matching using an n-gram Algorithm. Departmental Technical Report CSD-TR-91-16.
10. D.E. Knuth, J.H. Morris and V.R. Pratt: Fast Pattern Matching in strings. SIAM. J. Comput. **6** (1977) 323-350.
11. J. Shawe-Taylor: Fast String Matching in a Stationary Ergodic Source. Departmental Technical Report CSD-TR-633.
12. J. Shawe-Taylor and J.Y. Kim: Fast Two dimensional Pattern Matching. Departmental Technical Report CSD-TR-91-25.
13. D. Welsh: Codes and Cryptography. Oxford University Press (1988).
14. S. Wu and U. Manber: Fast Text Searching With Errors. Department of Computer Science TR91-11 Arizona University.

Heaviest Increasing/Common Subsequence Problems

Guy Jacobson and Kiem-Phong Vo

AT&T Bell Laboratories
600 Mountain Avenue
Murray Hill, NJ 07974

Abstract. In this paper, we define the *heaviest increasing subsequence* (HIS) and *heaviest common subsequence* (HCS) problems as natural generalizations of the well-studied *longest increasing subsequence* (LIS) and *longest common subsequence* (LCS) problems. We show how the famous Robinson-Schensted correspondence between permutations and pairs of Young tableaux can be extended to compute heaviest increasing subsequences. Then, we point out a simple weight-preserving correspondence between the HIS and HCS problems. ¿From this duality between the two problems, the Hunt-Szymanski LCS algorithm can be seen as a special case of the Robinson-Schensted algorithm. Our HIS algorithm immediately gives rise to a Hunt-Szymanski type of algorithm for HCS with the same time complexity. When weights are position-independent, we can exploit the structure inherent in the HIS-HCS correspondence to further refine the algorithm. This gives rise to a specialized HCS algorithm of the same type as the Apostolico-Guerra LCS algorithm.

1 Introduction

Given a sequence σ over some linearly ordered alphabet, the *longest increasing subsequence* (LIS) problem is to find a longest subsequence of σ that is strictly increasing. Given two sequences α and β over some general alphabet, the *longest common subsequence* (LCS) problem is to find a longest sequence γ that is a subsequence of both α and β.

Both the LIS and LCS problems have venerable histories. The LIS problem arises in the study of permutations, Young tableaux and plane partitions. These objects play central roles in the representation theory of the symmetric group initiated by Young[25] and MacMahon in the early part of the century. In 1938, Robinson[16] found an explicit correspondence between permutations and pairs of Young tableaux. This correspondence was rediscovered in 1961 by Schensted[18] who extended it to general integer sequences. There are many interesting results concerning the Robinson-Schensted correspondence. The reader is referred to other papers[20, 5, 22, 12] for more details. Schensted's main motivation was to compute an LIS from a given sequence of integers. This can be done by specializing the algorithm to compute only the left-most column of the Young tableau. Fredman[4] has shown that $O(n \log n)$ time is required to compute an LIS. Thus, the Robinson-Schensted algorithm is optimal for LIS. In a different guise, LIS can be used to compute a largest stable set for a permutation graph[6]. In this guise, LIS has a ready generalization: computing a heaviest stable set for a permutation graph whose nodes have non-uniform weights. This is a special case of the *heaviest increasing subsequence* (HIS) problem: Given a sequence over some linearly ordered alphabet and a weight function on the symbols and their positions in the sequence, find a subsequence with the heaviest sum of weights. We shall show how to generalize the Robinson-Schensted algorithm to compute an HIS in $O(n \log n)$ time.

The LCS problem was first studied in the context of the *string-to-string correction* problem. Wagner and Fischer[23] solved this problem using dynamic programming in quadratic time and space. Since then, LCS has found many practical applications: CRT screen updates, file differential comparison, data compression, spelling correction, and genetic sequencing[17]. It was the CRT screen update problem that led one of us (Vo) in 1983 to look into a weighted extension for LCS called the *minimal distance LCS* (MDLCS) problem. Here, a typical screen update involves two screens, the current one, and the desired one. The update algorithm must match the screen lines and issue hardware line insertion/deletion to align matched lines. To reduce screen disturbance, it is desirable that matched lines that are closely aligned be given preference over other matches. Thus, the minimal distance weight function assigns higher weights to closely aligned matched lines. The MDLCS problem is to find among all LCS's one that minimizes the total distances of all matched lines. Therefore, the MDLCS weight function combines the length of the common subsequence and the distance between their matches. This is an example of a class of general weight functions that assign values from some *ordered additive monoid* to common subsequences based on both matched symbols and their positions in the original sequences. The *heaviest common subsequence* (HCS) problem is to find common subsequences that maximize such weights. The dynamic programming algorithm is easily extended to solve HCS. This algorithm was implemented in the **curses** screen update library distributed with System V UNIX systems[21].

Aho, Hirschberg and Ullman[1] showed that quadratic time is needed to find the length of an LCS when the computation model allows only equal-unequal comparisons. On the other hand, in a more general computation model, Fredman's lower bound for the LIS problem gives a lower bound of $O(n \log n)$ for computing the LCS of two sequences of length n. This can be seen as follows. Let π be a permutation of the integers from 1 to n. A LCS between π and the sequence $1, 2, \ldots, n$ is also an LIS of π. The first general subquadratic algorithm for LCS was found by Masek and Patterson[13], who employed a "Four-Russians" approach to solve the problem in $O(n^2 / \log n)$ for finite alphabets and $O(n^2 \log \log n / \log n)$ for a general alphabet. The question of whether $O(n \log n)$ is a tight bound for the LCS problem is still open.

More recent work on the LCS problem focused on finding general algorithms whose efficiency is a function of certain characteristics of the problem instance.

Hunt and Szymanski[10, 11] gave an $O((r + n) \log n)$ algorithm where r is the total number of *matches*; a match is an ordered pair of positions (i, j) such that $\alpha_i = \beta_j$. This is efficient when the matches are sparse. Apostolico and Guerra[2] improved on the Hunt-Szymanski algorithm and described an $O(d \log n)$ algorithm where d is the number of *dominant matches*; a dominant match is an ordered pair (i, j) such that $\alpha_i = \beta_j$ and every LCS of the prefixes $\alpha_1 \cdots \alpha_i$ and $\beta_1 \cdots \beta_j$ has α_i as its final symbol. The quantity d is important because the number of dominant matches can be much smaller than the number of matches.

Another line of research into LCS seeks an algorithm that is fast when α and β are similar. This is practical, for example, if they are two versions of a text file. Myers[14] describes an $O(n\Delta)$ time algorithm, where Δ is the edit distance between the two strings (with unit cost insertion and deletion). Interested readers should see the work of Hirschberg[8], Nakatsu et al.[15], Hsu and Du[9], Wu et al,[24], and Chin and Poon[3] for other approaches.

The LIS and LCS problems are closely related. In fact, both the Hunt-Szymanski and Apostolico-Guerra algorithms are implicitly based on a correspondence between increasing and common subsequences. We shall state explicitly this bijective correspondence and note that it is weight-preserving. This allows us to map common subsequence problems to increasing subsequence problems and make use of the machinery in the Robinson-Schensted correspondence. It can be seen from this that the Hunt-Szymanski LCS algorithm is only a special case of the Robinson-Schensted algorithm. Then, it is easy to see how the inherent structure in the LCS problem can be exploited in the mapping to tune the Hunt-Szymanski algorithm. This refinement results in the Apostolico-Guerra LCS algorithm. In fact, this process of discovery uncovers a bug in the original description given by Apostolico and Guerra. Finally, applying our extension of the Robinson-Schensted algorithm for computing HIS, we derive fast algorithms for computing HCS.

2 Basic Definitions and Notations

To make the paper self-contained, in this section, we give all the basic definitions and specify conventions for how the algorithms will be presented.

2.1 Sequences and Subsequences

Let $\sigma = \sigma_1\sigma_2\cdots\sigma_p$ be a sequence over some alphabet A. We shall use σ_i to denote the ith symbol of σ, and $\sigma_{i...j}$ to denote the contiguous subsequence consisting of symbols in positions from i to j. A sequence τ is called a subsequence of σ if there is a sequence of integers $i_1 < i_2 < \cdots < i_l$ such that τ is equal to $\sigma_{i_1}\sigma_{i_2}\cdots\sigma_{i_l}$. If the alphabet A is linearly ordered, we say that τ is an increasing subsequence if $\tau_1 < \tau_2 < \cdots < \tau_l$. Given two sequences α and β, a sequence γ is called a common subsequence of α and β if it is both a subsequence of α and a subsequence of β. That is, there are two sequences of integers $i_1 < i_2 < \cdots < i_l$ and $j_1 < j_2 < \cdots < j_l$ such that γ is equal to $\alpha_{i_1}\alpha_{i_2}\cdots\alpha_{i_l}$ and $\beta_{j_1}\beta_{j_2}\cdots\beta_{j_l}$.

2.2 Dominant Matches vs. Edit Distance

A number of modern LCS algorithms have their complexity based on either the edit distance or the number of dominant matches. The edit distance between the strings α and β is the minimum number of character insertions and deletions required to transform α to β. Let the strings α and β have a LCS of length ρ and an edit distance of Δ. We will always have $|\alpha| + |\beta| = 2\rho + \Delta$, because each symbol in α or β but not in the LCS increases the edit distance by one.

Now, define a match (i, j) of the two letters $\alpha_i = \beta_j$ to be dominant if every LCS of $\alpha_{1...i}$ and $\beta_{1...j}$ must end at positions i and j. Following standard conventions, we denote the total number of matches by r and the number of dominant matches by d. Other authors[9, 2] have observed that d can be *much* smaller than r, especially when the two strings are very similar. We will now make this observation rigorous by proving a bound on d based on the edit distance Δ.

Let $\rho(i, j)$ denote the length of the LCS of prefixes $\alpha_{1...i}$ and $\beta_{1...j}$; similarly let $\Delta(i, j)$ denote the edit distance between the prefixes $\alpha_{1...i}$ and $\beta_{1...j}$. Say that a dominant match (i, j) is k-dominant if $\rho(i, j) = k$.

Theorem 1. *The number of dominant matches $d \leq \rho(\Delta + 1)$.*

Proof. Suppose there are d_k k-dominant matches. Sort them by increasing values of i: $\{(i_1, j_1), (i_2, j_2), \ldots, (i_{d_k}, j_{d_k})\}$ where $i_1 < i_2 < \ldots < i_{d_k}$ and $j_1 > j_2 > \cdots > j_{d_k}$. Now because the i's are strictly increasing integers and the j's are strictly decreasing, $i_l - i_1 \geq l - 1$ and $j_l - j_{d_k} \geq d_k - l$.
Now consider the edit distance $\Delta(i_l, j_l)$:

$$\Delta(i_l, j_l) = i_l + j_l - 2\rho(i_l, j_l) = i_l + j_l - 2k.$$

Because (i_1, j_1) is a k-dominant match, $i_1 \geq k$ and similarly, $j_{d_k} \geq k$. So

$$\Delta(i_l, j_l) \geq i_l + j_l - i_1 - j_{d_k}$$

rearranging:

$$\Delta(i_l, j_l) \geq (i_l - i_1) + (j_l - j_{d_k}).$$

Now we can use the inequalities derived earlier to get:

$$\Delta(i_l, j_l) \geq (l - 1) + (d_k - l) = d_k - 1.$$

Now consider the particular value of k with the largest number of k-dominant matches. A LCS of α and β can be constructed using only dominant matches, and then it must use one of these k-dominant matches, say (i_l, j_l). Now if (i_l, j_l) is a match that is used in the LCS of α and β, then $\Delta \geq \Delta(i_l, j_l)$. Therefore $\Delta + 1 \geq d_k$. Now since d_k is at least as great as any other d_l, for $1 \leq l \leq \rho$ then $\rho d_k \geq d$. Combining these two inequalities, we get $\rho(\Delta + 1) \geq d$.

A corollary of this theorem is that Apostolico and Guerra's $O(d \log n)$ LCS algorithm is also bounded by $O(n \log n\Delta)$ time, and so is never more than a log factor slower than Myer's $O(n\Delta)$ algorithm.

2.3 Ordered Additive Monoids as Weight Systems

As we have seen with the minimal distance LCS problem, the weight of a matched pair of symbols may not be just a simple value but can be made up from different components. For MDLCS, the components are the length of a common subsequence and a weight based on the distance among matched symbols as measured by their positions in the given sequences. Therefore, it is necessary to talk about more general weight systems. To this end, we define a class of objects called *ordered additive monoids*. An ordered additive monoid is a triple $(M, +, \leq)$ such that:

1. M is a set with a distinguished element 0.
2. For all $x \in M$, $x + 0 = x$.
3. For all $x, y \in M$, $x + y = y + x \in M$.
4. For all $x, y, z \in M$, $(x + y) + z = x + (y + z)$.
5. M is linearly ordered with respect to $\leq$.
6. For all $x, y \in M$, $x \leq x + y$ and $y \leq x + y$.

A simple example of an ordered additive monoid is the non-negative real numbers. A more nontrivial example is the monoid defined on the power set 2^S of a given set S. In this case, $+$ is set union, and $\leq$ can be taken as any linear extension of the partial order on 2^S defined by inclusion. As seen with the MDLCS problem, of interest to us is the fact that given two ordered additive monoids $(M, +_M, \leq_M)$ and $(N, +_N, \leq_N)$, we can construct a new ordered additive monoid on the set $M \times N$ by defining for all (u, v) and (x, y) in $M \times N$:

1. $(u, v) + (x, y) = (u +_M x, v +_N y)$.
2. $(u, v) \leq (x, y)$ if $u \leq_M x$ or $u = x$ and $v \leq_N y$.

As MDLCS suggested, we need to consider weight functions that depend on both symbols and their positions. Let N be the non-negative integers. Let A be an alphabet and M be an ordered additive monoid. Let M^+ be the set of non-zero elements of M. For increasing subsequence problems, a weight function is a map $\omega : N \times A \mapsto M^+$. The weight of a sequence σ over A is defined as $\sum_{1 \leq k \leq l} \omega(k, \sigma_k)$.

For common subsequence problems, there are two involved sequences over the alphabet A, α and β. A weight function is a function $\omega : N \times N \times A \mapsto M^+$ where M is again some ordered additive monoid. The weight of a common subsequence γ is defined as: $\sum_{1 \leq k \leq l} \omega(i_k, j_k, \gamma_k)$. Here i_k and j_k are the indices of matched pairs as they appear in the original sequences α and β.

It is not hard to see that the dynamic programming method for LCS can be extended to solve the HCS problem. Let Ω_{ij} be the weight of an HCS of $\alpha_{1...i}$ and $\beta_{1...j}$. Define w_{ij} as 0 if $\alpha_i \neq \beta_j$ and $\omega(i, j, \alpha_i)$ if the two symbols are the same. Below is the recursion to compute Ω_{ij}:

$$\Omega_{ij} = \max(\Omega_{i-1,j}, \Omega_{i,j-1}, \Omega_{i-1,j-1} + w_{ij})$$

2.4 Algorithm Presentation

We shall present algorithms in a pseudo-C syntax. Each algorithm is described with line numbers which are used in subsequent discussions. Frequently, we need to deal with ordered lists. Given a list L of objects of certain type, we shall require the following operations on L:

insert(L,o): insert the object o into the list L.
delete(L,o): delete the object o from the list L.
next(L,o): find the least element strictly larger than o in L.
prev(L,o): find the largest element strictly smaller than o in L
max(L): find the maximal element in L.
min(L): find the minimal element in L.

An important note on these list operations is that, using balanced tree structures, they can all be performed in $O(\log n)$ time where n is the number of objects involved. In practice, we use splay trees[19]. They are simple to implement, use less space, and work just as well as balanced trees. In the algorithms, ϕ will stand for some undefined object. C programmers may think of this as the NULL pointer. The operations next(), prev(), max() and min() when not defined will return ϕ. next() and prev() do not require that the argument object be already in L but it has to be of the right type. For convenience, next(L,ϕ) is equivalent to min(L). Similarly, prev(L,ϕ) is equivalent to max(L).

3 Computing a Heaviest Increasing Subsequence

The Robinson-Schensted algorithm computes a pair of tableaux from a sequence. For the purpose of computing an LIS, we don't need the entire algorithm, only the part that computes the left-most column of the left tableau. Figure 1 shows the simplified LIS algorithm.

```
 1. lis(σ₁σ₂···σₙ)
 2. {   L = φ;
 3.       for(i = 1; i <= n; i = i+1)
 4.       {   s = prev(L,σᵢ);
 5.           t = next(L,s);
 6.           if(t != φ)
 7.               delete(L,t);
 8.           insert(L,σᵢ);
 9.           node[σᵢ] = newnode(σᵢ, node[s]);
10.       }
11. }
```

Fig. 1. The Robinson-Schensted LIS algorithm

Remarks on Figure 1
2:   This line initializes the left-most column L of the Young tableau.
4:   This line computes an element s in L where the current symbol can be appended while maintaining the invariant that L is strictly increasing.
5-7: These lines replace the element after s with σ_i. In tableau parlance, t is *bumped* by σ_i.
8:   node is an auxiliary array that, for each element in L, contains a record of an element that precedes this element in an increasing subsequence. The function newnode() constructs such records and links them into a directed graph. At the end of the algorithm, we can search from the maximal element of L to recover an LIS of σ.

Note that in the lis() algorithm, at any given time, the length of an LIS of the prefix of σ considered thus far is kept implicitly as the height of the list L. For the weighted case, we must maintain the weight of an HIS of a prefix explicitly. Thus, the elements of L are pairs (s, w) with $s \in \sigma$ and w is the total weight of an HIS ending with s. We maintain the invariant that L is strictly increasing in both coordinates. Therefore, the ordering based on their first coordinate (the alphabet ordering) can be used to order L. Figure 2 shows the HIS algorithm.

```
 1. his(σ₁σ₂···σₙ, Ω)
 2. {   L = φ;
 3.     for(i = 1; i <= n; i = i+1)
 4.     {   (s,v) = prev(L,(σᵢ,0));
 5.         (t,w) = next(L,(s,v));
 6.         while((t,w) != φ)
 7.         {   if( v+Ω(i,σᵢ) < w )
 8.                 break;
 9.             delete(L,(t,w));
10.             (t,w) = next(L,(t,w));
11.         }
12.         if((t,w) == φ || σᵢ < t)
13.         {   insert(L,(σᵢ,v+Ω(i,σᵢ)));
14.             node[σᵢ] = newnode(σᵢ, node[s]);
15.         }
16.     }
17. }
```

Fig. 2. A $O(n \log n)$ HIS algorithm

Remarks on Figure 2

4:     prev() computes the largest element (s,v) in L such that s is strictly smaller than σ_i. This means that σ_i can be appended to any increasing subsequence ending at s to define a new increasing subsequence. If there is no such (s,v), we define v to be 0.

5-15: These lines replace lines 5–9 of the lis() algorithm. Bumping is done in the while() loop between lines 6–11. This ensures the invariant that the second coordinates of objects in L are strictly increasing. Line 12 tests to see if $(\sigma_i, v + \Omega(i, \sigma_i)))$ can be inserted into the list L while maintaining the invariant that the first coordinates of objects are strictly increasing. This test is needed because our weight function is also based on the indices of symbols. It can be omitted if the weight function only depends on the symbols. Line 13 does the actual insertion. Line 14 constructs a record so we can recover an actual HIS when the algorithm terminates.

To see how the algorithm runs, consider the sequence 9,2,6,1,1,2,5 in which all elements have their integral values as weights except that the first 1 has weight 2. Below is the progression of the list L as elements are processed:

9	2	6	1	1	2	5
9,9	2,2	2,2	1,2	1,2	1,2	1,2
	9,9	6,8	6,8	6,8	2,4	2,4
		9,9	9,9	9,9	6,8	5,9
					9,9	

Now, consider the list L after each iteration of the `for(;;)` loop. For convenience, we shall use L_i to denote the state of L after the ith iteration. Each element `(s,v)` on L defines an increasing subsequence ending at `s` by tracing the links created on line 14. We shall say that this sequence is defined by `s`.

We claim that for every increasing subsequence $s_1 s_2 \cdots s_k$ of $\sigma_{1\ldots i}$, there is an element $t \leq s_k$ in L_i that defines an increasing subsequence that is at least as heavy as $s_1 s_2 \cdots s_k$. From this, it follows that the maximal element of L_i defines an HIS of $\sigma_{1\ldots i}$. Thus, when the algorithm ends, the maximal element of L defines an HIS for the entire sequence σ.

We prove the claim by induction on `i`, the index variable of the `for(;;)` loop. The case `i=1` is clear. Now, assume the assertion for `i-1` and consider an increasing subsequence $s_1 s_2 \cdots s_k$ of $\sigma_1 \cdots i$. Consider the case when $s_k = \sigma_i$. By induction, there is an element $t \leq s_{k-1}$ in L_{i-1} that defines an increasing sequence that is at least as heavy as $s_1 \cdots s_{k-1}$. Since $t < \sigma_i$, t cannot be bumped off L on lines 6-11. After the ith iteration, either σ_i was inserted into L or it is already in L. Since L is strictly increasing in the weights, the sequence defined by σ_i satisfies the claim. So, assume that $s_k \neq \sigma_i$. Now there are two cases. The case $s_k < \sigma_i$ follows immediately since the part of L preceding σ_i is unchanged in the ith iteration. Assume $s_k > \sigma_i$, by the induction hypothesis, there is a sequence defined by some t in L_{i-1} that is at least as heavy as $s_1 \cdots s_k$. Now, either t is still in L_i and we are done, or t was bumped off L in the `while()` loop of lines 6-11. In this case, the `if()` statement of line 7 guarantees that the sequence defined by σ_i will be at least as heavy as the sequence defined by t (in the $i - 1$st step). This complete the proof of the claim. Therefore, the algorithm `his()` is correct.

To analyze the time complexity of `his()`, we observe that all operations in each iteration of the `for(;;)` take $O(\log n)$ time. Since the loop iterates n times, the total time is $O(n \log n)$. We have proved:

Theorem 2. *Let σ be a sequence over a linearly ordered alphabet A and Ω a weight function from A to some ordered additive monoid M. Algorithm* `his`(σ, Ω) *computes a heaviest increasing subsequence of σ in time $O(n \log n)$ where n is the length of σ.*

4   Computing a Heaviest Common Subsequence

Let N be the set of natural numbers. A *biletter* is an element of the set $N \times N$. Given an instance of a common subsequence problem, i.e., two sequences over some alphabet A, $\alpha = \alpha_1 \alpha_2 \cdots \alpha_m$ and $\beta = \beta_1 \beta_2 \cdots \beta_n$, we can construct from these sequences a corresponding *biword* (sequence of biletters) as in Figure 3.

For example, given the sequences `abac` and `baba`, `biword(abac,baba)` will construct the biword:

```
1 1 2 2 3 3
4 2 3 1 4 2
```

It is not hard to see by induction on i that every common subsequence of α and β maps to an increasing subsequence of the lower word of the biword. On the other hand, given an increasing subsequence of the lower word of the biword, it is easy to invert the indices and retrieve a common subsequence between α and β.

```
1. biword(α,β)
2. {   B = φ;
3.     for(i = 1; i <= m; i = i+1)
4.     {   Let P be the list of positions of αᵢ in β;
5.         for(k = max(P); k != φ; k = prev(P,k))
6.             append (i,k) to B;
7.     }
8. }
```

Fig. 3. The HCS-HIS correspondence

Let $\Omega : N \times N \times A \mapsto M$ be a weight function. We assign the weight of a matched pair of symbols to the lower part of the corresponding biletter. Then, `biword()` is a weight preserving function that maps bijectively every common subsequence of α and β to an increasing subsequence of the corresponding biword with the same weight.

Applying the `his()` algorithm from the last section to the lower part of the biword, we immediately have an algorithm for computing HCS. Of course, in practice there is no need to ever construct the biword explicitly. It can be generated from the lists of positions in β of the given symbols. The Hunt-Szymanski algorithm is essentially `lis()` where the biword is constructed on the fly. Figure 4 shows `hcs1()`, an algorithm for HCS. In the same way that `his()` is a generalization of the Robinson-Schensted algorithm `lis()`, this algorithm is a generalization of the Hunt-Szymanski algorithm.

The correctness of `hcs1()` follows immediately from that of `his()` and the discussions on the correspondence. The run time of `hcs1()` depends on r, the total number of matches between α and β, and the size of the list L which is less than $\min(n, m)$. Assume that $n < m$, we have:

Theorem 3. *Let α and β be two given sequences over an alphabet A. Let $\Omega : N \times N \times A \mapsto M$ be a weight function. Algorithm* `hcs1`(α, β, Ω) *computes a heaviest common subsequence between α and β in time* $O((r + m) \log n)$.

To see the algorithm working, consider the sequences `abca` and `aabd`. Let the weight of a matched symbol at positions `i` and `j` be the ordered pair $(1, 4 - |i - j|)$. For example, the weight of the match of the symbol `b` is $(1, 3)$. This is the MDLCS weight function. The corresponding biword is:

1 1 2 4 4
2 1 3 2 1

Below is the progress of the list L as the biletters are processed. The algorithm shows that the HCS is the sequence `ab` at positions 1,2 of `abca` and at positions 1,3 of `aabd`.

(1,2)	(1,1)	(2,3)	(4,2)	(4,1)
2,(1,3)	1,(1,4)	1,(1,4)	1,(1,4)	1,(1,4)
		3,(2,7)	3,(2,7)	3,(2,7)

```
1.  hcs1(α₁α₂···αₘ,β₁β₂···βₙ,Ω)
2.  {   for(i = 1; i <= n; i = i+1)
3.          insert(Position[βᵢ],i);
4.      L = φ;
5.      for(i = 1; i <= m; i = i+1)
6.      {   P = Position[αᵢ];
7.          for(j = max(P); j != φ; j = prev(P,j))
8.          {   (s,v) = prev(L,(j,0));
9.              (t,w) = next(L,(s,v));
10.             while((t,w) != φ)
11.             {   if(v+Ω(i,j,αᵢ) < w)
12.                     break;
13.                 delete(L,(t,w));
14.                 (t,w) = next(L,(t,w));
15.             }
16.             if((t,w) == φ || j < t)
17.                 insert(L,(j,v+Ω(i,j,αᵢ)));
18.         }
19.     }
20. }
```

Fig. 4. An $O(r \log n)$ HCS algorithm

Remarks on Figure 4

2-3: For each symbol in β, an ordered list of its positions is constructed.

4: This line initializes the list L as in algorithm his(). Each object to be stored in L compose from the lower part of a biletter (i.e., the matched index in β) and the weight of some corresponding common subsequence defined by this symbol.

5-7: Here, the two for(;;) loops essentially construct the corresponding biword on the fly. Note that the decreasing order processing of indices of matched symbols on line 7 is crucial for the correctness of the algorithm. This points out a bug in the original Apostolico-Guerra algorithm which traverses the lists of matched indices in increasing order.

8-17: These lines are straightforward translation of lines 4-15 in algorithm his(). For clarity, we omitted the construction of the linked list to retrieve an HCS. Note that on line 8, the function call prev(L,(j,0)) works because we are assuming that the implementation orders L by the matched indices only, not the weights. The while() loop on lines 10-15 ensures that the weights are strictly increasing on L.

5   Tuning the HCS Algorithm

The algorithm $hcs1(\alpha,\beta,\Omega)$ can be tuned further if we know more about the weight function Ω. This section considers a few main cases in which weights are known to follow some regular patterns. The complexity analyses of the tuned algorithms is based on weighted *dominant matches* which are defined as follows: A match $\alpha_i = \beta_j$ is dominant if every HIS of $\alpha_{1..i}$ and $\beta_{1..j}$ must end at α_i and β_j. Again, following standard conventions, we let d be the total number of weighted dominant matches.

Recall from section 2 that if Ω_{ij} is defined as the weight of a HIS between $\alpha_{1...i}$

and $\beta_{1\ldots j}$, then $\Omega_{ij} = max(\Omega_{i-1,j}, \Omega_{i,j-1}, w_{ij})$ where w_{ij} is $\Omega_{i-1,j-1}$ if $\alpha_i \neq \beta_j$ or $\Omega_{i-1,j-1} + \omega(i, j, \alpha_i)$ if they are equal. It can be seen easily by induction that a match is dominant whenever Ω_{ij} is defined by w_{ij}. That is, a match $\alpha_i = \beta_j$ is dominant when $w_{ij} > \Omega_{i-1,j}$ and $w_{ij} > \Omega_{i,j-1}$.

Assume an instance of the HCS problem with sequences α, β, and weight function Ω. We say that Ω is β-decreasing, if for every symbol in α, the weights of its matches are decreasing (but not necessarily strictly decreasing) as they appear from left to right in β. On the other hand, if for every symbol in α, the weights of its matches in β strictly increase from left to right, we say that Ω is β-increasing. We similarly define α-decreasing and α-increasing. The below result follows from a simple induction. It shows that algorithm hcs1() runs in $O(d \log n)$ for weight systems that are increasing.

Theorem 4. *If the weight function Ω is α-increasing and β-increasing, then every match is a dominant match.*

The rest of this section shows tunings of the algorithm based on whether or not the weight functions are α-decreasing, β-decreasing or both. We shall state conditions when the algorithms perform in $O(d \log n)$ time.

5.1 Computing HCS for β-Decreasing Weights

In this case, let s and j be defined as on lines 7–8 of algorithm hcs1(). The reverse order insertion of the sequence $s < j_1 < \cdots < j_k = j$ eventually amounts to the insertion of just j_1 since it is heaviest. This means that intermediate insertions can be avoided by directly computing j_1. Figure 5 shows the modified HCS algorithm. Line 10 is the new addition to algorithm hcs1().

Theorem 5. *If the weight function Ω is β-decreasing and α-increasing, then algorithm hcs2() runs in $O(d \log n)$ time.*

5.2 Computing HCS for α-Decreasing Weights

In this case, consider lines 8–9 of algorithm hcs1(). If j is currently on the list L, these lines will define t to be j. If the element immediately precedes j in L has not changed since j was inserted into L, then because the weight of the new j is less than the one already in L, lines 10–17 will leave L unchanged. This means that when an element j is inserted into L, we can delete it from its position list to avoid duplicate processing. However, we must insert it back into the position list if it gets removed from L or if its predecessor in L ever changes. Figure 6 shows the modified algorithm.

Theorem 6. *If the weight function Ω is α-decreasing and β-increasing, then algorithm hcs3() runs in $O(d \log n)$ time.*

```
 1. hcs2(α₁α₂···αₘ,β₁β₂···βₙ,Ω)
 2. {   for(i = 1; i <= n; i = i+1)
 3.         insert(Position[βᵢ],i);
 4.     L = φ;
 5.     for(i = 1; i <= m; i = i+1)
 6.     {   P = Position[αᵢ];
 7.         for(j = max(P); j != φ; j = prev(P,j))
 8.         {   (s,v) = prev(L,(j,0));
 9.             (t,w) = next(L,(s,v));
10.             j = next(P,s);
11.             while((t,w) != φ)
12.             {   if(v+Ω(i,j,αᵢ) < w)
13.                     break;
14.                 delete(L,(t,w));
15.                 (t,w) = next(L,(t,w));
16.             }
17.             if((t,w) == φ || j < t)
18.                 insert(L,(j,v+Ω(i,j,αᵢ)));
19.         }
20.     }
21. }
```

Fig. 5. An HCS algorithm for decreasing weights in β

5.3 Computing HCS When Weights are Position-Independent

An important special case that finds many practical applications is when the weights are dependent only on the symbols in the alphabet. In this case, both conditions of algorithms hcs2() and hcs3() apply. Further, the test on line 16 of hcs1() is not needed as we noted in the remarks following algorithm his(). Putting everything together, we have algorithm hcs4() (Figure 7) for computing an HCS when weights are position-independent. When all weights are constant, hcs4() reduces to the Apostolico-Guerra LCS algorithm.

5.4 Conclusions

In this paper, we defined the heaviest increasing subsequence and heaviest common subsequence problems as natural generalizations of the longest increasing and longest common subsequence problems. These problems are intimately related by a weight-preserving correspondence. We showed how to generalize the Robinson-Schensted LIS algorithm to solve the HIS problem in $O(n \log n)$ time. Then using the weight-preserving correspondence, we applied the new HIS algorithm to solve the HCS problem in $O(r \log n)$ time, where r is the number of matches. This algorithm is a generalization of the Hunt-Szymanski LCS algorithm. We showed through a sequence of simple refinements how to tune the HCS algorithm when weights followed certain regular patterns. In particular, when weights are position-independent, our HCS algorithm can be viewed as a generalization of the Apostolico-Guerra LCS algorithm. Typically, computing an HCS may require much fewer matches than the entire set of

```
 1.  hcs3(α₁α₂···αₘ,β₁β₂···βₙ,Ω)
 2.  {    for(i = 1; i <= n; i = i+1)
 3.             insert(Position[βᵢ],i);
 4.        L = φ;
 5.        for(i = 1; i <= m; i = i+1)
 6.        {    P = Position[αᵢ];
 7.             for(j = max(P); j != φ; j = prev(P,j))
 8.             {    (s,v) = prev(L,(j,0));
 9.                  (t,w) = next(L,(s,v));
10.                  while((t,w) != φ)
11.                  {    if(j < t)
12.                            insert(Position[βₜ],t);
13.                       if(v+Ω(i,j,αᵢ) < w)
14.                            break;
15.                       delete(L,(t,w));
16.                       (t,w) = next(L,(t,w));
17.                  }
18.                  if((t,w) == φ || j < t)
19.                  {    insert(L,(j,v+Ω(i,j,αᵢ)));
20.                       delete(P,j);
21.                  }
22.             }
23.        }
24.  }
```

Fig. 6. An HCS algorithm for decreasing weights in α

Remarks on Figure 6

11-12: These lines implement the condition that a position t on L must be reinserted into its Position list if it gets bumped off L or if the element preceding it in L changes.

20:　　 This line removes j from its Position list after it gets inserted into L so that redundant processing of j is avoided.

matches. We defined generalized *dominant matches*, and specified conditions under which all of our HCS algorithms would run in $O(d \log n)$ time where d is the number of dominant matches.

The Robinson-Schensted LIS algorithm is central in the combinatorial theory of tableaux and plane partitions. Our extension of the algorithm indicates that many of the interesting results in the theory may extend. Of even more interest is the weight-preserving correspondence between the HIS and HCS problems. In future work, we hope it will show new ways in using the machinery of tableau and plane partition theory to find out more about the structure of common subsequence problems.

```
1.  hcs4(α₁α₂ ··· αₘ,β₁β₂ ··· βₙ,Ω)
2.  {    for(i = 1; i <= n; i = i+1)
3.              insert(Position[βᵢ],i);
4.      L = φ;
5.      for(i = 1; i <= m; i = i+1)
6.      {   P = Position[αᵢ];
7.          for(j = max(P); j != φ; j = prev(P,j))
8.          {   (s,v) = prev(L,(j,0));
9.              (t,w) = next(L,(s,v));
10.             j = next(P,s);
11.             while((t,w) != φ)
12.             {   insert(Position[βₜ],t);
13.                 if(v+Ω(αᵢ) < w)
14.                     break;
15.                 delete(L,(t,w));
16.                 (t,w) = next(L,(t,w));
17.             }
18.             insert(L,(j,v+Ω(αᵢ)));
19.             delete(P,j);
20.         }
21.     }
22. }
```

Fig. 7. An HCS algorithm for position-independent weights

References

1. A. V. Aho, D. S. Hirschberg, and J. D. Ullman. Bounds on the complexity of the longest common subsequence problem. *JACM*, 23(1):1–12, 1976.
2. A. Apostolico and C Guerra. The longest common subsequence problem revisited. *Algorithmica*, 2:315–336, 1987.
3. Francis Y. L. Chin and C. K. Poon. A fast algorithm for computing longest common subsequences of small alphabet size. *Journal of Information Processing*, 13(4):463–469, 1990.
4. Michael L. Fredman. On computing the length of longest increasing subsequences. *Discrete Mathematics*, 11:29–35, 1975.
5. E. Gansner. *Matrix Correspondences and the Enumeration of Plane Partitions*. PhD thesis, MIT, Cambridge, MA, 1978.
6. M. Golumbic. *Algorithmic Graph Theory and Perfect Graphs*. Academic Press, 1980.
7. Daniel S. Hirschberg. A linear space algorithm for computing maximal common subsequences. *CACM*, 18(6):341–343, 1975.
8. Daniel S. Hirschberg. Algorithms for the longest common subsequence problem. *JACM*, 24(4):664–675, 1977.
9. W. J. Hsu and M. W. Du. New algorithms for the LCS problem. *JCSS*, 29:133–152, 1984.
10. J. W. Hunt and M. D. McIlroy. An algorithm for differential file comparison. Computer Science Technical Report 41, Bell Laboratories, 1975.
11. James W. Hunt and Thomas G. Szymanski. A fast algorithm for computing longest common subsequences. *CACM*, 20(5):350–353, 1977.

12. D. Knuth. *The Art of Computer Programming*, volume 3. Addison-Wesley, Reading, MA, 1973.
13. William J. Masek and Michael S. Patterson. A faster algorithm computing string edit distances. *JCSS*, 20:18–31, 1980.
14. Eugene W. Myers. An $O(ND)$ difference algorithm and its variations. *Algorithmica*, 1:251–266, 1986.
15. N. Nakatsu, Y. Kambayashi, and S. Yajima. A longest common subsequence algorithm suitable for similar text strings. *Acta Informatica*, 18:171–179, 1982.
16. G. De B. Robinson. On the representations of the symmetric group. *American J. Math.*, 60:745–760, 1938.
17. D. Sankoff and J.B. Kruskal. *Time Warps, String Edits and Macromolecules: The Theory and Practice of Sequence Comparisons*. Addison Wesley, Reading, MA, 1983.
18. C. Schensted. Largest increasing and decreasing subsequences. *Canadian J. Math.*, 13:179–191, 1961.
19. D. Sleator and R. Tarjan. Self-adjusting binary trees. *JACM*, 32:652–686, 1985.
20. R. Stanley. Theory and applications of plane partitions. *Stud. Applied Math.*, 50:259–279, 1971.
21. K.-P. Vo. More <curses>: the <screen> library. Technical report, AT&T Bell Laboratories, 1986.
22. K.-P. Vo and R. Whitney. Tableaux and matrix correspondences. *J. of Comb. Theory, Series A*, 35:323–359, 1983.
23. R. A. Wagner and M. J. Fischer. The string-to-string correction problem. *JACM*, 21(1):168–173, 1974.
24. Sun Wu, Udi Manber, Gene Myers, and Webb Miller. An $O(NP)$ sequence comparison algorithm. *Information Processing Letters*, 35(6):317–323, 1990.
25. A. Young. The collected papers of alfred young. *Math. Exp.*, 21, 1977.

Approximate Regular Expression Pattern Matching with Concave Gap Penalties*

James R. Knight   Eugene W. Myers

Department of Computer Science, University of Arizona
Tucson, AZ 85721

Abstract. Given a sequence A of length M and a regular expression R of length P, an approximate regular expression pattern matching algorithm computes the score of the best alignment between A and one of the sequences exactly matched by R. There are a variety of schemes for scoring alignments. In a concave gap-penalty scoring scheme, a function $\delta(a, b)$ gives the score of each aligned pair of symbols a and b, and a *concave* function $w(k)$ gives the score of a sequence of unaligned symbols, or gap, of length k. A function w is concave if and only if it has the property that for all $k > 1$, $w(k + 1) - w(k) \le w(k) - w(k-1)$. In this paper we present an $O(MP(\log M + \log^2 P))$ algorithm for approximate regular expression matching for an arbitrary δ and any concave w.

1 Introduction

The problem of approximately matching a regular expression with concave gap penalties falls into a family of *approximate pattern matching* problems that compute the score of an *optimal alignment* between a given query sequence and one of the sequences specified by the pattern. An alignment is simply a pairing of symbols between two sequences, $A = a_1 a_2 \ldots a_M$ and $B = b_1 b_2 \ldots b_N$ over alphabet Σ, such that the lines of the induced *trace* do not cross as illustrated below:

Alignment: $\begin{bmatrix} a \\ a \end{bmatrix} \begin{bmatrix} b \\ b \end{bmatrix} \begin{bmatrix} cad \\ \varepsilon \end{bmatrix} \begin{bmatrix} b \\ b \end{bmatrix} \begin{bmatrix} b \\ b \end{bmatrix} \begin{bmatrix} \varepsilon \\ a \end{bmatrix} \begin{bmatrix} c \\ b \end{bmatrix} \begin{bmatrix} c \\ c \end{bmatrix} \begin{bmatrix} c \\ c \end{bmatrix}$ Trace:

Gap of length 3   Gap of length 1

abcadbbccc
abbbabcc

Alignments are evaluated with a scoring scheme S that gives scores for each *aligned pair*, $[a_i, b_j]$, and each contiguous block of unaligned symbols or *gap*, $a_{i+1} a_{i+2} \cdots a_k$. The score of an alignment is the sum of the scores of each aligned pair and gap, and an optimal alignment is one of minimal score. An approximate pattern matching problem takes as input a sequence A, a pattern R, and an alignment scoring scheme S. It determines the best scoring alignment between A and one of the sequences B exactly matching R, where alignments are scored using scheme S.

In this paper we consider two scoring schemes, *symbol-based* and *concave gap penalty*. Both scoring schemes use an arbitrary function $\delta(a, b)$, for $a, b \in \Sigma$, to score the aligned pairs. The difference is in the scoring of gaps. In a symbol-based scheme,

---

* This work was supported partially by the National Institute of Health under Grant R01 LM04960 and the Aspen Center for Physics

δ is extended to be defined over an additional symbol ε not in Σ, and the score of an unaligned symbol a is given by $\delta(a, \varepsilon)$. The score of a gap $a_{i+1}a_{i+2}\ldots a_k$ is $\sum_{p=i+1}^{k} \delta(a_p, \varepsilon)$, the sum of the scores of the individual unaligned symbols. The score of gaps in B is defined symmetrically.

The concave gap penalty scheme is one of a number of gap-cost models where the cost of a gap is solely a function of its length. In such a scheme, an additional function $w(k)$ gives the cost of a gap of length k. A concave gap penalty scheme adds the requirement that w be *concave* in the sense that its first forward differences are non-increasing. Formally, w is concave if and only if $\Delta w(k) \geq \Delta w(k+1)$ for $k > 1$, where $\Delta w(k) \equiv w(k) - w(k-1)$. One example of such a function is the logarithmic function $w(k) = \alpha + \beta \log k$, for constants $\alpha, \beta > 0$.

The problem considered in this paper is a generalization of several earlier results. The basic sequence comparison problem, SEQ(A, B, $\{\delta\}$), finds the optimal alignment between A and B under symbol-based scoring scheme $S = \{\delta\}$. Several authors [10, 11, 13] independently discovered an $O(MN)$ algorithm for this problem where M and N are the lengths of A and B. In 1984, Waterman [14] generalized this classic problem by considering concave gap penalties, i.e. SEQ(A, B, $\{\delta, w\}$) where w is concave. A few years later, a number of authors arrived at an $O(MN(\log M + \log N))$ algorithm [4, 7, 2] using the concept of a *minimum envelope*. In another direction of generality, Myers and Miller [9] considered the problem of approximate matching regular expressions under symbol-dependent scoring schemes or RE(A, R, $\{\delta\}$). By observing that an automaton for R is a reducible graph, they devised a two sweep node listing algorithm requiring $O(MP)$ time where P is the length of R.

This paper presents an $O(MP(\log M + \log^2 P))$ algorithm for the problem, RE(A, R, $\{\delta, w\}$), of approximately matching A to regular expression R under a concave gap penalty scheme $S = \{\delta, w\}$. The best previous algorithm required $O(MP(M + P))$ or cubic time [9]. Our sub-cubic result builds on the earlier results above by combining the minimum envelope and two-sweep node listing ideas. However, the extension is not straightforward, requiring the use of *persistent* data structures [8] and collections of envelopes, some of which are organized as stacks.

2 Preliminaries

All the problems discussed in the introduction can be recast as problems of finding the cost of a shortest source-to-sink path in an *alignment graph* constructed from the sequence/pattern input to the problem. The reduction is such that each edge corresponds to a gap or aligned pair and is weighted according to the cost of that item. The graph edges are traditionally categorized into three types: *substitution edges* modeling aligned pairs; *deletion edges* modeling gaps in the input sequence A; and *insertion edges* modeling gaps in the pattern. The graph is constructed inductively so as to ensure that every path between two vertices models an alignment between corresponding substrings/subpatterns of the inputs. From these graphs, *dynamic programming recurrences* for computing the shortest path costs from the source to each vertex are easily derived. In all cases we seek the shortest path cost to a designated sink since every source-to-sink path models a complete alignment between the two inputs.

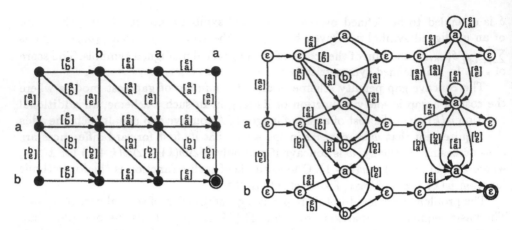

Fig. 1. Alignment graphs for SEQ(ab, baa, $\{\delta\}$) and RE(ab, (a|b)a*, $\{\delta\}$).

For the basic sequence comparison problem, $\text{SEQ}(A, B, \{\delta\})$, the resulting alignment graph takes the form of an acyclic graph whose vertices are layed out as an $(M+1) \times (N+1)$ matrix as in Figure 1a. The source vertex $(0,0)$ appears in the upper left corner of the matrix, and the sink vertex (M, N) is in the lower right corner. The set of paths from vertex (i, j) to (k, l) is in one-to-one correspondence with the alignments between $a_{i+1} a_{i+2} \ldots a_k$ and $b_{j+1} b_{j+2} \ldots b_l$ and their costs coincide. This graph provides the basis for the classic $O(MN)$ dynamic programming recurrence for computing the shortest path cost from the source vertex $(0,0)$ to vertex (i,j): $C_{i,j} = \min\{C_{i-1,j-1} + \delta(a_i, b_j), C_{i-1,j} + \delta(a_i, \varepsilon), C_{i,j-1} + \delta(\varepsilon, b_j)\}$. The more complex sequence comparison with gap penalties problem, $\text{SEQ}(A, B, \{\delta, w\})$, uses an alignment graph similar to that for the basic sequence comparison problem, but with additional edges to each vertex (i, j) from predecessors on row i and column j modeling the multi-symbol gaps. These extra edges are reflected in the insertion and deletion terms of the recurrence:

$$C_{i,j} = \min\{\ C_{i-1,j-1} + \delta(a_i, b_j),\ \min_{0 \le k < i}\{C_{k,j} + w(i-k)\},\ \min_{0 \le k < j}\{C_{i,k} + w(j-k)\}\ \} \quad (1)$$

This computation requires $O(MN(M+N))$ time as each application of the recurrence must consider all paths through the $O(M+N)$ incoming edges to vertex (i,j). In the next section on minimum envelopes, it will be revealed how the complexity can be reduced in the case where w is concave.

Generalizing to a regular expression pattern matching problem such as $\text{RE}(A, R, \{\delta\})$ involves an alignment graph where each "row" is formed from the states of an ε-NFA F constructed from R. Any regular expression R can be converted into an equivalent state-labeled, non-deterministic finite automaton with the inductive construction derived from [5], whose main construction steps are shown in Figure 2. An ε-NFA F constructed in this way contains $O(P)$ nodes and edges, and the in-degree and out-degree of every node is 2 or less. In addition, the structure of cycles in F has a special property. Term those edges introduced from ϕ_R to θ_R in the diagram of F_{R*}, *back edges*, and term the rest, *DAG edges*. Note that the graph restricted to the set of DAG edges is acyclic. Moreover, it can be shown that any cycle-free path

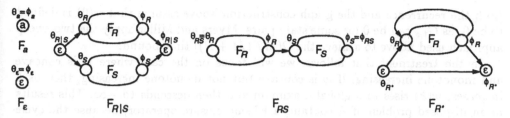

Fig. 2. Inductive ε-NFA construction.

in F has at most one back edge. Graphs with this property are commonly referred to as being *reducible* [1] or as having a *loop connectedness parameter* of 1 [3].

The alignment graph consists of $M+1$ copies of F with edges modeling the alignments between substrings of A and paths through F. An example is presented in Figure 1b. The dynamic programming recurrence is as follows:

$$C_{i,s} = \min\{ \min_{t \to s}\{C_{i-1,t} + \delta(a_i, \lambda_s)\},\ C_{i-1,s} + \delta(a_i, \varepsilon),\ \min_{t \to s}\{C_{i,t} + \delta(\varepsilon, \lambda_s)\} \}$$

where t and s are states in F, $t \to s$ denotes a transition in F and λ_s is the state label of state s. Note that cyclic dependencies can occur in this recurrence, because the underlying alignment graph can contain cycles of insertion edges along a row, corresponding to the cycles in F. Miller and Myers [7] used the above observations about the cycles in F to arrive at an $O(MP)$, row-based algorithm where the recurrence at each vertex is evaluated in two "topological" sweeps (according to the DAG edges of F) of each row in the alignment graph. The first sweep computes the recurrence at each vertex ignoring any insertion edges which correspond to back edges in F. The second sweep recomputes the recurrence, including those insertion edges ignored in the first sweep. The minimum of the two computed values gives the shortest path cost from the source vertex $(0, \theta)$, because a cycle free path between any two vertices cannot contain more than one insertion back edge on any row and all such paths are considered by the two sweeps on each row.

The introduction of a gap penalty scoring scheme affects the alignment graphs of Figure 1b similar to that in the sequence comparison case. The set of vertices remains unchanged, but extra edges must be added to represent the multi-symbol gaps. For the insertion gaps, the problem is more complex as there can be an infinite number of paths between two vertices in a row, each modeling a different sized gap. Due to this increased generality, it appears very difficult to treat the case of arbitrary w. The only result to date is an $O(MP(M+P))$ time algorithm by Myers and Miller [9] that treats the case where w is monotone increasing, i.e. $w(k) \le w(k+1)$. With this restriction, the path between vertices (i,t) and (i,s) modeling the least cost gap is a path from t to s in F with the fewest non-ε symbols. Let $G_{t,s}$, hereafter called the *gap distance* between t and s, be the number of non-ε labeled states on such a path. Thus, it suffices to add a single insertion edge from (i,t) to (i,s), of cost $w(G_{t,s})$, for every pair of vertices where a path from t to s in F, denoted $t \xrightarrow{*} s$, exists. This results in the following recurrence:

$$C_{i,s} = \min\{ \min_{t \to s}\{C_{i-1,t} + \delta(a_i, \lambda_s)\},\ \min_{0 \le k < i}\{C_{k,s} + w(i-k)\},\ \min_{\forall t: t \xrightarrow{*} s} \{C_{i,t} + w(G_{t,s})\} \}$$

Both the recurrence and the graph construction above assume that $w(0)$ is defined to be 0, as $G_{t,s}$ can be 0 for some state pairs. Myers and Miller apply the two-sweep approach used above to achieve the $O(MP(M+P))$ time bound.

In the treatment that follows, we will focus on the case where w is concave and monotone increasing. If w is concave but not monotone increasing, then as k increases, $w(k)$ rises to a global maximum and then descends to $-\infty$. This results in an ill-posed problem if R contains a Kleene closure operator, because the cycle allows infinitely sized insertion gaps. The alternate case where R contains no Kleene closures is well posed and can be solved in $O(MP(\log M + \log^2 P))$ time, but is not discussed here.

3 Minimum Envelopes

The inefficiency of the dynamic programming algorithm for SEQ(A, B, $\{\delta, w\}$) is that, for each i and j, it takes $O(M)$ and $O(N)$ time to compute the deletion and insertion terms. This is the best one can do by considering the computation of $C_{i,j}$ in isolation. However, considering the deletion terms over a whole column of the alignment graph produces the following "one-dimensional" characterization:

$$D_i = \min_{0 \le k < i}\{V_k + w(i-k)\} \tag{2}$$

where V_k is $C_{k,j}$ and D_i is the value of the deletion term for $C_{i,j}$. The insertion terms along a row can be characterized similarly.

The key to a faster algorithm for this one-dimensional problem is to capture the future contribution of the terms in each minimum. In other words, at position i, identify the values to be contributed from the first $i-1$ *candidates* to the computations of D_{i+x}, where $x \ge 0$. This is expressed as a *minimum envelope*, $E_i(x) = \min_{0 \le k < i}\{V_k + w((i-k)+x)\}$, over the domain $x \in [0, M-i]$. Each candidate k captures the future contribution of the kth term in D_{i+x}, and the envelope E_i captures the contribution of the first $i-1$ candidates at $i+x$. Simple algebra reveals that $D_i = E_i(0)$. Thus the problem becomes one of incrementally computing each E_i in increasing order. That is, given a data structure modeling E_{i-1}, construct a data structure modeling $E_i(x) = \min\{E_{i-1}(x+1), V_{i-1} + w(1+x)\}$.

When w is a concave function, the candidates' contribution to future D_i becomes that of Figure 3a. Each candidate's contribution takes the form of a translated image of w, $\alpha + w(\beta + x)$, where $\alpha = V_k$ and $\beta = i - k$. Waterman [14] shows that translated versions of the same concave curve intersect each other at most once, so a given candidate can be minimal over a single interval of x values, if at all. Call the candidates with non-empty minimal intervals *active candidates*. These candidates partition the envelope's domain, and so E_i can be modeled using a list of these candidates, ordered in increasing order of the right endpoints of their intervals. This list is called a *candidate list*. Each list element contains three fields $<\alpha, \beta, x>$, the α and β values of the candidate and the right endpoint of the candidate's interval. Note that in such a list, the candidates' β values are also in increasing order, since curves with small β's (and hence larger first forward differences of w) rise more quickly than those with larger β's.

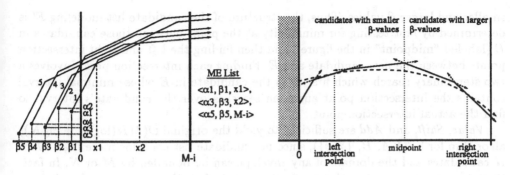

Fig. 3. a) A minimum envelope and its candidate list. b) Depiction of the *Add* operation.

The equations for D_i and E_{i+1} above suggest three operations for manipulating candidate lists,

(1) *Value*(E, x): returns the value of E at x, or $E(x)$, when $x \geq 0$
(2) *Shift*(E, Δ): returns a candidate list $E'(x) = E(x + \Delta)$ when $\Delta \geq 0$
(3) *Add*(E, α, β): returns $E'(x) = \min\{E(x), \alpha + w(\beta + x)\}$ when $\beta > 0$

and the following algorithm for solving the one-dimensional problem of Equation 2,

```
E ← [ ]
for i ← 1 to M do
{   E ← Add(Shift(E, 1), V_{i-1}, 1)
    D_i ← Value(E, 0).
}
```

The remainder of the section shows how the three operations can be realized in logarithmic time, giving an $O(M \log M)$ bound for this algorithm.

The candidate lists are implemented as *applicative* or *persistent* height-balanced trees. Myers [8] develops an implementation which permits element access, sublist selection, list concatenation and binary search in time logarithmic to the length of the lists. It uses the standard height-balanced tree operations, but makes copies of the nodes normally altered, thus preserving the logarithmic time bound at the expense of logarithmic space. In addition, the β- and x-fields of candidates in the tree are stored as offsets to the β and x values of the parent. At the root, the β- and x-fields hold the actual β and x values of the root's candidate. With this "trick", described most notably in [12], all of the β and x values in the tree can be incremented or decremented in $O(1)$ time, simply by changing the values at the root.

Operation *Value* is realized using a binary search over E to find the active candidate at x and then computing that candidate's value at x. Operation *Shift* transforms E into E' by adding Δ to the β values of all of the active candidates in E, subtracting Δ from the x values (since the candidates' intervals also shift by Δ), and finally deleting the now inactive candidates in the list, i.e. those candidates which are minimal in E only in the range $[0, \Delta]$. Operation *Add* involves the possible replacement of an interior section of E with the new candidate, as depicted in Figure 3b. The new candidate can be minimal only over an interval between the candidates with smaller and larger β values. Thus, the structure of the candidate list modeling E' is determined by first testing for minimality at the point between those candidates in E (labeled "midpoint" in the figure), and then finding the left and right intersection points between the new candidate and E. Finding each intersection point involves a two step binary search which first finds the candidate in E whose minimal interval contains the intersection point and then searches over the candidate's interval to find the actual intersection point.

Value, *Shift*, and *Add* are sufficient to yield the original $O(MN(\log M + \log N))$ algorithm for SEQ(A, B, $\{\delta, w\}$), since no candidate list contains more than M or N candidates and the domain of any envelope can be bounded by M or N. In fact, they are more general than those actually needed, since the parameters β and Δ are 1 in every call to *Add* and *Shift*. The algorithms in [4, 7, 2] hinge critically on these special circumstances, which permits E to be implemented as a simple stack and dramatically simplifies the algorithms needed for *Add* and *Shift*. However, our goal is an algorithm for the more complex sequence versus regular expression problem, and the more general operations are needed for that algorithm.

4 Regular Expression Pattern Matching

As in the sequence vs. sequence problem, the deletion term $\min_{0 \le k < i}\{C_{k,s} + w(i - k)\}$ and the insertion term $\min_{\forall t : t \overset{*}{\to} s}\{C_{i,t} + w(G_{t,s})\}$ of the recurrence for RE(A, R, $\{\delta, w\}$) give the cubic time bound for the Myers-Miller two-sweep algorithm. The previous section shows that the problem of delivering the deletion terms in a column constitutes a "one-dimensional" problem solvable in $O(M \log M)$ time. This section formulates the problem of delivering the insertion terms in a row as a more complex one-dimensional problem and solves it in $O(P \log^2 P)$ time. These insertion terms are computed in two sweeps as in the algorithm in Section 2. Combining these one-dimensional solutions with the Myers-Miller algorithm yields an algorithm that computes $C_{M,\phi}$ in $O(MP(\log M + \log^2 P))$ time.

This one-dimensional problem is cast in terms of envelopes. What is desired at each state s is the envelope $E_s(x) = \min_{\forall t : t \overset{*}{\to} s}\{V_t + w(G_{t,s} + x)\}$, where V_t is the minimum of the deletion and substitution branches of $C_{i,t}$'s recurrence. Then, $C_{i,s}$ is simply $E_s(0)$. Note that there is at most a single candidate in E_s for each state in F. Because states and candidates correspond, candidates in an envelope will often be referred to in terms of their originating states.

The E_s values are actually arrived at in two topological sweeps. In a first sweep, the envelopes $E1_s$ consider only insertion gaps whose underlying paths in F are restricted to the DAG edges. The second sweep envelopes $E2_s$ consider gaps whose corresponding paths in F have exactly one back edge. Because only cycle free paths must be considered, $E_s(x) = \min\{E1_s(x), E2_s(x)\}$. As will be seen, the envelopes $E1$ and $E2$ are modeled by up to $O(\log P)$ candidate lists.

4.1 The First Sweep Algorithm

The data structures modeling $E1_s$ are constructed incrementally from s's predecessor states. The difficulty lies in 1) developing the construction so that each candidate in the data structures at s occurs in the *frame* of s, i.e. a candidate from t has

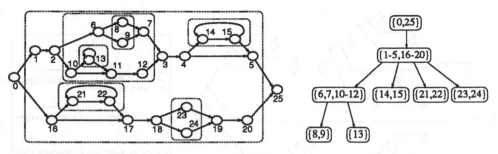

Fig. 4. The nesting tree for an example ε-NFA.

$\beta = G_{t,s}$, and 2) keeping within the $O(P \log^2 P)$ time bound. This is accomplished by partitioning the candidates in $E1_s$ into two groups, called the up predecessors and the down predecessors. The partition is based on the relative position of states t and s in the nesting of alternation and Kleene closure sub-automaton in F. To capture this positioning, a structure called a *nesting tree* is introduced. Each node in the nesting tree corresponds to an alternation or Kleene closure sub-automaton (henceforth, simply referred to as a *sub-machine*) and the edges represent the nesting of the sub-machines. In addition, each node is labeled with a *node set* containing the set of states which appear inside the corresponding sub-machine and outside all nested sub-machines. An example is given in Figure 4. Observe that the relative position of two states in the nesting of sub-machines is mirrored in the relative position of the states in the tree and that the node sets of the tree partition the states in the ε-NFA. For a state s, let N_s denote the node whose node set contains it.

The up predecessors at a state s consist of the states t where $t \xrightarrow{*} s$ and N_s is "up" in the nesting tree from N_t, i.e. $N_s \xrightarrow{*} N_t$. The other predecessors of s are down predecessors. This corresponds to those states where N_s is "up and then down" from N_t in the nesting tree (strictly "down" paths are considered "up and then down" paths). Since the node sets in the nesting tree contain all of the states in the ε-NFA, the up and down predecessors make up the set of states with candidates in $E1_s$.

The incremental construction of the up predecessors data structure, called the *up list*, is shown in graphical terms in Figure 5. The up list at each state s consists of a single candidate list which models $EU1_s$, the envelope of all up predecessors of s. The edges in the construction graph form an *up tree* that specifies the construction of the up list at each node. For the up tree nodes at state s with a candidate insertion edge and possibly an up tree edge, the construction takes the up list at the tail of the up tree edge, uses operation *Shift* to adjust the β values of the candidates if necessary, and uses operation *Add* to insert the candidate from s, $V_s + w(x)$. For those nodes in the tree with two incoming up tree edges, operation *Shift* is first used to adjust the β values, then an operation called *Merge* is used to combine the two lists. *Merge* takes two candidate lists and constructs the candidate list modeling the intersection of the two envelopes. It uses operation *Add* to insert the candidates from the shorter list into the longer list, thus constructing a list modeling the combined minimum envelope. This construction creates a list containing all of the candidates from states nested as deeply as s which are predecessors of s. Exactly those predecessors appear in the nesting sub-tree rooted at N_s.

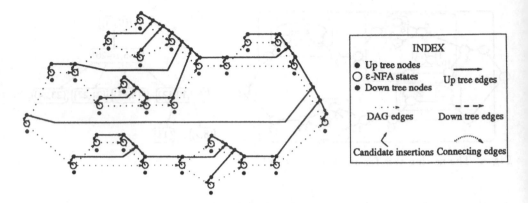

Fig. 5. The up list construction graph over an example ε-NFA F.

The construction, except for the *Merge* operations, takes $O(P \log P)$ time using $O(P)$ *Adds* and *Shifts*. At the nodes with two incoming up tree edges, a particular *Merge* may require $O(P)$ *Adds* to form the combined list, but over the course of the algorithm, only $O(P \log P)$ *Add* operations are needed. This is true because the up tree, with the candidate insertion edges, forms a binary tree with a single candidate added per leaf. An induction over any binary tree with P leaves can show that if the time spent at each node is proportional to the number of leaves of the smaller child sub-tree, the total time spent is bounded by $O(P \log P)$. Since merging according to the smaller list takes even fewer *Add* operations as the larger sub-tree can have the smaller list, $O(P \log P)$ *Adds* and $O(P \log^2 P)$ time is needed.

The construction of the down states data structure, or *down list*, is given in Figure 6. The down list consists of up to $\log P + 1$ candidate lists, denoted $H_{0,s}$, $H_{1,s}, \ldots, H_{k_s,s}$, and a value k_s holding the number of candidate lists needed at s. The down predecessors at s consists of those states which go "up and then down" to s. Since the up predecessors are those states which go "up" to a state, this suggests an algorithm which merges or *incorporates* the up lists from each sub-machine's start state as the topological sweep moves "down" into the sub-machine. This idea is reflected in the connecting edges shown in the Figure 6 graph.

The simple algorithm for this construction uses a single envelope and operation *Merge* to incorporate each of the incoming up lists. This algorithm, however, is $O(P^2)$, because the size of too many lists being merged can be $O(P)$. Our algorithm merges only some of the up lists into a designated candidate list, H_0, and keeps the rest unmerged in a stack of unmerged up lists, $H_1, H_2, \ldots, H_{k_s}$. By performing a "balancing act" between the cost of merging into H_0 and the cost of constructing and evaluating each envelope in the stack, the overall time bound of the construction can be kept under $O(P \log^2 P)$. The construction algorithm must guarantee that 1) no more than $\log P + 1$ unmerged up lists are pushed on the stack at any state, and 2) the *Merge* operations used in the construction of H_0 require no more than $O(P \log P)$ *Add* operations.

The decisions of how to incorporate the up lists are called *copy decisions*, and the problem of making copy decisions which ensure the two conditions above is called the

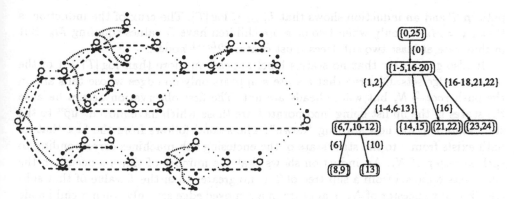

Fig. 6. The down tree construction graph and edge labeled nesting tree.

copy decision problem. A copy decision is made at the start state of each alternation and Kleene closure sub-automata. Term these sub-automata of F *sub-machines.* The procedure making the copy decisions uses the nesting tree T constructed from F. This tree has two useful properties in characterizing these decisions, 1) each edge corresponds to the copy decisions made at the start state of the sub-machine corresponding to the edge, and 2) the copy decisions made on paths through the down tree correspond to the edges along paths through the nesting tree. If each edge in T is labeled with the candidates in the up list being incorporated at the corresponding start state (see Figure 6), then the following *label removal problem* characterizes the copy decision problem:

> Remove a subset of edge sets s.t. (1) no path in T is labeled with $\geq \log |T|$ edge sets and (2) no state $s \in F$ appears in $\geq \log |T| + 1$ of the removed edge sets.

The fate of each edge set models the copy decision at the corresponding sub-machine. To wit, an edge set's removal represents the decision to merge the up list into H_0. Conversely, retaining an edge set represents the decision to push the up list onto the stack. An argument shows that any label removal procedure solving the label removal problem also solves the copy decision problem under this correspondence.

The procedure used to solve the label removal problem [copy decision problem] is as follows: Let L_b be the number of edge sets left on the path from a node b with the most edge sets. Let $c_1, c_2, \ldots, c_h$ always denote the children of a node b, and let $M_b = \max\{L_{c_1}, L_{c_2}, \ldots, L_{c_h}\}$. Compute L_b bottom up for nodes b in T and determine which edge sets to remove as follows:

1. $h = 0$ (a leaf). Set $L_b = 0$.
2. $h > 1$ and $\exists i, j : [i \neq j \ \& \ L_{c_i} = L_{c_j} = M_b]$ (two maximal children). Set $L_b = L_{c_i} + 1$ and leave all edge sets [Push the up list onto the stack].
3. $h \geq 1$ and $\exists i$ s.t. $[L_{c_i} = M_b \ \& \ \forall j \neq i : L_{c_i} > L_{c_j}]$ (a unique maximal child). Set $L_b = L_{c_i}$, remove the edge set on $b \to c_i$ [Merge into H_0] and leave the other edge sets [Push onto the stack].

This procedure guarantees that no more than $\log |T|$ edge sets occur on any path in the resulting tree, because L_{root} equals the maximum number of edge sets on any path in T and an induction shows that $L_{root} \leq \log |T|$. The crux of the induction is that $L_b = M_b + 1$ only when two or more children have L values equaling M_b. But in that case at least two sub-trees must contain 2^{L_b-1} nodes.

It also guarantees that no state's label appears in more than $\log |T| + 1$ of the removed edge sets. Observe that a state s appears only on edges whose tails are on the path $root \overset{*}{\rightarrow} N_s$ but whose heads are not. The first observation occurs because the states in the up list being incorporated are those which have moved "up" to the start state and are now starting "down". The second follows from the fact that no path exists from s to the start state of the enclosing sub-machines corresponding to each ancestor of N_s. An induction shows that the number of times s occurs in the edges sets removed from a sub-tree of T is no greater than the L value of that sub-tree. For any ancestor of N_s, s appears in a removed edge set only when a child node which is *not* an ancestor of N_s has the unique maximal L value. Thus, whenever the number of removals of edges sets containing s increases, moving up the path from N_s to $root$, the L value at the corresponding ancestor also increases.

4.2 The Second Sweep Algorithm

The construction algorithm for the second sweep is similar to that of the first sweep down list. The first sweep's up list is incorporated into the second sweep data structures at each back edge in F. For a path from t to s containing one back edge to be cycle free, t and s must appear "inside" that back edge in F. The up list at each back edge's tail contains the candidates from all states inside the back edge, so incorporating those up lists constructs a data structure modeling $E2_s$. The construction is not straightforward however, requiring two additional data structures and an additional computation during the first sweep. The details are given in [6].

References

1. Allen, F. E.: Control Flow Analysis. *SIGPLAN Notices* 5 (1970) 1-19.
2. Galil, Z., Giancarlo, R.: Speeding Up Dynamic Programming with Applications to Molecular Biology. *Theo. Comp. Sci.* 64 (1989) 107-118.
3. Hecht, M. S., Ullman, J. D.: A Simple Algorithm for Global Dataflow Analysis Programs. *SIAM J. Comp.* 4(4) (1975) 519-532.
4. Hirschberg, D. S., Larmore, L. L.: The Least Weight Subsequence Problem. *SIAM J. Comp.* 16(4) (1987) 628-638.
5. Hopcroft, J. E., Ullman, J. D.: Introduction to Automata Theory, Languages, and Computation, Chapter 2. Reading: Addison-Wesley, 1979.
6. Knight, J. R., Myers, E. W.: Approximate Regular Expression Pattern Matching with Concave Gap Penalties. TR 92-12, Dept. of CS, Univ. of Arizona, Tucson, AZ, 1992.
7. Miller, W., Myers, E. W.: Sequence Comparison with Concave Weighting Functions. *Bull. Math. Bio.* 50(2) (1988) 97-120.
8. Myers, E. W.: Efficient Applicative Data Types. *Proc. 11th Symp. POPL* (1984) 66-75.
9. Myers, E. W., Miller, W.: Approximate Matching of Regular Expressions. *Bull. Math. Bio.* 51(1) (1989) 33-56.
10. Needleman, S. B., Wunsch, C. D.: A General Method Applicable to the Search for Similarities in the Amino Acid Sequence of Two Proteins. *J. Mole. Bio.* 48 (1970) 443-453.

11. Sankoff, D.: Matching Sequences Under Deletion/Insertion Constraints. *Proc. Nat. Acad. Sci. U. S. A.* **69** (1972) 4-6.
12. Sleator, D. D., Tarjan, R. E.: Self-Adjusting Binary Search Trees. *J. ACM* **32(3)** (1985) 652-686.
13. Wagner, R. A., Fischer, M. J.: The String-to-String Correction Problem. *J. ACM* **21(1)** (1974) 168-173.
14. Waterman, M. S.: General Methods of Sequence Comparison. *Bull. Math. Bio.* **46** (1984) 473-501.

Matrix Longest Common Subsequence Problem, Duality and Hilbert Bases

Pavel A. Pevzner, Michael S. Waterman

Department of Mathematics and of Molecular Biology
University of Southern California
Los Angeles, California, 90089-1113

Abstract. Although a number of efficient algorithms for the longest common subsequence (LCS) problem have been suggested since the 1970's, there is no duality theorem for the LCS problem. In the present paper a simple duality theorem is proved for the LCS problem and for a wide class of partial orders generalizing the notion of common subsequence. An algorithm for finding generalized LCS is suggested which has the classical dynamic programming algorithm as a special case. It is shown that the generalized LCS problem is closely associated with the minimal Hilbert basis problem. The Jeroslav-Schrijver characterization of minimal Hilbert bases gives an $O(n)$ estimation for the number of elementary edit operations for generalized LCS.

1 Introduction

Biological molecules can be represented as long strings of letters from a finite alphabet, 4 letters for DNA and 20 letters for proteins. Currently a large effort is being expended in the experimental determination of these genetic sequences from various organisms. Biologists ask which known sequences are evolutionary related to a newly determined sequence. The primary events in sequence evolution are *substitution*, when one letter is replaced by another, and *insertion* or *deletion* of a letter. These are the edit operations in the *minimum edit distance* problem. In this paper, we explore duality theorems and primal-dual algorithms for minimum edit distance problem. In particular we demonstrate that some advanced algorithms for the minimum edit distance problem are different implementations of the primal-dual algorithm.

The simplest and most often studied minimum edit distance problem in computer science is the *longest common subsequence* (LCS) problem, which is to find a longest subsequence common to two sequences. The LCS problem is equivalent to the problem of finding the minimum number of inserted or deleted letters to transform one sequence into the other. We give a matrix generalization of LCS, A-LCS, for 2×2 matrices A. Several alignment problems of biological interest are included in this family of A-LCS problems.

A-LCS is a path in a comparability graph for a partial order. The classical Needleman-Wunsch ([NW70]) dynamic programming algorithm decomposes each long arc in this graph into short arcs, thereby achieving its efficiency. We study similar reductions for A-LCS and apply the Jeroslav-Schrijver characterization of Hilbert bases to this problem. We give a geometric interpretation of elementary edit operations for A-LCS and demonstrate that the number of elementary edit operations equals the size of the minimum Hilbert basis in the corresponding cone. A paper treating these and related topics in more detail will appear in [PW92].

2 Examples and Definitions

A *partially ordered set* or briefly a *poset* is a pair $(P, \prec)$ such that P is a set and $\prec$ is a transitive and irreflexive binary relation on P, i.e. $p \prec q$ and $q \prec r$ imply $p \prec r$. A *chain* is a subset of P where any two elements are comparable, and an *antichain* is a subset where no two elements are comparable. A *sequence* in a poset is an ordered chain $p_1 \prec p_2 \ldots \prec p_t$. Partial orders $\prec$ and $\prec^*$ are called *conjugate* ([KT82]) if for any two distinct $p_1, p_2 \epsilon P$ the following condition holds:

$$p_1 \text{ and } p_2 \text{ are } \prec\text{-comparable} \iff p_1 \text{ and } p_2 \text{ are } \prec^*\text{-incomparable}$$

Let w be an arbitrary non-negative integer valued function on P:

$$w : P \longrightarrow Z^+$$

For a partial order $\prec$, a sequence $p_1 p_2 \ldots p_k$ in P, maximizing

$$\sum_{i=1}^{k} w(p_i), \tag{1}$$

is called a *longest $\prec$-sequence*.

Let $I = \{1, 2 \ldots, n\}$ and $J = \{1, 2 \ldots, m\}$. As discussed in the introduction, our interest is in the comparison of two sequences $s = s_1 s_2 \ldots s_n$ and $t = t_1 t_2 \ldots t_m$. For this reason we study $P \subseteq I \times J$ (often $p = (i, j) \epsilon P$ denotes $s_i = t_j$). Let $p_1 = (i_1, j_1)$ and $p_2 = (i_2, j_2)$ be two arbitrary elements in $I \times J$. Denote

$$\Delta i = \Delta i(p_1, p_2) = i_2 - i_1,$$
$$\Delta j = \Delta j(p_1, p_2) = j_2 - j_1,$$
$$\Delta = \Delta(p_1, p_2) = (\Delta i, \Delta j)$$

Consider a few examples of partial orders on $I \times J$ (Fig.1).

(i) Common subsequences(CS): $p_1 \prec_1 p_2 \iff \Delta i > 0, \Delta j > 0$
(ii) Common forests(CF): $p_1 \prec_2 p_2 \iff \Delta i \geq 0, \Delta j \geq 0$
(iii) Common inverted subsequences(CIS): $p_1 \prec_3 p_2 \iff \Delta i > 0, \Delta j < 0$
(iv) Common inverted forests(CIF): $p_1 \prec_4 p_2 \iff \Delta i \geq 0, \Delta j \leq 0$

Partial orders $\prec_1$ and $\prec_3$ are particular cases of a partial order defined by an arbitrary 2×2 matrix $A = (a_{ij})$:

$$p_1 \prec_A p_2 \iff A\Delta^T > 0. \tag{2}$$

For $A = \begin{pmatrix} 1 & 0 \\ 0 & 1 \end{pmatrix}$ we have $\prec_1$, and for $A = \begin{pmatrix} 1 & 0 \\ 0 & -1 \end{pmatrix}$ we have $\prec_3$. A partial order defined by A is called an *A-order*, and a sequence in A-order is called an *A-sequence*. Similarly, define $\overline{A}$-order and $\overline{A}$-sequence by the inequality

$$p_1 \prec_{\overline{A}} p_2 \iff A\Delta^T \geq 0. \tag{3}$$

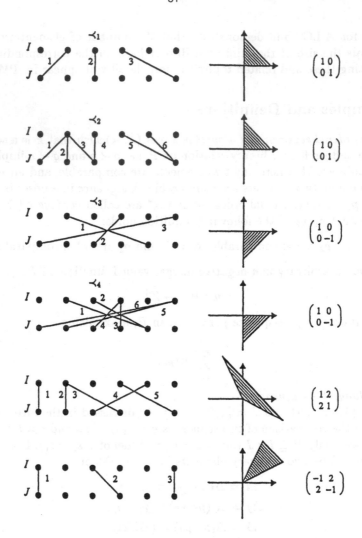

Fig. 1. Examples of sequences and corresponding cones for various partial orders.

The matrix A determines a *cone* in R^2 (Fig.1). The set of vectors Δ fulfilling (2) is designated $cone(A)$, while the set of vectors Δ fulfilling (3) is denoted $cone(\overline{A})$. For partial order A, an A-sequence $p_1 p_2 \ldots p_k$ in P maximizing (1) is called a *longest common sequence* for A or A-LCS ($\overline{A}$-LCS is defined similarly). For $P = I \times J$, $A = \begin{pmatrix} 1 & 0 \\ 0 & 1 \end{pmatrix}$ and w defined for sequences $\mathbf{s}$ and $\mathbf{t}$ according to the rule:

$$w(p) = w(i, j) = \begin{cases} 1, & s_i = t_j \\ 0, & \text{otherwise} \end{cases},$$

the problem (1) coincides with the longest common subsequence problem.

Let $\mathcal{C} = \{C\}$ be a family of subsets of a set P. $C' \subseteq \mathcal{C}$ is called a *cover* of a function w if:

$\forall p \epsilon P$ there exist at least $w(p)$ subsets in family C' containing an element p

For $w \equiv 1$ on P, C' is a cover if and only if each $p \epsilon P$ is contained in at least one of subsets $C \epsilon C'$. The number of elements in C' is called the *size* of the cover C' and a cover of minimum size is called a *minimum cover* of w by $\mathcal{C}$.

3 Duality for Longest $\prec$-sequence Problems

The following lemma is proven by an easy application of Dilworth's theorem [D50].

Lemma 1. *Let $\prec$ and $\prec^*$ be conjugate partial orders on P. Then the length of a longest $\prec$-sequence in P equals the size of a minimum cover of w by $\prec^*$-sequences.*

The binary relation on P defined by $p_1 \sqsubset p_2 \Longleftrightarrow p_1 \prec p_2$ or $p_1 \prec^* p_2$ can easily be shown to be linear

Lemma 2. $\sqsubset$ *is a linear order on P.*

Let $\mathcal{P} = p_1 p_2 \ldots p_l$ be an arbitrary sequence of the members of P, and $\mathcal{P}_i = p_1 p_2 \ldots p_i$. Let $\mathcal{C}_i = \{C_1, C_2, \ldots, C_j\}$ be a cover of $\mathcal{P}_i$ by $\prec^*$-sequences and let $p_1^{max}, p_2^{max}, \ldots, p_j^{max}$ be the $\prec^*$-maximum elements in $C_1, C_2, \ldots, C_j$ correspondingly. Consider an algorithm for constructing a cover $\mathcal{C}_{i+1}$ from $\mathcal{C}_i$.
Algorithm 1.
Let k be the minimum index $(1 \leq k \leq j)$ fulfilling the condition (Fig.2)

$$p_k^{max} \prec^* p_{i+1}, \tag{4}$$

and if the condition (4) fails for all k assume $k = j + 1$.
   If $k < j + 1$, add p_{i+1} to C_k and define

$$\mathcal{C}_{i+1} = \{C_1, C_2, \ldots, C_{k-1}, C_k \bigcup \{p_{i+1}\}, C_{k+1}, \ldots, C_j\}.$$

If $k = j + 1$ add $\{p_{i+1}\}$ as a new $\prec^*$-sequence to the cover $\mathcal{C}_{i+1}$:

$$\mathcal{C}_{i+1} = \{C_1, C_2, \ldots, C_j, C_{j+1} = \{p_{i+1}\}\,\}.$$

Define also a *reference* $ref(p)$ for p_{i+1} by

$$ref(p_{i+1}) = \begin{cases} p_{k-1}^{max} & \text{if } k > 1 \\ \emptyset & \text{otherwise} \end{cases}$$

$\square$

Assume $\mathcal{C}_1$ consists only of the set $C_1 = \{p_1\}$ and $ref(p_1) = \emptyset$. Applying algorithm 1 $|P| - 1$ times, we will construct a cover $\mathcal{C}_l$ of P and a set of references $ref(p)$ for each $p \epsilon P$. The size of the cover $\mathcal{C}_l$ depends on the choice of the ordering of $\mathcal{P}$. The following proposition shows that if $\mathcal{P}$ is the ordering of P in $\sqsubset$, then algorithm 1 gives a primal-dual algorithm for simultaneous solutions of (i) the longest $\prec$-sequence problem and (ii)the minimum $\prec^*$-cover problem (we suppose for simplicity that $w \equiv 1$).

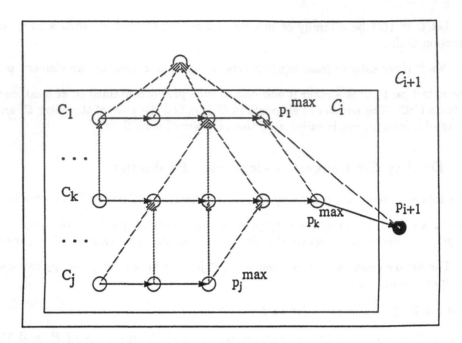

Fig. 2. Addition of p_{i+1} to the $\prec^*$-sequence C_k in the cover $\mathcal{C}_i = \{C_1, C_2, \ldots, C_j\}$. The references $ref(p) = q$ correspond to the (dashed) arcs (p, q).

Proposition 3. *If $\mathcal{P} = p_1 p_2 \ldots p_l$ is the ordering of P in $\sqsubset$, then algorithm 1 constructs a minimum cover $\mathcal{C}_l = \{C_1, C_2, \ldots, C_t\}$ of P by $\prec^*$-sequences. A traceback of references $ref(p)$ defines a longest $\prec$-sequence of length t for each $p\epsilon C_t$.*

Proof. We show that for each i ($1 \leq i \leq l$) the cover $\mathcal{C}_i = \{C_1, C_2, \ldots, C_j\}$ satisfies the condition

$$\forall k > 1, \forall p\epsilon C_k : \quad ref(p) \prec p. \tag{5}$$

Trivially this condition holds for C_1. We suppose that it holds for $\mathcal{C}_i$ and prove it for $\mathcal{C}_{i+1}$. Consider two following cases:

1) $k < j + 1$ (see condition (4), algorithm 1).
   In this case $ref(p_{i+1}) = p_{k-1}^{max}$. Since $\mathcal{P}$ is the $\sqsubset$-ordering then $p_{k-1}^{max} \sqsubset p_{i+1}$ and therefore either $p_{k-1}^{max} \prec p_{i+1}$ or $p_{k-1}^{max} \prec^* p_{i+1}$. Since k is the minimum index fulfilling $p_k^{max} \prec^* p_{i+1}$, then $p_{k-1}^{max} \prec p_{i+1}$ and therefore condition (5) holds for $\mathcal{C}_{i+1}$.
2) $k = j + 1$.
   In this case $ref(p_{i+1}) = p_j^{max}$. Since $\mathcal{P}$ is the $\sqsubset$-ordering then either $p_j^{max} \prec p_{i+1}$ or $p_j^{max} \prec^* p_{i+1}$. Since $k = j + 1$, condition (4) fails for each $k \leq j$. Therefore $p_j^{max} \prec p_{i+1}$ and condition (5) holds for $\mathcal{C}_{i+1}$.

Obviously each cover $\mathcal{C}_l = C_1 C_2 \ldots C_t$ fulfilling condition (5) determines (through the traceback procedure) a $\prec$-sequence of length t for each $p\epsilon C_t$. According to lemma 1 each such sequence is a $\prec$-longest sequence, and $\mathcal{C}_l$ is a minimum cover of P by $\prec^*$-sequences. $\quad\square$

We remark that algorithm 1 is a modification of the maximum path algorithm for the *comparability graph* of partial order $\prec$. According to the Dushnik-Miller theorem ([DM41]), partial order $\prec$ has a conjugate partial order $\prec^*$ if and only if the *dimension* of $\prec$ is ≤ 2. The linear order $\sqsubset$ together with the linear order $\sqsubset'$ defined by the rule

$$p_1 \sqsubset' p_2 \Longleftrightarrow p_1 \prec p_2 \text{ or } p_2 \prec^* p_1$$

yield a 2-dimensional representation of the partial order $\prec$. Notice also that lemma 1 is closely related to 'minimal antichain cover-longest chain' version of Dilworth's theorem and Robacker's theorem ([Fulk71]) about 'maximum cut packing-minimum path length' in graphs.

4 Duality for A-LCS Problems

Lemma 4. *Let* $A = \begin{pmatrix} a_{11} & a_{12} \\ a_{21} & a_{22} \end{pmatrix}$ *be an arbitrary* 2×2 *matrix and* $A^* = \begin{pmatrix} a_{11} & a_{12} \\ -a_{21} & -a_{22} \end{pmatrix}$.
Then the partial orders A and $\overline{A^*}$ are conjugate.

The next lemma is an immediate corollary of Lemmas 1 and 4:

Lemma 5. *Let* $A = \begin{pmatrix} a_{11} & a_{12} \\ a_{21} & a_{22} \end{pmatrix}$ *be a* 2×2 *matrix and* $A^* = \begin{pmatrix} a_{11} & a_{12} \\ -a_{21} & -a_{22} \end{pmatrix}$. *Then*

(i) *the length of A-LCS equals the size of a minimum cover of w by $\overline{A^*}$ -sequences, and*

(ii) *the length of $\overline{A^*}$-LCS equals the size of a minimum cover of w by A-sequences .*

Applying lemma 5 to the matrix $A = \begin{pmatrix} 1 & 0 \\ 0 & 1 \end{pmatrix}$ we derive the following theorem:

Theorem 6. .

- *The length of a longest CS equals the size of a minimum cover by CIF.*
- *The length of a longest CF equals the size of a minimum cover by CIS.*
- *The length of a longest CIS equals the size of a minimum cover by CF.*
- *The length of a longest CIF equals the size of a minimum cover by CS.*

For the LCS problem and a fixed length alphabet, algorithm 1 can be implemented in $O(nL)$ time where L is the length of longest common sequence or in $O((r + n)\log n)$ time where r is the total number of matches between the two input sequences. These improvements of the Needleman-Wunsch algorithm have been suggested by Hirschberg ([H77]) and Hunt and Szymanski ([HS77]). $\prec^*$-chains in algorithm 1 correspond to the *k-candidates* in Hirschberg's algorithm. Maximal elements of $\prec^*$-chains in algorithm 1 correspond to the *dominant matches* in Apostolico's improvement ([A86]) of Hunt-Szymanski's algorithm. Further improvements of algorithm 1 for the LCS problem can be found in ([A86], [AG87], [EGGI90]). The relationships between algorithm 1, advanced LCS algorithms and Robinson-Schensted-Knuth algorithm for Young tableaux ([Saga91]) will be considered in detail elsewhere.

5 Maximum Paths in Graphs and A-LCS Problem

A longest $\prec$-sequence problem can be reformulated as a 'maximum path' problem in the weighted directed graph $G(P, E, w)$, where E and w are defined by the rule

$$(p_1, p_2)\epsilon E \Longleftrightarrow p_1 \prec p_2$$

$$w(p_1, p_2) = w(p_2).$$

Sankoff [S72] first proposed an $O(n^2)$ algorithm for the LCS-problem in computer science, but a few authors developed closely related algorithms even earlier in molecular biology and speech processing. As a matter of fact the contribution of these authors is concerned with transformations of $G(P, E)$ to reduce computational complexity. They increased $|P|$ with a simultaneous significant decrease of $|E|$ by 'decomposition' of each 'long' arc into short arcs.

The classical Needleman-Wunsch algorithm for the LCS problem has running time $O(n^2)$ due to the special *arrangement* (systolic schedule, [CPHW91]) of vertices of G. Arranging vertices allows implementation of the maximum path algorithm for acyclic graphs in $O(|E|)$ time and gives $O(n^2)$ running time for the LCS-problem. Unfortunately the Needleman-Wunsch transformation of $G(P, E)$ into the graph with $O(n^2)$ arcs is not valid for an arbitrary A-order. Below we describe a transformation of $G(P, E)$ that decreases the number of arcs significantly, where Needleman-Wunsch transformation is a special case.

6 Algorithms for A-LCS Problems

Consider the A-LCS problem and let L_1 and L_2 be the lines $a_{11}x + a_{12}y = 0$ and $a_{21}x + a_{22}y = 0$, respectively. Without loss of generality below we suppose that $r = (i_1, j_1)$ and $s = (i_2, j_2)$ are the first integer points on the lines L_1 and L_2 fulfilling the conditions

$$|x| \leq n, |y| \leq m, \tag{6}$$

and $A = \begin{pmatrix} j_1 & -i_1 \\ -j_2 & i_2 \end{pmatrix}$.

Let $v_1, v_2 \ldots, v_k$ be the set V of all non-zero integer vectors (or vertices) of the parallelogram Π, defined by points $0, r, s, r + s$. The number of elements of V equals $\|V\| = |i_1 j_2 - i_2 j_1| + 2 = |det(A)| + 2$

Consider a graph $G^*(I \times J, E^*)$ with vertex set $I \times J$ and arc set E^* determined by V

$$(p_1, p_2)\epsilon E^* \Longleftrightarrow (p_2 - p_1)\epsilon V \tag{7}$$

Define weighting functions w and $\overline{w}$ on E^* according to the rule:

$$w(p_1, p_2) = \begin{cases} w(p_2), & \text{if } p_2\epsilon P, \text{ and } (p_2 - p_1) \neq r, s \\ 0, & \text{otherwise} \end{cases} \tag{8}$$

$$\overline{w}(p_1, p_2) = \begin{cases} w(p_2), & \text{if } p_2\epsilon P \\ 0, & \text{otherwise} \end{cases} \tag{9}$$

Theorems 7 and 8 below reduce A-LCS and $\overline{A}$-LCS problems to longest path problems.

Theorem 7. *The length of a $\overline{A}$-LCS coincides with the length of a $\overline{w}$- longest path in G^*.*

Proof. Obviously, each path $p_1, p_2, ..., p_t$ in $G^*(I \times J, E^*, \overline{w})$ corresponds to an $\overline{A}$-sequence of the same length, since the vertices of this path with $w(p_k) > 0$ correspond to elements of an $\overline{A}$-sequence. To prove the theorem, it is sufficient to prove that each $\overline{A}$-sequence $p_1, p_2, ..., p_t$ has a corresponding path in the G^*-graph of at least the same length.

Let p_{k-1}, p_k be two arbitrary sequential elements of an $\overline{A}$-sequence. Since $\Delta = \Delta_k(p_{k-1}, p_k)$ belongs to $cone(\overline{A})$, then

$$\Delta = xr + ys = \lfloor x \rfloor r + \lfloor y \rfloor s + \langle x \rangle r + \langle y \rangle s$$

($\lfloor x \rfloor$ is the integer part of x and $\langle x \rangle$ is the fractional part of x). The vector $p = \langle x \rangle r + \langle y \rangle s$ belongs to the parallelogram Π and is integer ($p = \Delta - \lfloor x \rfloor r + \lfloor y \rfloor s$). Therefore Δ is decomposed as a sum of $\lfloor x \rfloor + \lfloor y \rfloor + 1$ (or $\lfloor x \rfloor + \lfloor y \rfloor$ if $p = 0$) vectors defined by vertices from V. The decomposition for each pair p_{k-1}, p_k determines a path in G^* that visits vertices $p_1, p_2, ..., p_t$ and therefore has at least the same length as the $\overline{A}$-sequence $p_1 p_2 ... p_t$.   □

Theorem 8. *The length of an A-LCS coincides with the length of a w-longest path in G^*.*

Proof. Let $\mathcal{P} = p_1 p_2 ... p_t$ be an arbitrary path in G^* . Consider a subsequence of $\mathcal{P}$ defined by vertices: $\mathcal{P}' = \{p_k : p_k \epsilon P, (p_k - p_{k-1}) \neq r, s\}$. Observe that $\mathcal{P}'$ is an A-sequence and according to (8) the length of this A-sequence coincides with the w-length of P:

$$\sum_{k=2}^{t} w(p_{k-1}, p_k) = \sum_{p_k \epsilon \mathcal{P}'} w(p_{k-1}, p_k) = \sum_{p_k \epsilon \mathcal{P}'} w(p_k).$$

To prove the theorem, it is sufficient to prove that each A-sequence $p_1, p_2, ..., p_t$ has a corresponding path in graph G^* with the same w-length. Let p_{k-1}, p_k be two arbitrary sequential elements of the A-sequence. As was proved in Theorem 7, $\Delta = \Delta_k(p_{k-1}, p_k) = \lfloor x \rfloor r + \lfloor y \rfloor s + p$. If $p \neq 0$, define $\mathcal{P}_k$ to be the path consisting of $\lfloor x \rfloor$ arcs r, $\lfloor y \rfloor$ arcs s and ending with p. According to (8) only the last arc of this path has positive weight, equal to $w(p_k)$. If $p = 0$, then $\lfloor x \rfloor > 0$, $\lfloor y \rfloor > 0$ since otherwise p_{k-1} and p_k would be incomparable. Let $\mathcal{P}_k$ be the path consisting of $\lfloor x \rfloor - 1$ arcs r, $\lfloor y \rfloor - 1$ arcs s and the arc $r + s$ at the end. According to (8) only the last arc of this path has positive weight, equal to $w(p_k)$. Thus each pair p_{k-1}, p_k determines a path $\mathcal{P}_k$ in $G^*(I \times J, E^*, w)$, and the only last arc of this path has positive weight $w(p_k)$. Therefore the length of the path $\mathcal{P}_1, \mathcal{P}_2, ..., \mathcal{P}_t$ equals the length of the A-sequence $p_1, p_2, ..., p_t$.   □

7   A-LCS Problems and Hilbert Bases

According to theorems 7 and 8, finding longest A and $\overline{A}$ sequences requires about kn^2 operations, where k is the maximum vertex degree in G^* ($k = |det(A)| + 2$). For the classical LCS problem, and for the longest CF, CIS, CIF problems, $\det(A)=1$ and $k = 3$ (Fig.3a,b) as in the usual Needleman-Wunsch algorithm. For $A = \begin{pmatrix} 1 & 2 \\ 2 & 1 \end{pmatrix}$ and $A = \begin{pmatrix} -1 & 2 \\ 2 & -1 \end{pmatrix}$, $\det(A) = -3$ and $k = 5$ (Fig.3c,d), but we can further decrease k as some points in Π are non-negative integer combinations of others.

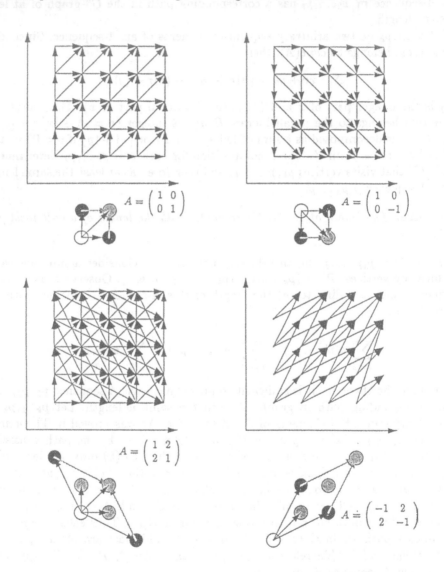

Fig. 3. Examples of parallelograms Π for various matrices A and the corresponding graphs G^*. The dark nodes in Π correspond to r and s.

We describe a procedure to eliminate arcs in the graphs G^* for A- and $\overline{A}$-LCS problems. A set H of integer vectors in the cone $K = \{x : Ax \geq 0\}$ is called a *Hilbert* *basis* of K if each integer vector in K is a non-negative integer linear combination of vectors in H. Let H be a Hilbert basis of the cone $\{x : Ax \geq 0\}$, and let $H(A)$ be the intersection of H and Π, where Π is the parallelogram defined by A.

Observe that theorem 7 still holds even if we define E^* by

$$(p_1, p_2) \epsilon E^* \iff (p_2 - p_1) \epsilon H(A)$$

instead of by (7). Similarly theorem 8 still holds even if we define E^* by

$$(p_1, p_2) \epsilon E^* \iff (p_2 - p_1) \epsilon H(A) \bigcup \{r + s\}$$

instead of by (7). These observations allow further transformations of G^* excluding arcs which do not belong to the Hilbert basis.

For example when $A = \begin{pmatrix} -1 & 2 \\ 2 & -1 \end{pmatrix}$, $H(A) = \{(1, 2), (1, 1), (2, 1)\}$. Therefore we can reduce the maximum degree of G^* for this matrix to 3. Unfortunately we can't guarantee that $H(A)$ contains $O(1)$ integer points for an arbitrary A-matrix. Some interesting questions arise:

1. Given A, find $H(A)$ of minimal size.
2. Find the maximum size of a minimal Hilbert basis $H(A)$ for matrix $A = (a_{ij})$ with $|a_{ij}| \leq n$.

Fortunately using the following characterization of Hilbert bases ([S81]), we can find $H(A)$ in $O(n)$ time and prove that the maximum size of minimal Hilbert basis of $H(A)$ is $O(n)$.

Theorem 9. *A set of all integer vectors in K which are not non-negative integer linear combinations of other integer vectors in K is the minimal Hilbert basis in K.*

The theorem implies a sufficient condition for a point (i,j) to lie outside the minimal Hilbert basis of the plane lattice Π : if all points in 1-neighbourhood of (i, j) (the set $\{(i', j') : |i' - i| \leq 1, |j' - j| \leq 1\}$) belong to Π, then (i, j) does not belong to the minimal Hilbert basis of Π. Therefore (i, j) can belong to the minimal Hilbert basis only if its 1-neighbourhood intersects the boundary of Π. This implies that cardinality of the minimal Hilbert basis of Π is at most $O(n + m)$ and that it is easy to find the minimal Hilbert basis in $O(n)$ time. Therefore the number of arcs in G can be reduced to $O(n^3)$, and this yields $O(n^3)$ A- and $\overline{A}$-LCS algorithms for an arbitrary A-matrix.

The LCS problem is often discussed in the terms of two elementary edit operations insertions and deletions. Generalized LCS problems require at least $|H(A)|$ elementary edit operations; each operation corresponds to a vector from the Hilbert basis. It is worth noting that, although the number of elementary edit operations can be as large as $O(n)$, an integer analog of the Caratheodory theorem ([CFS86]) implies that each arc between comparable elements can be decomposed as a sum of only 4 elementary edit operations.

For arbitrary 2×2 non-singular matrices A and B each A-LCS problem can be reduced to a B-LCS by the transformation $\begin{pmatrix} i' \\ j' \end{pmatrix} = B^{-1} \cdot A \begin{pmatrix} i \\ j \end{pmatrix}$. For arbitrary

A and $B = \begin{pmatrix} 1 & 0 \\ 0 & 1 \end{pmatrix}$ this transformation reduces the A-LCS problem for n-letter words to the classical LCS problem for $O(n^2)$ words and yields a $O(n^4)$ algorithm for the A-LCS problem. Nevertheless, the Hunt-Szymanski algorithm applied after such transformation yields A-LCS algorithm with running time at most $O(n^2 \log n)$.

8 Acknowledgements

The research was supported in part by the National Science Foundation (DMS 90-05833) and the National Institute of Health (GM-36230). We are grateful to Gian-Carlo Rota, Anatoly Rubinov and Martin Vingron for helpful discussions as well as Andras Sebo and Alexander Vainshtein for useful comments on Hilbert bases.

References

[A86] Apostolico A.: Improving the worst-case perfomance of the Hunt-Szymanski strategy for the longest common subsequence of two strings. Inform. Process. Lett. **23** (1986) 63-69

[AG87] Apostolico A., Guerra C.: The longest common subsequence problem revisited. Algoritmica 2(1987) 315-336

[CPHW91] Chow E.T., Hunkapiller T., Peterson J.C., Zimmerman B.A., Waterman M.S.: A systolic array processor for biological information signal processing. Proc. of International Conference on Supercomputing (ICS-91) June 17-21,1991 (to appear)

[CFS86] Cook W.,Fonlupt J.,Schrijver A.: An integer analogue of Caratheodory's theorem. J. of Combinatorial Theory (B) **40** (1986) 63-70

[D50] Dilworth R.P.: A decomposition theorem for partially ordered sets Ann. Math. **51** (1950) 161-165

[DM41] Dushnik B., Miller E.W.: Partially ordered sets. Am. J. Math. **63** (1941) 600-610

[EGGI90] Eppstein D., Galil Z., Giancarlo R., Italiano G. F. Sparse dynamic programming; Extended Abstract *Proc. first ACM-SIAM SODA* (1990) 513-522

[Fulk71] Fulkerson D.R.: Blocking and antiblocking polyhedra. Mathematical programming. **1** (1971) 168-194

[H77] Hirscberg D.S. Algorithms for the longest common subsequence problem. J. ACM **24** (1977) 664-675

[HS77] Hunt J.W., Szymanski T.G.: A fast algorithm for computing longest common subsequences. Comm. ACM **20** (1977) 350-353

[KT82] Kelly D., Trotter W.T.: Dimension theory for ordered sets. In I.Rival (ed.) Ordered sets Reidel, Dordrecht/Boston (1982)

[NW70] Needleman S.B., Wunsch C.D.: A general method applicable to the search for similarities in the amino acid sequence of two proteins. J.Mol.Biol. **48** (1970) 443-453

[PW92] Pevzner P., Waterman M.: Generalized sequence alignment and duality. Adv. in Appl. Math. (1992) (in press)

[Saga91] Sagan B.E.: The symmetric group. Representations, combinatorial algorithms and symmetric functions. Wadsworth and Brooks/Cole (1991)

[S72] Sankoff D.: Matching sequences under deletion-insertion constraints. Proc. Nat. Acad. Sci. USA **69** (1972) 4-6

[S81] Schrijver A.: On total dual integrality. Linear algebra and its applications. **38** (1981) 27-32

From Regular Expressions to DFA's Using Compressed NFA's*

Chia-Hsiang Chang and Robert Paige

New York University/Courant Institute
251 Mercer St. New York, NY 10012
email: changch@cs.nyu.edu, paige@cs.nyu.edu

Abstract. We show how to turn a regular expression R of length r into an $O(s)$ space representation of McNaughton and Yamada's NFA, where s is the number of occurrences of alphabet symbols in R, and $s + 1$ is the number of NFA states. The standard adjacency list representation of McNaughton and Yamada's NFA takes up $s + s^2$ space in the worst case. The adjacency list representation of the NFA produced by Thompson takes up between $2r$ and $6r$ space, where r can be arbitrarily larger than s. Given any set V of NFA states, our representation can be used to compute the set U of states one transition away from the states in V in optimal time $O(|V| + |U|)$. McNaughton and Yamada's NFA requires $\Theta(|V| \times |U|)$ time in the worst case. Using Thompson's NFA, the equivalent calculation requires $\Theta(r)$ time in the worst case.

An implementation of our NFA representation confirms that it takes up an order of magnitude less space than McNaughton and Yamada's machine. An implementation to produce a DFA from our NFA representation by subset construction shows linear and quadratic speedups over subset construction starting from both Thompson's and McNaughton and Yamada's NFA's. It also shows that the DFA produced from our NFA is as much as one order of magnitude smaller than DFA's constructed from the two other NFA's.

Throughout this paper the importance of syntax is stressed in the design of our algorithms. In particular, we exploit a method of program improvement in which costly repeated calculations can be avoided by establishing and maintaining program invariants. This method of symbolic finite differencing has been used previously by Douglas Smith to derive efficient functional programs.

1 Introduction

The growing importance of regular languages and their associated computational problems in languages and compilers is underscored by the granting of the Turing Award to Rabin and Scott in 1976, in part, for their ground breaking logical and algorithmic work in regular languages [17]. Of special significance was their construction of the canonical minimum state DFA that had been described non-constructively in the proof of the Myhill-Nerode Theorem[15,16]. Rabin and Scott's work, which was motivated by theoretical considerations, has gained in importance

---

* This research was partially supported by Office of Naval Research Grant No. N00014-90-J-1890 and Air Force Office of Scientific Research Grant No. AFOSR-91-0308.

as the number of practical applications has grown. In particular, the construction of finite automata from regular expressions is of central importance to the compilation of communicating processes[4], string pattern matching[3], model checking[10], lexical scanning[2], and VLSI layout design[22]; unit-time incremental acceptance testing in a DFA is also a crucial step in LR_k parsing[13]; algorithms for acceptance testing and DFA construction from regular expressions are implemented in the UNIX operating system[18].

Throughout this paper our model of computation is a uniform cost sequential RAM [1]. We report the following four results.

1. Recently Berry and Sethi[5] used results of Brzozowski[7] to formally derive and improve McNaughton and Yamada's algorithm[14] for turning regular expressions into NFA's. NFA's produced by this algorithm have fewer states than NFA's produced by Thompson's algorithm[21], and are believed to outperform Thompson's NFA's for acceptance testing. Berry and Sethi's algorithm has two passes and can easily be implemented to run in time $\Theta(m)$ and auxiliary space $\Theta(r)$, where r is the length of the regular expression, and m is the number of edges in the NFA produced. More recently, Brüggemann-Klein[6] presents a two-pass algorithm to compute McNaughton and Yamada's NFA using the same resource bounds as Berry and Sethi. We present an algorithm that computes the same NFA in a single left-to-right scan over the regular expression. It runs in the same asymptotic time $\Theta(m)$ as Berry and Sethi, but it improves the auxiliary space (i.e. space in addition to the r units of space needed to store the input) to $\Theta(s)$, where s is the number of occurrences of alphabet symbols appearing in the regular expression.

2. One disadvantage of McNaughton and Yamada's NFA is that its worst case number of edges is $m = \Theta(s^2)$. Thompson's NFA only has between r and $2r$ states and between r and $4r$ edges. We introduce a new compressed data structure, called the CNNFA, that uses only $\Theta(s)$ space to represent McNaughton and Yamada's NFA. The CNNFA can be constructed from a regular expression R in $\Theta(r)$ time and $O(s)$ auxiliary space. It supports acceptance testing in worst-case time $O(s|x|)$ for arbitrary string x, and a promising new way to construct DFA's faster than the classical subset construction of Rabin and Scott.

3. Our main theoretical result is a proof that the CNNFA can be used to compute the set of states U one edge away from an arbitrary set of states V in McNaughton and Yamada's NFA in optimal time $O(|V| + |U|)$. The previous best worst-case time is $\Theta(|V| \times |U|)$. This is the essential idea that explains the superior performance of the CNNFA in both acceptance testing and DFA construction.

4. We give empirical evidence that our algorithm for NFA acceptance testing using the CNNFA outperforms competing algorithms using either Thompson's or McNaughton and Yamada's NFA. We give more dramatic empirical evidence that constructing a DFA from the CNNFA can be achieved in time one order of magnitude faster than the classical Rabin and Scott subset construction (cf. Chapter 3 of [2]) starting from either Thompson's NFA or McNaughton and Yamada's NFA. Our benchmarks also indicate better performance using Thompson's NFA over McNaughton and Yamada's NFA for acceptance testing

and subset construction. This observation runs counter to the judgment of those using McNaughton and Yamada's NFA throughout UNIX.

The next section presents standard terminology and background material, and can be skipped by anyone who knows Chapter 3 of [2]. Section 3 reformulates McNaughton and Yamada's algorithm from an automata theoretic point of view. Section 4 describes a new algorithm to turn a regular expression into a McNaughton and Yamada's NFA. In Section 5 we show how to construct the CNNFA. Analysis of the CNNFA is presented in Theorem 11, which is our main theoretical result. In section 6, we show how to further compress the CNNFA. Section 7 discusses experimental results showing how the CNNFA compares with other NFA's in solving acceptance testing and DFA construction. Section 8 mentions future research.

2 Terminology and Background

With few exceptions the following basic definitions and terminology can be found in [2,11]. By an *alphabet* we mean a finite nonempty set of symbols. If Σ is an alphabet, then Σ^* denotes the set of all finite strings of symbols in Σ. The empty string is denoted by λ. If x and y are two strings, then xy denotes the concatenation of x and y. Any subset of Σ^* is a *language* over Σ.

Definition 1. Let L, L_1, L_2 be languages over Σ. The following expressions can be used to define new languages.

- $\emptyset$ denotes the empty set
- $L_1 L_2 = \{xy : x \in L_1, y \in L_2\}$ denotes *product*
- $L^0 = \{\lambda\}$ if $L \neq \emptyset$; $\emptyset^0 = \emptyset$
- $L^{i+1} = LL^i$, where $i \geq 0$
- $L^* = \cup_{i=0}^{\infty} L^i$
- $L^T = \{x : ax \in L | a \in \Sigma\}$ denotes the *tail* of L

In later discussions we will make use of the identities below, which follow directly from the preceding definition.

$$L\{\lambda\} = \{\lambda\}L = L \tag{1}$$

$$L\emptyset = \emptyset L = \emptyset \tag{2}$$

$$(L_1 \cup L_2)^T = L_1^T \cup L_2^T \tag{3}$$

$$(L_1 L_2)^T = L_1^T L_2 \text{ if } \lambda \notin L_1; \text{ otherwise, } (L_1 L_2)^T = L_1^T L_2 \cup L_2^T \tag{4}$$

$$(L^*)^T = L^T L^* \tag{5}$$

Kleene [12] characterized a subclass of languages called *regular languages* in terms of *regular expressions*.

Definition 2. The regular expressions over alphabet Σ and the languages they denote are defined inductively as follows.

- $\emptyset$ is a regular expression that denotes the empty set
- λ is a regular expression that denotes set $\{\lambda\}$

– a is a regular expression that denotes $\{a\}$, where $a \in \Sigma$

If J and K are regular expressions that represent languages L_J and L_K, then the following are also regular expressions:

– $J|K$ (alternation) represents $L_J \cup L_K$
– JK (product) represents $L_J L_K$
– J^* (star) represents $\cup_{i=0}^{\infty} L_J^i$

By convention star has higher precedence than product, which has higher precedence than alternation. Both product and alternation are left associative. Parentheses are used to override precedence. Without loss of generality, we will assume throughout this paper that regular expressions have no occurrences of $\emptyset$.

Regular expressions have been used in a variety of practical applications to specify regular languages in a perspicuous way. The problem of deciding whether a given string belongs to the language denoted by a particular regular expression can be implemented efficiently using finite automata defined below.

Definition 3. A *nondeterministic finite automata* (abbr. NFA) M is a 5-tuple $(\Sigma, Q, I, F, \delta)$, where Σ is an alphabet, Q is a set of states, $I \subseteq Q$ is a set of initial states, $F \subseteq Q$ is a set of final states, and $\delta \subseteq Q \times (\Sigma \times Q)$ is a state transition map. It is useful to view NFA M as a labeled directed graph with states as vertices and an edge labeled a connecting state q to state p for every pair $[q, [a, p]]$ belonging to δ. For all $q \in Q$ and $a \in \Sigma$ we use the notation $\delta(q, a)$ to denote the set $\{p : [q, [a, p]] \in \delta\}$ of all states reachable from state q by a single edge labeled a. It is helpful to extend the notation for transition map δ in the following way. If $V \subseteq Q$, $x \in \Sigma^*$, and $B \subseteq \Sigma^*$, then we define,

- $\delta(V, a) = \cup_{q \in V} \delta(q, a)$ • $\delta(q, ax) = \delta(\delta(q, a), x)$
- $\delta(V, x) = \cup_{q \in V} \delta(q, x)$ • $\delta(V, B) = \cup_{b \in B} \delta(V, b)$

The language accepted by M, denoted by L_M, is defined by the rule, $x \in L_M$ if and only if $\delta(I, x) \cap F \neq \emptyset$. In other words, $L_M = \{x \in \Sigma^* | \delta(I, x) \cap F \neq \emptyset\}$. NFA M is a *deterministic finite automata* (abbr. DFA) if transition map δ has no more than one edge with the same label leading out from each state, and if I contains exactly one state.

Kleene also characterized the regular languages in terms of languages accepted by DFA's. Rabin and Scott [17] showed that NFA's also characterize the regular languages, and their work led to algorithms to decide whether an arbitrary string is accepted by an NFA. Regular expressions and NFA's that represent the same regular language are said to be *equivalent*.

There are two main approaches for turning regular expressions into equivalent NFA's. One, due to Thompson [21], constructs an NFA (augmented with λ edges) in which the number of states n is somewhere between the length r of the regular expression and $2r$, and the outdegree of any state is no greater than 2. The number of edges m is between r and $4r$. Thompson's construction is a simple, bottom-up method that processes the regular expression as it is parsed. The time and space is linear in r.

Another approach, based on Berry and Sethi's [5] improvement to McNaughton and Yamada [14], constructs an NFA in which the number of states n is precisely one plus the number s of occurrences of alphabet symbols appearing in the regular expression. In general, s can be arbitrarily smaller than r. However, the number of edges in McNaughton and Yamada's NFA is $m = \Omega(s^2)$ in the worst case, which can be one order of magnitude larger than the bound for Thompson's NFA. Berry and Sethi's construction scans the regular expression twice, and, with only a little effort, both passes can be made to run in linear time and auxiliary space with respect to r plus the size of the NFA.

There is one main approach for turning NFA's (constructed by either of the two methods above) into DFA's. This is by Rabin and Scott's subset construction [17].

3 McNaughton and Yamada's NFA

It is convenient to reformulate McNaughton and Yamada's transformation from regular expressions to NFA's[14] in the following way.

Definition 4. A *normal NFA* (abbr. NNFA) is an NFA in which all edges leading into the same state have the same label. Thus, it is convenient to label states instead of edges, and we represent an NNFA M as a 6-tuple $(\Sigma, Q, \delta, I, F, A)$, where Σ is an alphabet, Q is a set of states, $\delta \subseteq Q \times Q$ is a set of (unlabeled) edges, $I \subseteq Q$ is a set of initial states, $F \subseteq Q$ is a set of final states, and $A : Q \to \Sigma$ maps states $x \in Q$ into labels $A(x)$ belonging to alphabet Σ. The language L_M accepted by NNFA M is the set of strings $x \in \Sigma^*$ formed from concatenating labels on all but the first state of a path from a state in I to a state in F. A *McNaughton/Yamada NNFA* (abbr. MYNNFA) is an NNFA with one initial state of zero in-degree (see Fig. 1).

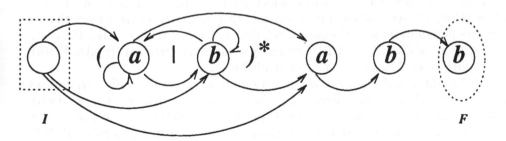

Fig. 1. An MYNNFA equivalent to regular expression $(a|b)^*abb$

We sometimes omit Σ in NNFA specifications when it is obvious. It is useful (and completely harmless) to sometimes allow the label map A to be undefined on states with zero in-degree. For example, we will not define A on the initial state of an MYNNFA.

Definition 5. The *tail* of an MYNNFA $M = (\Sigma, Q, \delta, I = \{q_0\}, F, A)$ is an NNFA $M^T = (\Sigma^T, Q^T, \delta^T, I^T, F^T, A^T)$, where $\Sigma^T = \Sigma$, $Q^T = Q - \{q_0\}$, $\delta^T = \{[x, y] \in$

$\delta|x \neq q_0\}$, $I^T = \{y : [q_0, y] \in \delta\}$, $F^T = F - \{q_0\}$, and $A^T = A$. Fig. 2 shows the tail of the MYNNFA given in Fig. 1.

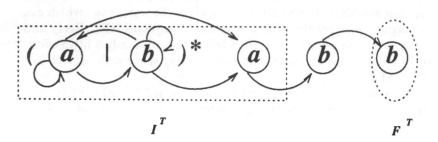

Fig. 2. The tail of an MYNNFA equivalent to regular expression $(a|b)^* abb$

If MYNNFA M accepts language L_M, then the identity $L_M^T = (L_M)^T$ holds; that is, the language accepted by tail machine M^T is the same as the tail of the language accepted by M. However, given the tail of some MYNNFA M, we can not compute an MYNNFA equivalent to M without also knowing whether $\lambda \in L_M$. Let $null_M = \{\lambda\}$ if $\lambda \in L_M$; otherwise, let $null_M = \emptyset$. Now we can reconstruct an MYNNFA $M = (\Sigma, Q, \delta, I, F, A)$ from its tail $M^T = (\Sigma^T, Q^T, \delta^T, I^T, F^T, A^T)$ and from $null_M$ using the equations

$$\Sigma = \Sigma^T, \; Q = Q^T \cup \{q_0\}, \; \delta = \delta^T \cup \{[q_0, y] : y \in I^T\}, \; I = \{q_0\},$$
$$F = F^T \cup \{q_0\}null_M, \; A = A^T, \tag{6}$$

where q_0 is a new state.

It is a desirable and obvious fact (which follows immediately from the definition of an MYNNFA) that when A is one-to-one, then no state can have more than one edge leading to states with the same label. Hence, such an MYNNFA is a DFA. More generally, an MYNNFA is a DFA if and only if the binary relation $\{[x, y] \in \delta | A(y) = a\}$ is single-valued for every alphabet symbol $a \in \Sigma$.

McNaughton and Yamada's algorithm inputs a regular expression R, and computes an MYNNFA M that accepts L_R. Their algorithm can be implemented within a left-to-right parse of R without actually producing a parse tree. To explain how the construction is done, we use the notational convention that M_R denotes an MYNNFA equivalent to regular expression R. Each time a subexpression J of R is reduced during parsing, $null_J$ and M_J^T are computed, where M_J is an MYNNFA equivalent to J. The last step computes an MYNNFA M_R from M_R^T and $null_R$ by equations (6).

Theorem 6. *(McNaughton and Yamada) Given any regular expression R with s occurrences of alphabet symbols from Σ, an MYNNFA M_R with $s + 1$ states can be constructed.*

Proof. The proof uses structural induction to show that for any regular expression R, we can always compute $null_R$ and M_R^T for some MYNNFA M_R. Then equations

(6) can be used to obtain M_R. We assume a fixed alphabet Σ. There are two base cases, which are easily verified.

$$M_\lambda^T = (Q_\lambda^T = \emptyset, \delta_\lambda^T = \emptyset, I_\lambda^T = \emptyset, F_\lambda^T = \emptyset, A_\lambda^T = \emptyset), null_\lambda = \{\lambda\} \tag{7}$$
$$M_a^T = (Q_a^T = \{q\}, \delta_a^T = \emptyset, I_a^T = \{q\}, F_a^T = \{q\}, A_a^T = \{[q,a]\}), null_a = \emptyset, \tag{8}$$
$$where\ a \in \Sigma, and\ q\ is\ a\ distinct\ state$$

To use induction, we assume that J and K are two arbitrary regular expressions equivalent respectively to MYNNFA's M_J and M_K with $M_J^T = (Q_J^T, I_J^T, F_J^T, \delta_J^T, A_J^T)$ and $M_K^T = (Q_K^T, I_K^T, F_K^T, \delta_K^T, A_K^T)$, where Q_J^T and Q_K^T are disjoint. Then we can use (3), (4), and (5) to verify that

$$M_{J|K}^T = (Q_{J|K}^T = Q_J^T \cup Q_K^T, \delta_{J|K}^T = \delta_J^T \cup \delta_K^T, I_{J|K}^T = I_J^T \cup I_K^T, F_{J|K}^T = F_J^T \cup F_K^T,$$
$$A_{J|K}^T = A_J^T \cup A_K^T), null_{J|K} = null_J \cup null_K \tag{9}$$
$$M_{JK}^T = (Q_{JK}^T = Q_J^T \cup Q_K^T, \delta_{JK}^T = \delta_J^T \cup \delta_K^T \cup F_J^T I_K^T, I_{JK}^T = I_J^T \cup null_J I_K^T,$$
$$F_{JK}^T = F_K^T \cup null_K F_J^T, A_{JK}^T = A_J^T \cup A_K^T), null_{JK} = null_J null_K \tag{10}$$
$$M_{J\cdot}^T = (Q_{J\cdot}^T = Q_J^T, \delta_{J\cdot}^T = \delta_J^T \cup F_J^T I_J^T, I_{J\cdot}^T = I_J^T, F_{J\cdot}^T = F_J^T, A_{J\cdot}^T = A_J^T),$$
$$null_{J\cdot} = \{\lambda\} \tag{11}$$

The preceding formulas are illustrated in Fig. 3.

Disjointness of the unions used to form the set of states for the cases $J|K$ and JK proves the assertion about the number of states. The validity of the disjointness assumption follows from the fact that new states can only be obtained from rule (8), and each new state is distinct. We can convert M_R^T into M_R using equations (6). □

The proof of Theorem 6 leads to McNaughton and Yamada's algorithm. The construction of label function A shows that when all of the occurrences of alphabet symbols appearing in the regular expression are distinct, then A is one-to-one. In this case, a DFA would be produced.

Analysis determines that this algorithm falls short of optimal performance, because the operation $\delta_J^T \cup F_J^T I_J^T$ within formula (11) for $M_{J\cdot}^T$ is not disjoint; all other unions are disjoint and can be implemented in unit time. In particular, this overlapping union makes McNaughton and Yamada's algorithm use time $\theta(s^3 \log s)$ to transform regular expression $((a_1|\lambda)(\cdots((a_{s-1}|\lambda)(a_s|\lambda)^*)^* \cdots)^*)^*$ into an MYNNFA with $s+1$ states and s^2 edges.

This redundancy is made explicit in two examples. After two applications of rule (11), we obtain the expansion

$$\delta_{J\cdot\cdot}^T = \delta_{J\cdot}^T \cup F_{J\cdot}^T \cdot I_{J\cdot}^T$$
$$= \delta_J^T \cup F_J^T I_J^T \cup F_J^T I_J^T,$$

in which that product $F_J^T I_J^T$ is redundant. If $null_J = null_K = \{\lambda\}$, then application of rules (11) and (10) gives us the expansion

$$\delta_{(JK)\cdot}^T = \delta_{JK}^T \cup F_{JK}^T I_{JK}^T$$
$$= \delta_J^T \cup \delta_K^T \cup F_J^T I_K^T \cup (F_J^T \cup F_K^T)(I_J^T \cup I_K^T),$$

in which product $F_J^T I_K^T$ is redundant.

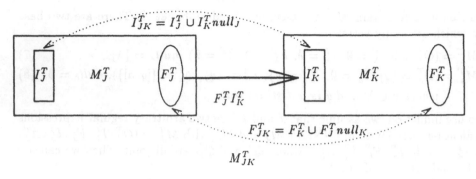

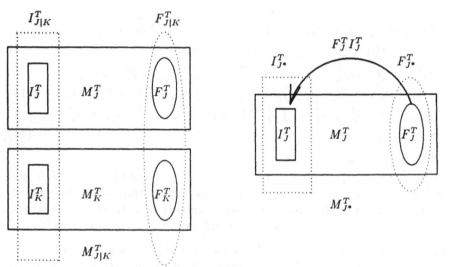

Fig. 3. Tail machine construction.

4 Faster NFA Construction

By recognizing the overlapping union $\delta_J^T \cup F_J^T I_J^T$ within formula (11) for $M_{J\bullet}^T$ as the source of inefficiency, we can maintain invariant $nred_J = F_J^T I_J^T - \delta_J^T$ in order to replace the overlapping union by the equivalent disjoint union $\delta_J^T \cup nred_J$. In order to maintain $nred_R$ as a component of the *tail* NNFA computation given above, we can use the following recursive definition, obtained by simplifying expression $F_R^T I_R^T - \delta_R^T$ and using the rules from the proof of Theorem 6.

$$nred_\lambda = \emptyset \tag{12}$$

$$nred_a = F_a^T I_a^T, \, where \, a \in \Sigma \tag{13}$$

$$nred_{J|K} = nred_J \cup nred_K \cup F_J^T I_K^T \cup F_K^T I_J^T \tag{14}$$

$$nred_{JK} = F_K^T I_J^T \cup null_K \, nred_J \cup null_J \, nred_K \tag{15}$$

$$nred_{J\bullet} = \emptyset \tag{16}$$

Rules (12), (13) and (16) are trivial. Rule (14) follows from applying distributive laws to simplify formula

$$nred_{J|K} = (F_J^T \cup F_K^T)(I_J^T \cup I_K^T) - (\delta_J^T \cup \delta_K^T)$$

Rule (15) is obtained by applying distributed laws to simplify formula,

$$nred_{JK} = (F_K^T \cup null_K F_J^T)(I_J^T \cup null_J I_K^T) - (\delta_J^T \cup \delta_K^T \cup F_J^T I_K^T)$$

The preceding idea embodies a general method of symbolic finite differencing for deriving efficient functional programs. This method has been mechanized and used extensively by Douglas Smith within his program transformation system called KIDS (see for example [19]).

Each union operation in the preceding rules is disjoint and, hence, $O(1)$ time implementable. However our solution creates a new problem. Potentially costly edges resulting from product operations occurring in rules (14) and (15) may be useless, because they are never incorporated into δ. These edge may be useless for two reasons – (1) if the regular expression is star free, and (2) if the edges are eliminated by rule (15).

To overcome this problem we will use lazy evaluation to compute products only when they actually contribute edges to the NNFA. Thus, instead of maintaining a union $nred_R$ of products, we will maintain a set $lazynred_R$ of pairs of sets. Consequently, the overlapping union $\delta_J^T \cup F_J^T I_J^T$ within formula (11) for $M_{J^\bullet}^T$ can be replaced by

$$\delta_J^T \cup (\cup_{[A,B] \in lazynred_J} AB) \tag{17}$$

However, this solution creates another problem: the sets forming F^T and I^T, which are computed by the rules to construct the *tail* of an NNFA, must be persistent in the following sense. Let the sets in the sequence forming F^T (respectively I^T) be called *F-sets* (respectively *I-sets*). Each *F-set* (respectively *I-set*) could be stored as a first (respectively second) component of a pair belonging to *lazynred*. Given any such pair, we need to iterate through the *I-set* G stored in the second component of the pair in $O(|G|)$ time.

The sequence of *F-sets* (respectively *I-sets*) are formed by two operations: 1. create a new singleton set; and 2. form a new set by taking the disjoint union of two previous sets in the sequence. Clearly, each of these sequences can be stored as a binary forest in which each subtree in the forest represents a set in the sequence, where the elements of the set are stored in the frontier. By construction each internal node in the forest has two children.

We call the forest storing the *F-sets* (respectively *I-sets*) the *F-forest* (respectively *I-forest*). For each node n belonging to the *F-forest* (respectively *I-forest*), let $Fset(n)$ (respectively $Iset(n)$) denote the *F-set* (respectively *I-set*) represented by n.

Each node in the *F-forest* and *I-forest* except the roots stores a parent pointer. Each node n in the *I-forest* also stores a pointer to the leftmost leaf of the subtree rooted in n and a pointer to the rightmost leaf of the subtree rooted n. The frontier nodes of the *I-forest* are linked.

This data structure preserves the unit-time disjoint union for *F-sets* and *I-sets*, and supports linear time iteration through the frontier of any node in the *I-forest*. Since all the *F-sets* and *I-sets* are subsets of the NNFA states Q, the *F-forest* and *I-forest* each is stored in $O(|Q|)$ space.

Theorem 7. *For any regular expression R we can compute $lazynred_R$ in time $O(r)$ and auxiliary space $O(s)$, where r is the size of regular expression R, and s is the number of occurrences of alphabet symbols appearing in R.*

Proof. If G and H are two sets, let $pair(G, H) = \{[G, H]\}$ if both G and H are nonempty; otherwise, let $pair(G, H) = \emptyset$. The proof makes use of the following recursive definition of $lazynred_R$ obtained from the recursive definition of $nred_R$.

$$lazynred_\lambda = \emptyset \tag{18}$$

$$lazynred_a = pair(F_a^T, I_a^T), \text{where } a \in \Sigma \tag{19}$$

$$lazynred_{J|K} = lazynred_J \cup lazynred_K \cup pair(F_J^T, I_K^T) \cup pair(F_K^T, I_J^T) \tag{20}$$

$$lazynred_{JK} = pair(F_K^T, I_J^T) \cup null_K lazynred_J \cup null_J lazynred_K \tag{21}$$

$$lazynred_{J^*} = \emptyset \tag{22}$$

Operation $pair(G, H)$ takes unit time and space. Each union operation occurring in the rules above is disjoint and, hence, implementable in unit time. Rule (19) contributes unit time and space for each alphabet symbol occurring in R, or $O(s)$ time and space overall. Rule (20) contributes unit time for each alternation operator appearing in R or $O(r)$ time overall. It contributes two units of space for each alternation operator both of whose alternands contain at least one alphabet symbol. Hence, the overall space contributed by this rule is less than $2s$. By a similar argument, Rule (21) contributes $O(r)$ time and less than s space overall. The other two rules contribute no more than $O(r)$ time overall. Hence, the time and space needed to compute $lazynred_R$ is $O(r)$ and $O(s)$ respectively. □

By Theorems 6 and 7, and by the fact that $nred_R$ can be computed from $lazynred_R$ in $O(|nred_R|)$ time using formula (17), we have our first theoretical result.

Theorem 8. *For any regular expression R we can compute an equivalent $MYNNFA$ with $s + 1$ states in time $O(r + m)$ and auxiliary space $O(s)$, where r is the size of regular expression R, m is the number of edges in the $MYNNFA$, and s is the number of occurrences of alphabet symbols appearing in R.*

5 Improving Space for McNaughton and Yamada's NFA

Theorem 8 leads to a new algorithm that computes the adjacency form of the MYNNFA M_R in a single left-to-right shift/reduce parse of the regular expression R. Although this improves upon the algorithm of Berry and Sethi, McNaughton and Yamada's NFA has certain theoretical disadvantages over Thompson's simpler NFA. Recall that for regular expression $(((((a_1^*|a_2)^*|a_3)^*...a_k)^*$ the number of edges in McNaughton and Yamada's NFA is the square of the number of edges in Thompson's NFA.

Nevertheless, we can modify the algorithm just given so that in $O(r)$ time it produces an $O(s)$ space compressed NNFA that encodes McNaughton and Yamada's NFA, and that supports acceptance testing in $O(s|x|)$ time. In the same way that $nred_R$ was represented more compactly as $lazynred_R$, we can represent δ_R, which is a union of cartesian products, as a set $lazy\delta_R$ of pairs of set-valued arguments of these products. If M_R is the compressed NNFA equivalent to regular expression R, then the rules for M_R^T are given just below:

$$lazy\delta_\lambda^T = \emptyset \tag{23}$$
$$lazy\delta_a = \emptyset \tag{24}$$
$$lazy\delta_{J|K}^T = lazy\delta_J^T \cup lazy\delta_K^T \tag{25}$$
$$lazy\delta_{JK}^T = pair(F_J^T, I_K^T) \cup lazy\delta_J^T \cup lazy\delta_K^T \tag{26}$$
$$lazy\delta_{J*}^T = lazy\delta_J^T \cup lazynred_J \tag{27}$$

After the preceding rules are processed we can obtain a representation for M_R by introducing a new state q_0 and by adding the pair $[q_0, I_R^T]$ to $lazy\delta_R^T$ in accordance with equations (6).

We now show how to use $lazy\delta_R$ to simulate δ_R. If V is a subset of the MYNNFA states Q, then we can compute the collection of states $\delta(V, a)$ for all of the alphabet symbols $a \in \Sigma$ as follows. First we compute

$$finddomain(V) = \{X : [X, Y] \in lazy\delta | V \cap X \neq \emptyset\}$$

which is used to find the set of next states

$$next\_states(V) = \{Y : [X, Y] \in lazy\delta | X \in finddomain(V)\}$$

Finally, for each alphabet symbol $a \in \Sigma$, we see that

$$\delta(V, a) = \{q : Y \in next\_states(V), q \in Y | A(q) = a\}$$

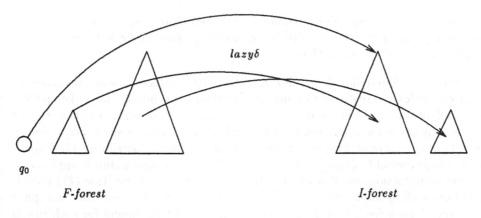

Fig. 4. Compressed NNFA organization.

In order to explain how $lazy\delta$ is implemented, we will use some additional terminology. For each F-set G represented by node n in the F-forest, n stores a pointer to a list of nodes in the I-forest representing set $\{Y : [G, Y] \in lazy\delta\}$. Furthermore, the F-forest and I-forest are compressed to only store nodes representing sets that appear as the first or second components of a pair $[X, Y] \in lazy\delta$. In other words, we make *lazyδ* a total onto binary relation. This can be achieved on-line as the *F-forest* and *I-forest* are constructed by a kind of path compression that affects the preprocessing time and space by no more than a small constant factor. Refer to Fig. 4 to see how the compressed NFA's are organized. Fig. 5 illustrates a compressed NNFA equivalent to regular expression $(a|b)^*abb$. Thus, we have

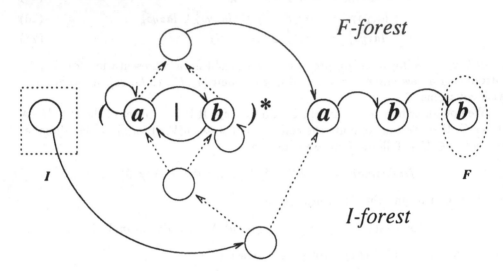

Fig. 5. A Compressed NNFA equivalent to regular expression $(a|b)^*abb$

Theorem 9. *For any regular expression R, its equivalent compressed NNFA, consisting of F-forest, I-forest and lazyδ, takes up $O(s)$ space and can be computed in time $O(r)$ and auxiliary space $O(s)$.*

Proof. Since each internal node in the *F-forest* and *I-forest* have at least two children, and since their leaves are distinct occurrences of alphabet symbols, they take up $O(s)$ space. Each of the unions in the rules to compute $lazyδ^T$ is disjoint, and hence takes unit time. By the same argument used to analyze the overall space contributed by Rule (21) in the proof of Theorem 7, we see that Rule (26) contributes $O(s)$ space and $O(r)$ time overall to $lazyδ_R^T$. By Rule (22), Theorem 7, and a simple application of structural induction, we also see that the space contributed by Rule (27) (which results from adding *lazynred* to $lazyδ^T$) overall is $O(s)$. It takes unit time and space to construct $lazyδ_R$ from $lazyδ_R^T$ and $null_R$. The overall time bound for each rule is easily seen to be $O(r)$. □

The compressed NNFA also supports an efficient evaluation of the three preceding queries in order to simulate transition map $δ$. The best previous worst case time bound for giving a subset V of states and computing the collection of sets $δ(V, a)$ for all of the alphabet symbols $a \in \Sigma$ is $\Theta(|V| \times |δ(V, \Sigma)|)$ using an adjacency list implementation of McNaughton and Yamada's NFA, or $\Theta(r)$ using Thompson's NFA.

In Theorem 11 we improve this bound, and obtain, essentially, optimal asymptotic time without exceeding $O(s)$ space. This is our main theoretical result. It explains the apparent superior performance of acceptance testing using the compressed NNFA over Thompson's. It explains more convincingly why constructing a DFA starting from the compressed NNFA is at least one order of magnitude faster than when we start from either Thompson's or McNaughton and Yamada's NFA. These empirical results are presented in section 7.

Before proving the theorem, we will first prove the following technical lemma.

Lemma 10. *Let V be a set of states in the compressed NNFA built from regular expression R, and let $lazy\delta_V = \{[X, Y] : [X, Y] \in lazy\delta | X \cap V \neq \emptyset\}$. Then $|lazy\delta_V| = O(|V| + |\delta(V, \Sigma)|)$.*

Proof. The result follows from proving that $O(|V| + |\delta(V, \Sigma)|)$ is a bound for each of the subsets of $lazy\delta_V$ contributed by rules (19), (20), (21), and (26) respectively. The bound holds for subsets contributed by rules (19), (20), and (26), because they form one-to-one maps.

The proof for the subset contributed by (21) is split into two cases. For convenience, let V_J denote the set of states in V such that their corresponding symbol occurrences appear in regular expression J, where J is a subexpression of R. First, consider the set A of pairs $[F_K^T, I_J^T] \in lazy\delta_V$ for subexpressions JK, where $V_J = \emptyset$. We claim that these edges form a one-to-many map, which implies the bound. Suppose this were not the case. Then we would have a subexpression JK, and a subexpression J_1J_2 of J such that $I_J^T = I_{J_1}^T$ and pairs $[F_K^T, I_J^T]$ and $[F_{J_2}^T, I_{J_1}^T]$ belonging to A. However, since J contains no occurrence of an alphabet symbol in V, then J_2 does not either. Hence, the pair $[F_{J_2}^T, I_{J_1}^T]$ cannot belong to A. Hence, the claim holds.

Next, consider the set B of pairs $[F_K^T, I_J^T] \in lazy\delta_V$ for subexpressions JK, where $V_J \neq \emptyset$. Proceeding from inner-most to outer-most subexpression JK, we charge each pair $[F_K^T, I_J^T] \in B$ to an uncharged state in V_J. A simple structural induction would show that V_J contains at least one uncharged state. Let J_1J_2 be an inner-most subexpression in R such that V_{J_1} is nonempty, and $[F_{J_2}^T, I_{J_1}^T] \in lazy\delta_V$. Then both V_{J_1} and V_{J_2} contains at least one uncharged state. After an uncharged state in V_{J_1} is charged, $V_{J_1J_2}$ still contains an uncharged state from V_{J_2}. The inductive step is similar. The result follows. $\square$

Theorem 11. *Given any subset V of the compressed NNFA states, we can compute all of the sets $\delta(V, a)$ for every alphabet symbols $a \in \Sigma$ in time $O(|V| + |\delta(V, \Sigma)|)$.*

Proof. The sets belonging to $finddomain(V)$ are represented by all the nodes P_V along the paths from the states belonging to V to the roots of the *F-forest*. These nodes P_V can be found in $O(|V| + |P_V|)$ time by a marked traversal of parent pointers in the forest. Observe that $|P_V|$ can be much larger than $|V|$.

Computing $next\_states(V)$ involves two steps. First, for each node $n \in P_V$, we traverse a nonempty list of nodes in the *I-forest* representing $\{Y : [Fset(n), Y] \in lazy\delta\}$. This step takes time linear in the sum of the lengths of these lists. (Observe that this number can be much larger than $|P_V|$.) Second, if D_V is the set of all nodes

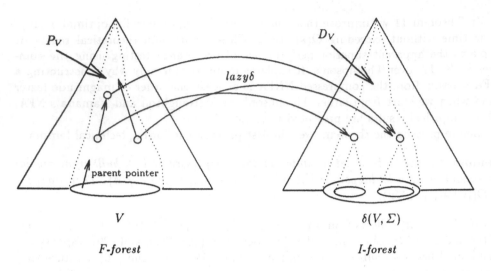

Fig. 6. To compute $\delta(V, a)$ in a compressed NNFA

in the *I-forest* belonging to these lists, then $next\_states(V) = \{Iset(n) : n \in D_V\}$. We can compute the set $next\_states(V)$ in $O(|\{[Fset(n), Y] : n \in P_V, [Fset(n), Y] \in lazy\delta\}|) = O(|V| + |\delta(V, \Sigma)|)$ time by Lemma 10.

Calculating $\delta(V, \Sigma)$ involves computing the union of the sets belonging to $next\_states(V)$. This is achieved in $O(|\delta(V, \Sigma)|)$ time using the left and right descendant pointers stored in each node belonging to D_V, traversing the unmarked leaves in the frontier, and marking leaves as they are traversed. Multiset discrimination [8] can be used to separate out all of the sets $\{q \in \delta(V, \Sigma)|A(q) = a\}$ for each $a \in \Sigma$ in time $O(|\delta(V, \Sigma)|)$. See Fig. 6 for a illustration of $\delta(V, a)$ computation. □

Consider an NFA constructed from the following regular expression:

$$(\overbrace{\lambda|(\lambda|(\cdots(\lambda|a)^*)^*)\cdots)^*}^{k \ *'s})^n$$

In order to follow transitions labeled 'a', we have to examine $\Theta(n^2)$ edges and $\Theta(n)$ states in $\Theta(n^2)$ time for McNaughton and Yamada's NFA, $\Theta(kn)$ states and edges in $\Theta(kn)$ time for Thompson's machine, and $\Theta(n)$ states and edges in $\Theta(n)$ time for the compressed NNFA.

6 Further Optimization

In this section, we introduce three simple transformations, *packing*, *path compression* and *tree contraction*, which can greatly improve the compressed NNFA representation. If $lazy\delta$ contains both $[F_1, I]$ and $[F_2, I]$, and if there exists an *F-set* $F = F_1 \cup F_2$, then the *F-set promotion* transformation packs $[F_1, I]$ and $[F_2, I]$ within $lazy\delta$ into a single pair $[F, I]$ (see Fig. 7). Similarly, if $lazy\delta$ contains both $[F, I_1]$ and $[F, I_2]$, and if there exists an *I-set* $I = I_1 \cup I_2$, then the *I-set promotion* transformation packs $[F, I_1]$

and $[F, I_2]$ within $lazy\delta$ into a single pair $[F, I]$ (see Fig. 7). In a single linear time bottom up traversal of the F-forest performing I-set and F-set promotion, the packing transformation can dramatically simplify the representation of $lazy\delta$.[9] In the case of regular expression $((a_1|\lambda)(\cdots((a_{s-1}|\lambda)(a_s|\lambda)^*)^* \cdots)^*)^*$ packing can simplify $lazy\delta$ from $3s - 1$ pairs into two pairs. The original compressed NNFA equivalent to $((a_1|\lambda)(\cdots((a_{s-1}|\lambda)(a_s|\lambda)^*)^* \cdots)^*)^*$ is shown in Fig. 8; the compressed NNFA resulting from packing transformation is illustrated in Fig. 9.

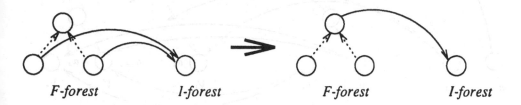

F-forest   I-forest   F-forest   I-forest

F-set Promotion

F-forest   I-forest   F-forest   I-forest

I-set Promotion

Fig. 7. *F-set* and *I-set* promotions

At the same time, we can carry out the same kind of path compression described in Section 6, so that the *F-forest* and *I-forest* only contain nodes in the domain (respectively range) of $lazy\delta$. However, whereas previously the forest leaves (corresponding to NFA states) were unaffected by compression, the packing transformation can remove leaves in the *F-forest* and *I-forest* from the domain and (respectively) range of $lazy\delta$. When path compression eliminates leaves, we need to turn the symbol assignment map A into a multi-valued mapping; that is, whenever leaves q1,...,qk are replaced by leaf q, we take the following steps;

- remove the old leaves q1,...,qk from the domain of A;
- assign the set of symbols $\{y : x \in \{q1, ..., qk\}, [x, y] \in A\}$ to A at q.

As an example of this, consider the expression $((a_1|\lambda)(\cdots((a_{s-1}|\lambda)(a_s|\lambda)^*)^* \cdots)^*)^*$ once again. Path compression will turn the data structure into the one depicted in Fig. 10. The original compressed NNFA has $4s - 1$ states, and after path compression it has only two states. In using our compressed representation to simulate an NFA,

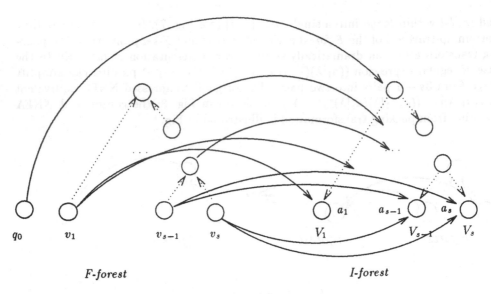

Fig. 8. A compressed NNFA equivalent $((a_1|\lambda)(\cdots((a_{s-1}|\lambda)(a_s|\lambda)^*)^*\cdots)^*)^*$ without optimization.

the transition edge t can be taken only if the current transition symbol belongs to $\{a_1, a_2, \cdots a_s\}$ which labels node C_1.

The reduced number of NFA states resulting from packing and path compression partly explains the superior performance of the compressed NNFA in both acceptance testing and DFA construction. Fig. 11 illustrates a compressed NFA resulting from applying packing and path compression to Fig. 5.

The tree contraction transformation is like the inverse of packing. It works as follows: (1) when an internal *F-forest* node n has k_1 outgoing edges and k_2 incoming edges, and if $k_1 k_2 \leq k_1 + k_2$, then we can replace node n and the $k_1 + k_2$ edges incident to n by $k_1 k_2$ edges (see Fig. 12); and (2) when an internal *I-forest* node n has k_1 incoming edges and k_2 outgoing edges, and if $k_1 k_2 \leq k_1 + k_2$, then we can replace node n and the $k_1 + k_2$ edges incident to n by $k_1 k_2$ edges (see Fig. 12). After applying the tree contraction transformation to the compressed NNFA of Fig. 11, one *I-forest* node is eliminated (see Fig. 13). In the remainder of this paper, we call the optimized compressed NNFA representation the CNNFA. Fig. 13 illustrates an CNNFA equivalent to regular expression $(a|b)^* abb$. It contains 5 states and 6 edges in contrast to the 9 states and 14 edges found in the MYNNFA of Fig. 5.

7 Performance Benchmark

Experiments to benchmark the performance of the CNNFA have been carried out for a range of regular expression patterns against a number of machines including Thompson's NFA, an optimized form of Thompson's NFA, and McNaughton and Yamada's NFA[14]. We build Thompson's NFA according to the construction rules described in [2]. Thompson's NFA usually contains redundant states and λ-edges. However, to our knowledge there is no obvious/efficient algorithm to optimize

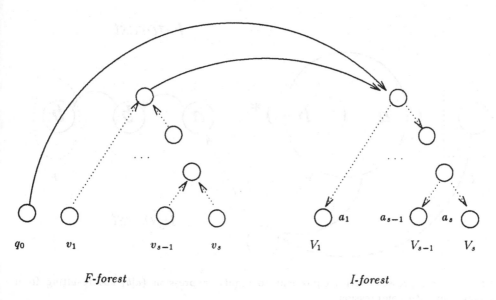

Fig. 9. A compressed NNFA equivalent to $((a_1|\lambda)(\cdots((a_{s-1}|\lambda)(a_s|\lambda)^*)^*\cdots)^*)^*$ resulting from Packing.

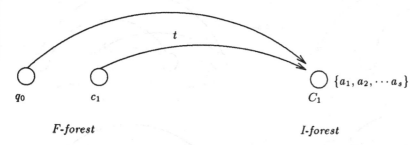

Fig. 10. A compressed NNFA equivalent to $((a_1|\lambda)(\cdots((a_{s-1}|\lambda)(a_s|\lambda)^*)^*\cdots)^*)^*$ resulting from packing and path compression.

Thompson's NFA without blowing up the linear space constraint. We therefore devise some simple but effective transformations that eliminate redundant states and edges in most of the test cases.

Our acceptance testing experiments show that the CNNFA outperforms Thompson's NFA, Thompson's NFA optimized, and McNaughton and Yamada's NFA. See Fig. 14 for an acceptance testing benchmark summary. The benchmark summary indicates that the CNNFA is slower than all other machines for $(abc\cdots)$ and $(abc\cdots)^*$ patterns. This is an anomalous shortcoming of our current implementation, which will be eliminated in the next version.

The benchmark for subset construction is more favorable. The CNNFA outperforms the other machines not only in DFA construction time but also in constructed machine size. Subset construction is compared among the following five starting machines: the CNNFA, Thompson's NFA, Thompson's NFA optimized, Thompson's NFA using kernel items heuristic[2], and McNaughton and Yamada's NFA. See Fig.

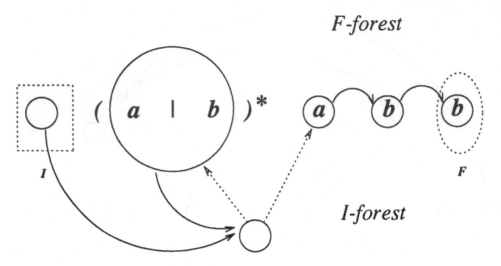

Fig. 11. A compressed NNFA equivalent to regular expression $(a|b)^*abb$ resulting from packing and path compression

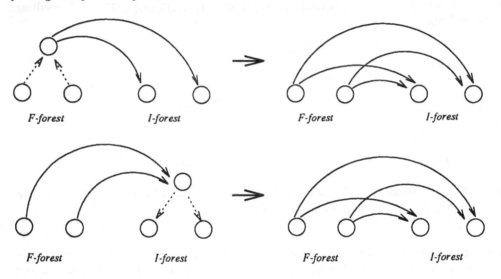

Fig. 12. Tree contraction

15 for a high level specification of the classical Rabin and Scott subset construction for producing a DFA σ from an NFA δ.

We implemented the subset construction specification tailored to the CNNFA and other machines. The only differences in these implementations is in the calculation of $\delta(V, \Sigma)$, where we use the efficient procedure described by Theorem 11, and in the ϵ-closure step, which is performed only by Thompson's NFA and Thompson's NFA optimized. The CNNFA achieves linear speedup and constructs a linearly smaller DFA in many of the test cases. See Fig. 16 and 17 for a benchmark summary. The raw timing data are found in [?]. All the tests described in this paper are performed

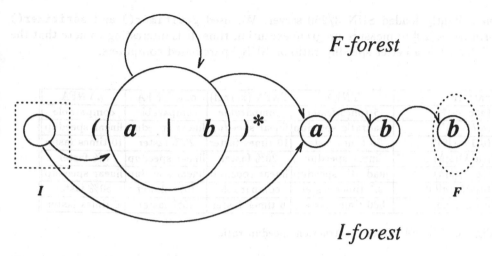

Fig. 13. An CNNFA equivalent to regular expression $(a|b)^*abb$

pattern	TNFA	opt. TNFA	MYNFA
$(abc\cdots)$	75% slower	55% slower	75% slower
$(a|b|\cdots)^*$	12 times faster	2 times faster	50% faster
$((a|\lambda)(b|\lambda)\cdots-)^*$	2 times faster	25% faster	80% slower
$((a|\lambda)(b|\lambda)\cdots)^*$	16 times faster	8 times faster	50% faster
$((a|\lambda)^n-)^*$	comparable	50% slower	linearly faster
programming language	7 times faster	60% faster	2 times faster

Fig. 14. CNNFA acceptance test speedup ratio

```
σ := ∅
workset := {{q₀}}
while ∃V ∈ workset do
        workset := workset - {V}
        for each symbol a ∈ Σ and set of states B = {x ∈ δ(V, Σ)|A(x) = a},
                where B ≠ ∅ do
            σ(V, a) := B
            B := ε-closure(B)
            if B does not belong to the domain of σ then
                    workset := workset ∪{B}
            end if
        end for
end while
```

Fig. 15. Rabin and Scott's subset construction

on a lightly loaded SUN 3/250 server. We used `getitimer()` and `setitimer()`
primitives [20] to measure program execution time. It is interesting to note that the
CNNFA has a better speedup ratio on SUN Sparc based computers.

pattern	TNFA	TNFA (kernel)	opt. TNFA	MYNFA
$(abc\cdots)^*$	5 times faster	comparable	comparable	comparable
$(a\|b\|\cdots)^*$	quadratic speedup	linear speedup	linear speedup	linear speedup
$(0\|1\cdots\|9)^n$	70 times faster	10 times faster	20% faster	10 times faster
$((a\|\lambda)(b\|\lambda)\cdots-)^*$	linear speedup	20% faster	linear speedup	5% faster
$((a\|\lambda)(b\|\lambda)\cdots)^*$	quadratic speedup	linear speedup	linear speedup	linear speedup
$(a\|b)^*a(a\|b)^n$	2.5 times faster	comparable	10% slower	50% faster
prog. lang.	800 times faster	6 times faster	40% faster	6 times faster

Fig. 16. CNNFA subset construction speedup ratio

pattern	TNFA	TNFA (kernel)	opt. TNFA	MYNFA
$(abc\cdots)^*$	comparable	comparable	comparable	comparable
$(a\|b\|c\cdots)^*$	linearly smaller	linearly smaller	comparable	linearly smaller
$(0\|1\cdots9)^n$	200 times smaller	10 times smaller	comparable	10 times smaller
$((a\|\lambda)(b\|\lambda)\cdots-)^*$	3 times smaller	comparable	comparable	comparable
$((a\|\lambda)(b\|\lambda)\cdots)^*$	linearly smaller	linearly smaller	comparable	linearly smaller
$(a\|b)^*a(a\|b)^n$	4 times smaller	comparable	comparable	comparable
prog. lang.	10 times smaller	5 times smaller	comparable	5 times smaller

Fig. 17. DFA size improvement ratio starting from the CNNFA

Recently at Columbia University's Theory Day, Aho reported a highly efficient
heuristic for deciding whether a given string belongs to the language denoted by
a regular expression, i.e. both string and regular expression are dynamic(cf. page
128 of [2]). This problem is needed for UNIX tools such as egrep. Aho's heuristic
constructs a McNaughton and Yamada's NFA first, and subsequently builds a DFA
piecemeal as the input string is scanned from left to right. Benchmarks showing
substantial computational improvement in adopting the CNNFA to Aho's heuristic
are found in [9].

8 Conclusion

Theoretical analysis and confirming empirical evidence demonstrates that our proposed CNNFA leads to a substantially more efficient way of turning regular expressions into DFA's (and minimum state DFA's in particular) than other NFA's in
current use. It would be interesting future research to analyze the effect of packing and path compression on the CNNFA. It would also be worthwhile to obtain a
sharper analysis of the constant factors in comparing the CNNFA with other NFA's.

9 Acknowledgement

We are grateful for helpful comments from Anne Brüggemann-Klein.

References

1. Aho, A., Hopcroft, J. and Ullman J., "Design and Analysis of Computer Algorithms", *Reading*, Addison-Wesley, 1974.
2. Aho, A., Sethi, R. and Ullman, J., "Compilers Principles, Techniques, and Tools", *Reading*, Addison-Wesley, 1986.
3. Aho, A., "Pattern Matching in Strings", in *Formal Language Theory*, ed. R. V. Book, Academic Press, Inc. 1980.
4. Berry, G. and Cosserat, L., "The Esterel synchronous programming language and its mathematical semantics" in *Seminar in Concurrency*, S. D. Brookes, A. W. Roscoe, and G. Winskel, eds., LNCS 197, Springer-Verlag, 1985.
5. Berry, G. and Sethi, R., "From Regular Expressions to Deterministic Automata" *Theoretical Computer Science*, 48 (1986), pp. 117-126.
6. Brüggemann-Klein, A., "Regular Expressions into Finite Automata", To appear in *Theoretical Computer Science*, 1992.
7. Brzozowski, J., "Derivatives of Regular Expressions", *JACM*, Vol. 11, No. 4., Oct. 1964, pp. 481-494.
8. Cai, J. and Paige, R., "Look Ma, No Hashing, And No Arrays Neither", *ACM POPL*, Jan. 1991, pp. 143 - 154.
9. Chang, C., Ph. D. Thesis, *To Appear*, 1992.
10. Emerson, E. and Lei, C., "Model Checking in the Propositional Mu-Calculus", *Proc. IEEE Conf. on Logic in Computer Science*, 1986, pp. 86 - 106.
11. Hopcroft, J. and Ullman, J., "Formal Languages and Their Relation to Automata",*Reading*, Addison-Wesley, 1969.
12. Kleene, S., "Representation of events in nerve nets and finite automata", in *Automata Studies, Ann. Math. Studies No. 34*, Princeton U. Press, 1956, pp. 3 - 41.
13. Knuth, D., "On the translation of languages from left to right", *Information and Control*, Vol. 8, Num. 6, 1965, pp. 607 - 639.
14. McNaughton, R. and Yamada, H. "Regular Expressions and State Graphs for Automata", *IRA Trans. on Electronic Computers*, Vol. EC-9, No. 1, Mar. 1960, pp 39-47.
15. Myhill, J., "Finite automata and representation of events," WADC, Tech. Rep. 57-624, 1957.
16. Nerode, A., "Linear automaton transformations," *Proc. Amer. Math Soc.*, Vol. 9, pp. 541 - 544, 1958.
17. Rabin, M. and Scott, D., "Finite automata and their decision problems" *IBM J. Res. Develop.*, Vol. 3, No. 2, Apr., 1959, pp. 114 - 125.
18. Ritchie, D. and Thompson, K. "The UNIX Time-Sharing System" *Communication ACM*, Vol. 17, No. 7, Jul., 1974, pp. 365 - 375.
19. Smith, D., "KIDS - A Knowledge-Based Software Development System", in Proc. Workshop on Automating Software Design, AAAI-88, 1988.
20. "SunOS Reference Manual VOL. II", *Programmer's Manual*, SUN microsystems, 1989.
21. Thompson, K., "Regular Expression search Algorithm", *Communication ACM* 11:6 (1968), pp. 419-422.
22. Ullman, J., "Computational Aspects of VLSI", *Computer Science Press*, 1984.

Identifying Periodic Occurrences of a Template with Applications to Protein Structure

Vincent A. Fischetti[1], Gad M. Landau[2]*, Jeanette P. Schmidt[2]**, and Peter H. Sellers[1]

[1] Rockefeller University, 1230 York Avenue, New York, NY 10021.
[2] Polytechnic University, 333 Jay Street, Brooklyn, NY 11201.

Abstract. We consider a string matching problem where the pattern is a template that matches many different strings with various degrees of perfection. The quality of a match is given by a penalty matrix that assigns each pair of characters a score that characterizes how well the characters match. Superfluous characters in the text and superfluous characters in the pattern may also occur and the respective penalties for such gaps in the alignment are also given by the penalty matrix. For a text T of length n, and a template P of length m, we wish to find the best alignment of T with P^n, which is the concatenation of n copies of P, (m will typically be much smaller than n). Such an alignment can simply be obtained by solving a dynamic programming problem of size $O(n^2 m)$, and ignoring the periodic character of P^n. We show that the structure of P^n can be exploited and the problem reduced to essentially solving a dynamic programming of size $O(mn)$. If the complexity of computing gap penalties is $O(1)$, (which is frequently the case), our algorithm runs in $O(mn)$ time. The problem was motivated by a protein structure problem.

1 Introduction

String matching and its many generalizations is a widely studied problem in computer science. One possible generalization that has been researched is *approximate string matching* - finding occurrences of a pattern in a text where *differences* (insertions and deletions) are allowed and *matches* may be defined by a function, with values in some range $(-r, r)$, which specifies how well a character from the pattern "matches" a given character in the text. (Positive values indicate "favorable matches", while negative values indicate "unfavorable matches".)

Given a text T of length n, and a pattern P of length m, in the exact string matching problem one finds all the locations (t_i) in the text such that $P = t_i t_{i+1} \ldots t_{i+m-1}$. When differences are allowed, however, every location in the text matches the pattern with some differences. A clarification of the definition of a "match" of the pattern is therefore needed. Most known algorithms for approximate string matching have two basic steps. In the first step each substring of the text receives a score, which reflects the quality of the match between the pattern and the given substring. In

* Partially supported by NSF grant CCR-8908286.
** Partially supported by NSF grant CCR-9110255 and the New York State Science and Technology Foundation Center for Advanced Technology.

the second step some locations are chosen as sufficiently "good" matches, worthy to be listed as "matches" in the output. The decision of what constitutes a "good" match can be made in many ways: [U-83], [LV-89] and [GP-90] use the edit distance between the matched strings as a measure (i.e the number of insertions, deletions and mismatches, which have to be performed to obtain the matched string from the pattern), and declare every match whose edit distance is below a predetermined threshold a "good" match. Many other measures have been defined in the literature [SK-83]. Note that the measure of the quality of a match between the pattern and a substring of the text depends on the application. In many applications in molecular biology a penalty table is given. This table gives the penalty value for the deletion and insertion of each letter of the alphabet, as well as the value for matching any pair of characters. In the simplest case the score of a match is simply the sum of the corresponding values in the penalty matrix. In some cases however gaps (successive insertions or deletions) get penalties that are different from the sum of the penalties of each insertion (deletion). A discussion on this subject is given in [GG-89].

In the approximate string matching problem discussed here, the pattern is not uniquely defined. Given a pattern P, we would like to detect whether P (or a cyclic rotation of P) is periodically repeated in the text. We define $\bar{P} = P^n$, such that P^n is built by concatenating n copies of P. Our goal can be restated as finding the best match between a substring of $\bar{P}$ and a substring of the text. Notice that the substrings of $\bar{P}$ can be characterized by a suffix of P followed by i copies of P followed by a prefix of P. Since P^n is of length mn, it is very likely that the best match with T will be attained with a much shorter subsequence of $\bar{P}$, presumably of length no more than n, unless the penalty matrix gives very high rewards for matches and very minor penalties to gaps. Our algorithm is particularly useful if m is much smaller than n, and is designed to find many repetitions of P.

Section 2 contains a short description of the motivation of the problem, which comes from protein structure determination. The exact definition of the problem is given in Section 3. Section 4 describes a simple dynamic programming algorithm that solves the problem by computing $O(mn^2)$ values. Section 5 describes the new algorithm.

2 Protein structure motivation

The secondary structure of proteins is critical for their proper function. For the majority of proteins, the secondary structure can not simply be resolved by examining the sequence of the amino acids that define the protein. While x-ray crystallography is the only reliable means available today by which to solve the structure of a protein molecule, it can not always be used. Some algorithms have been designed to help estimate the conformation of proteins based on the amino acid sequence, employing the chemical characteristics of the amino acids and their position within the sequence ([CF-74], [GOR-78]). Structural factors of a molecule are important in the overall protein conformation. The α-helical structure is the most common because of its ability in stabilizing proteins through short regions of helix-helix packing, referred to as a coiled-coil interaction, ([CP-86], [FP-92]).

The ability to form an α-helical coiled-coil structure is based on the presence of a seven-residue repeat pattern denoted by (a-b-c-d-e-f-g). The first and the fourth position, (a and d), are generally apolar or hydrophobic amino acids which, in the context of the α-helix, form an inclined stripe around the axis of the helix [CP-86]. Discontinuities in the heptad pattern, such as changes in frame have been seen in coiled-coil molecules. To date, the method used to determine the presence of a heptad pattern in protein molecules is by inspection.

Note that in the above application the alphabet of the pattern and that of the text are disjoint. The pattern consists of the seven positions (a-b-c-d-e-f-g), while the alphabet of the text consists of the usual alphabet that represents proteins: the 20 amino acids. A protein features a coiled-coil structure if it contains some (relatively long) 7-residue periodicity.

Based on the algorithm presented in this manuscript a program was developed which determines whether a protein exhibits the 7-residue periodicity, [FP-92]. To determine this periodicity a penalty matrix based on the knowledge of the typical sequence structure of coiled-coil protein was developed. The method used to construct the penalty matrix is a type of profile analysis, ([GLE-87], [LLE-91]).

The quality of the fit of a given amino acid to a given position in the heptad was based on data tabulated in [CP-90] and [LD-91], for four known classes of coiled-coil proteins, for which the 7-residue periodicity had been determined by inspection. For each amino acid i the average frequency p_{ij} of occurring in position j, (in the four classes of known coiled-coil molecules), was divided by the overall frequency P_i of amino acid i in GENBANK. To avoid values of $-\infty$, in case p_{ij} is zero, we used $\ln((p_{ij} + \epsilon)/(P_i + \epsilon))$, (for some small $\epsilon > 0$), as the penalty for placing amino acid i in position j. ϵ was chosen of the order of the rounding error used in the computation of p_{ij} and P_i.

High positive values therefore correspond to a very good fit, while low negative numbers indicate a bad fit. Gap penalties were chosen empirically, and took into account that a given sequence may contain several coiled-coil regions, separated by blocks that do not have a coiled-coil conformation and do therefore not exhibit the 7-residue periodicity.

Inspired by the above application, this paper addresses the problem of determining a repeat structure in a text for a pattern of arbitrary length and an arbitrarily specified penalty matrix.

3   The problem

We are given:
(i) a text $T = t_1 \ldots t_n$ over an alphabet Σ_1;
(ii) a pattern $P = p_1 \ldots p_m$ over an alphabet Σ_2;
(iii) a penalty matrix M defined over $\Sigma_1 \times \Sigma_2$;
(iv) a gap function G, defined over all pairs (τ, ρ), where τ is a substring of T and ρ is a substring of any cyclic rotation of P.

Let $\bar{P} = P^n$, where P^n is built by concatenating n copies of P. For convenience of notation we also define $p_0 = p_m$, so that each character $\bar{p}_j$ in $\bar{P}$ is equal to $p_{j \bmod m}$.

The penalty matrix M gives the value of every possible alignment of a character of the pattern and a character of the text. As indicated earlier, negative values in this matrix indicate a bad match, while positive values indicate a good match.

The gap function G has four parameters, two indices from the text and two indices from the pattern. $G(i_1, i_2, j_1, j_2)$ is defined for $0 < i_1 \leq i_2 \leq n$ and $0 \leq j_1, j_2 \leq m$. $G(i_1, i_2, j_1, j_2)$ is the penalty for simultaneously deleting the substring $t_{i_1+1} t_{i_1+2} \ldots t_{i_2}$ from the text and the substring of length $(j_2 - j_1) \pmod m$ in the periodic pattern which starts after character p_{j_1}. For convenience of notation we will denote the above substring simply by $p_{(j_1+1)} \ldots p_{j_2}$, (i.e. if $j_2 < j_1 < m$, $p_{(j_1+1)} \ldots p_{j_2}$ stands for $p_{j_1+1} \ldots p_m p_1 \ldots p_{j_2}$; if $j_2 < j_1 = m$ it stands for $p_1 \ldots p_{j_2}$ and if $j_2 = j_1$ it denotes the empty substring). Note that:

(1) the string $p_{(j_1+1)} \ldots p_{j_2}$ represents all substrings $\bar{p}_{(x_1+1)} \bar{p}_{(x_1+2)} \ldots \bar{p}_{x_2}$ of the same length in $\bar{P}$, for which $x_1 = j_1 \pmod m$ and $x_2 = j_2 \pmod m$;

(2) substrings of length m or more in $\bar{P}$ do not have a representative, since there could not possibly be a gain of "jumping" over an entire period of the pattern, and therefore the corresponding gap penalty has not been defined.

We also require G to obey the following properties:

$$G(i, i, j, j) = 0 \qquad (a)$$
$$\text{for } i_1 < i_2 \text{ or } j_1 \neq j_2 \quad G(i_1, i_2, j_1, j_2) < 0 \qquad (b)$$

as well as the triangle inequality:

$$\forall i_1 \leq i \leq i_2, \text{ and } \forall j_1, j, j_2 \ |G(i_1, i_2, j_1, j_2)| \leq |G(i_1, i, j_1, j)| + |G(i, i_2, j, j_2)| \ (c)$$

A simple and quite frequently considered case occurs when the gap function G is additive, and there is a function g defined on $\Sigma_1 \cup \Sigma_2$ so that:

$$G(i_1, i_2, j_1, j_2) = \sum_{i_1 < i \leq i_2} g(t_i) + \sum_{j = (j_1+1) \ldots (j_2) \pmod m} g(p_j), \text{ (note that } p_0 = p_m).$$

Definition: An alignment between the substring $t_i t_{i+1} \ldots t_\ell$ of T and the substring $\bar{p}_j \bar{p}_{j+1} \ldots \bar{p}_s$ of $\bar{P}$ is specified by two strictly increasing sequences of indices ($i \leq i_1, i_2, \ldots i_w \leq \ell$) and ($j \leq j_1, j_2, \ldots j_w \leq s$), with the interpretation that for $h = 1 \cdots w$, t_{i_h} is matched to $p_{j_h \bmod m}$, and all other characters in the substrings are unmatched. The score of the alignment is specified by the penalty matrix M and the gap function G, and is given by the following expression, where $i_{w+1} = \ell + 1$, $j_{w+1} = s + 1$, $i_0 = i - 1$, $j_0 = j - 1$:

$$\sum_{h=1}^{w} M(t_{i_h}, p_{j_h \bmod m}) + (\sum_{h=0}^{w} G(i_h, i_{h+1} - 1, j_h \pmod m), (j_{h+1} - 1) \pmod m))).$$

Definition: An optimal alignment between $\bar{P}$ and T is an alignment of a *substring* of $\bar{P}$ and a *substring* of T that yields the maximum score.

The aim of our algorithm is to produce an optimal alignment between T and $\bar{P}$.

This entails not simply finding the best alignment of two given substrings, but also choosing the best substrings. Note that the alignment of the *chosen* substrings will always start and end with a match that carries a positive score.

In general, in order to find the alignment with the highest score efficiently one needs to make some assumptions on the "civilized behavior" of the gap penalties. In the present article we are not overly concerned how gap penalties are chosen, since the main contribution of the present work is a reduction from a large size dynamic programming problem to a much smaller scale dynamic programming program. Any assumptions on the gap penalties can be carried from one solution to the other.

4   A simple but inefficient algorithm

Dynamic programming is a well known technique that is frequently used to solve approximate string matching problems ([NW-69], [S-74], [SK-83]). In this section we show how a dynamic programming problem of size $O(mn^2)$ can be used to solve our problem. We construct a $n \times nm$ matrix D whose entries $D[\ell, j]$ contain the score of the best alignment of a suffix of $t_1 t_2 \ldots t_\ell$ with any suffix of $\bar{p}_1 \bar{p}_2 \ldots \bar{p}_j$, or 0 if this score turns out to be negative. Assigning 0 to $D[\ell, j]$ allows an alignment to start with the match of character $t_{\ell+1}$ and $\bar{p}_{j+1}$. The maximum value in D can be traced back to obtain the desired alignment.

The following algorithm computes the matrix D, when G obeys properties (a), (b) and (c) from the previous section.

Notation: We denote by $\bar{G}$ the gap function defined on substrings of size less than m of $\bar{P}$: $\forall i_1 \leq i_2, j_1 \leq j_2$, $\bar{G}(i_1, i_2, j_1, j_2) = G(i_1, i_2, j_1 \pmod{m}, j_2 \pmod{m})$.

> for $i := 0$ to n do
>   for $j := 0$ to mn do
> $D[i, j] := 0$
> for $i := 1$ to n do
>   for $j := 1$ to mn do
> $(*)D[i, j] := \max\{D[i-1, j-1] + M[t_i, p_{j \bmod m}];$
> $\displaystyle \max_{\substack{i_0 \leq i \\ j-m < j_0 \leq j}} D[i_0, j_0] + \bar{G}(i_0, i, j_0, j)\}$
>
>   ($D[i, j]$ is the maximum of up to $im + 1$ numbers. These numbers are 0 (the current value of $D[i, j]$), the value obtained from its predecessors on its diagonal followed by a match, and the $im - 1$ predecessors in the upper left rectangle of width m respectively.)
>   od
> od

Remark: Gap functions are frequently quite simple functions and generally the maximum in $(*)$ can be computed much faster. As is well known, when the gap function G is additive, the maximum in $(*)$ taken in the above program can be computed in $O(1)$ time by considering only the three neighbors of $D[i, j]$, $(D[i-1, j-1]$, $D[i-1, j]$ and $D[i, j-1])$. If G is additive for small gaps and then remains constant, (which is a very suitable gap function for our problem), $D[i, j]$ may also be computed in $O(1)$ time, by considering four previously computed values, (the three values mentioned before and the maximal score computed so far).

Example: Let $t_1 \ldots t_9$ be *cabaabddc*, let P be *ab*, and assume for simplicity that all deletions and insertions of i characters carry a penalty of $-i$, and that $M(x, y) = 1$ if $x = y$ and -1 otherwise. The matrix D that corresponds to these strings is given below:

		0	1	2	3	4	5	6	7	8	
			a	b	a	b	a	b	a	b	...
0		0	0	0	0	0	0	0	0	0	...
1	c	0	0	0	0	0	0	0	0	0	...
2	a	0	1	0	1	0	1	0	1	0	...
3	b	0	0	2	1	2	1	2	1	2	...
4	a	0	1	1	3	2	3	2	3	2	...
5	a	0	1	0	2	2	3	2	3	2	...
6	b	0	0	2	1	3	2	4	3	4	...
7	d	0	0	1	1	2	2	3	3	3	...
8	d	0	0	0	0	1	1	2	2	2	...
9	c	0	0	0	0	0	0	1	1	1	...

Notice that the two optimal alignments found correspond actually to the same alignment.

5   The new algorithm

The new algorithm computes a matrix $\hat{D}$ of size $O(nm)$. $\hat{D}[i, j]$, (for $j = 1 \ldots m$), will hold the score of the best alignment of a substring of T that ends with t_i and a substring of $\bar{P}$ that ends with any $\bar{p}_k$ for which $j = k \pmod m$, (i.e. p_j). The above example illustrates why it seems sufficient to compute and keep only m locations in each row. Each row in D eventually becomes periodic, and $D(i, j) = D(i, j + m)$, for sufficiently large j. In row 2 in our example, the period is 1 0, while in row 7 the period is 3 3. For any j, the values $D[i, j], D[i, j + m], D[i, j + 2m] \ldots$ clearly form an increasing sequence of scores. This follows directly from the fact that any suffix of $\bar{P}_j = \bar{p}_1, \ldots \bar{p}_j$ is also a suffix of $\bar{P}_{j+m}$ and therefore the set of alignments competing for the maximal score when aligning with $\bar{P}_j$ is included in the corresponding set for $\bar{P}_{j+m}$. In addition, the largest number of pattern characters one might delete per character of text is $m - 1$. It is easy to see that the values in row i will therefore be periodic starting from entry $D[i, im]$, (at the latest). If $D[i, \ell]$, for $\ell \geq im$, corresponds to an alignment starting with a match at $\bar{p}_j$, then $D[i, \ell + m]$ will correspond to the same alignment starting with a match at $\bar{p}_{j+m}$.

The matrix $\hat{D}$ will therefore consist of the values in the above mentioned periods and contain the maximum value of D. We are going to show that each entry in $\hat{D}$ can be computed as efficiently as an entry in D. The maximum value in $\hat{D}$ can be easily traced back to obtain the desired alignment. Since we will do two passes through each row, an auxiliary column 0, which is identical to the last column m, will be used for wraparound.

We shall now describe the computation of $\hat{D}$.

In the dynamic programming computation of D, an entry $D[i,j] > 0$ is obtained from the score computed at some entry $D[i_0, j_0]$, for $i_0 \leq i$ and $j_0 < j$. If $D[i,j]$ is in the periodic region, then $\hat{D}[i, j \bmod m]$ will be computed from the score $\hat{D}[i'_0, j'_0]$, with $i'_0 = i_0$ and $j'_0 = j_0 \bmod m$. As a result a score $\hat{D}[i,j]$, may depend on a score $\hat{D}[i, j_0]$, (in the same row), with $j_0 > j$. To overcome this difficulty the computation of each of the rows of $\hat{D}$ will be done in two rounds. In the first round we compute each of the entries $\hat{D}[i,j]$, $(j > 0)$, using all previously computed values in $\hat{D}$, assuming for the moment that all the entries in row i are correct. In the second round we feed the value $\hat{D}[i, m]$ back to $\hat{D}[i, 0]$ and correct those values $\hat{D}[i, j]$, which depend on a score $\hat{D}[i, j_0]$, for $j_0 > j$. We shall prove that all entries in $\hat{D}[i; 0 \ldots m]$ will have been correctly computed by the end of this second round.

Assume that the submatrix $\hat{D}[1 \ldots i-1; 1 \ldots m]$ has been correctly computed. To compute row i we recall that the entry $\hat{D}[i, j]$ is the maximum score of an alignment of any suffix of $t_1 \ldots t_i$ and any substring of $\bar{P}$ that ends with p_j. This alignment may either end with the match of p_j and t_i or with a gap. We distinguish between two type of gaps. In the first type, the gap is in the text, and possibly also in the pattern, i.e. the last match in the alignment is between p_k and t_ℓ, for $\ell < i$ and any k. In the second type, the gap is in the pattern only, i.e. the last match in the alignment is between p_k and t_i, for some $k \neq j$, followed by the deletion of $p_{k+1} \ldots p_j$.

The three cases listed above correspond to the following computations:

(a)   If the alignment is obtained from an alignment of a suffix of $t_1 \ldots t_{i-1}$, and a substring of $\bar{P}$ that ends with p_{j-1}, followed by the match of t_i and p_j, then $\hat{D}[i, j] := \hat{D}[i-1, j-1] + M[t_i, p_j]$.

(b)   If the alignment is obtained from an alignment of of $t_1 \ldots t_\ell$, $\ell < i$ and a substring of $\bar{P}$ for which t_ℓ is matched to p_k (for any k), the subsequent deletion of both $t_{\ell+1} \ldots t_i$ and $p_{k+1} \ldots p_j$, then $\hat{D}[i, j] := \hat{D}[\ell, k] + G(\ell, i, k, j)$. (If $k = j$ no characters from the pattern are deleted.)

(c)   If the alignment is obtained from an alignment of a suffix of $t_1 \ldots t_i$, and a substring of $\bar{P}$ that ends with a match of t_i and p_k (for any $k \neq j$) and the deletion of the substring $p_{k+1} \ldots p_j$ in the pattern, then $\hat{D}[i, j] := \hat{D}[i, k] + G(i, i, k, j)$.

In the first step we compute the scores for alignments corresponding to cases (a) and (b) above, since theses scores depend only on previously correctly computed scores. In the second step we consider case (c). Suppose that $\hat{D}[i, j]$ gets a higher score in the second step than in the first step. It follows that $\hat{D}[i, j]$ was obtained from the score at $\hat{D}[i, k]$, followed by the deletion of $p_{k+1} \ldots p_j$. Since G obeys the triangle inequality, and $\hat{D}[i, j]$ could not get its best score directly by case (b), the alignment corresponding to $\hat{D}[i, k]$ must have ended with a match, and was therefore correctly computed in the first step. After the second round all entries will therefore be correct, and the time to compute an entry in $\hat{D}$ is at most twice the one to compute an entry in D.

If the full size dynamic programming problem could be solved in $O(gn^2m)$ time, where g accounts for the complexity of computing the gap function, then our reduced dynamic programming problem can be solved in $O(gnm)$ time.

Remark: As mentioned before when the gap function is additive, (and hence $g = O(1)$), we need only consider three neighbors for each entry. Therefore, in step 1 $\hat{D}[i, 1]$ is computed by considering its neighbors on the diagonal and above. $\hat{D}[i, j]$,

$(j > 1)$ is computed by considering its three neighbors, (possibly using a temporary value for $\hat{D}[i, j-1]$ that is too low). When $\hat{D}[i, m]$ has been computed in the first step, its value gets assigned to $\hat{D}[0, m]$ and the second step is started. Since gaps can only reduce the score, the maximal score in $\hat{D}[i, 1 \ldots m]$, will always have been correctly computed in the first step. Moreover, because G is additive, all scores to the right of the maximum will also be correct. In particular $\hat{D}[0, m]$ will have the right score after the first pass and therefore in the second pass the remaining scores (if any) will be adjusted. The above argument also shows that, (when gap values are additive), as soon as a value computed in the first pass equals a value computed in the second pass, the second pass can be terminated.

By the previous claim the following algorithm computes the values of the matrix $\hat{D}$ correctly. The auxiliary column 0, which is identical to the last column m, is used for wraparound.

Notation: The symbol $\preceq$ denotes the lexicographic order or pairs, i.e. $(i, j) \preceq (i_1, j_1)$ if $(i < i_1)$ or $((i = i_1)$ and $(j \leq j_1))$.

```
for i := 0 to n do
    for j := 0 to m do
        D̂[i, j] := 0
for i := 1 to n do
    D̂[i − 1, 0] := D̂[i − 1, m]
    for j := 1 to m do
        D̂[i, j] := max{D̂[i − 1, j − 1] + M[tᵢ, pⱼ];
                          max      D̂[i₀, j₀] + G(i₀, i, j₀, j)}
                     (i₀,j₀)⪯(i,j)
    od
    for j := 1 to m do
        D̂[i, j] := max{D̂[i, j];   max   D̂[i, j₀] + G(i, i, j₀, j)}
                               0≤j₀≤m
    od
od
```

Example Let $t_1 \ldots t_9$ be *cabaabddc* and let P be *ab*. The matrix $\hat{D}$ corresponding to this pattern matching problem is given below:

		0	1	2
		(b)	*a*	*b*
0		0	0	0
1	*c*	0	0	0
2	*a*	0	1	0
3	*b*	→2	1	2→
4	*a*	→2	3→	2→
5	*a*	2	3	2
6	*b*	4	3	4
7	*d*	3	3	3
8	*d*	2	2	2
9	*c*	1	1	1

Complexity: We compute a matrix of size $n(m+1)$, if the gap functions permit it, each value is computed in constant time, which gives total time $\Theta(nm)$.

6 Conclusions

The algorithmic problem addressed in this manuscript was motivated by a problem in protein structure determination. A program based on the presented algorithm was implemented and is currently in use at Rockefeller University, [FP-92]. The program is also being used as a tool to study some of the surface characteristics of several coiled-coil proteins.

References

[CF-74] P.Y. Chou and G.D. Fasman, "Prediction of protein conformation," *Biochemistry*, Vol. 13, 1974, pp. 222-245.

[CP-86] C. Cohen, and D.A.D. Parry, "Alpha-helical coiled coils – a widespread motif in proteins," *T.I.B.S.*, Vol. 11, 1986, pp. 245-248.

[CP-90] J. F. Conway and D. A. D. Parry, "Structural features in the heptad substructure and longer range repeats of two-stranded *alpha*-fibrous proteins," *Int. J. Biol. Macromol.*, Vol. 4, 1990, pp. 328-333.

[FP-92] V. A. Fischetti, V. Pancholi, P. Sellers, J. Schmidt, G. Landau, X. Xu, O. Schneewind, Streptococcal M protein: A common Structural Motif Used by Gram-positive Bacteria for Biological Active Surface Molecules, to appear *Molecular Recognition in Host-Parasite Interactions: Mechanisms in viral, bacterial and parasite infections. Published by Plenum Publishing.*

[GG-89] Z. Galil and R. Giancarlo, "Speeding up dynamic programming with applications to molecular biology," *Theoretical Computer Science*, Vol. 64, 1989, pp. 107-118.

[GLE-87] M. Gribskov, A. D. McLachlan, and D. Eisenberg, "Profile analysis: Detection of distantly related proteins," *Proc. Natl. Acad. Sci.*, Vol. 84, 1987, pp. 4355-4358.

[GOR-78] J. Garnier, D.J. Osguthorpe, and B. Robson, "Analysis of the accuracy and implications of simple methods for predicting the secondary structure of globular proteins," *J. Molecular Biology*, Vol. 120, 1978, pp. 97-120.

[GP-90] Z. Galil and K. Park, "An Improved Algorithm for Approximate String Matching," *SIAM J. Comp.*, Vol. 19, 1990, pp. 989-999.

[LD-91] A. Lupas, M. Van Dyke, J. Stock, "Predicting Coiled Coil from Protein Sequences, *Science* Vol. 252, 1990, pp. 1162-1164.

[LLE-91] R. Lüthy, A. D. McLachlan, and D. Eisenberg Secondary Structure-Based Profiles: Use of Structure-Conserving Scoring Tables in Searching Protein Sequence Databases for Structural Similarities'" *Proteins*, Vol. 10, 1991, pp. 229-239.

[LV-89] G.M. Landau and U. Vishkin, "Fast parallel and serial approximate string matching," *Journal of Algorithms*, Vol. 10, No. 2, June 1989, pp. 157-169.

[NW-69] S.B. Needleman and C.D. Wunsch, "A general method applicable to the search for similarities in the amino acid sequences of two proteins," *J. Molecular Biology*, Vol. 48, 1969, pp. 443-453.

[S-74]   P.H. Sellers, "On the theory and computation of evolutionary distance," *SIAM J. Appl. Math*, Vol. 26, No. 4, 1974, pp. 787-793.

[SK-83] D. Sankoff and J.B. Kruskal (editors), *Time Warps, String Edits, and Macromolecules: the Theory and Practice of Sequence Comparison*, Addison-Wesley, Reading, MA, 1983.

[U-83]   E. Ukkonen, "On approximate string matching," *Proc. Int. Conf. Found. Comp. Theor.*, Lecture Notes in Computer Science 158, Springer-Verlag, 1983, pp. 487-495.

Edit Distance for Genome Comparison Based on Non-Local Operations *

David Sankoff

Centre de recherches mathématiques, Université de Montréal
C.P. 6128, succursale "A", Montréal, Québec H3C 3J7
(sankoff@ere.umontreal.ca)

Abstract. Detailed knowledge of gene maps and and even complete nucleotide sequences for small genomes has led to the feasibility of evolutionary inference based on the macrostructure of entire genomes, rather than on the traditional comparison of homologous versions of a single gene in different organisms. In this paper, we define a number of measures of gene order rearrangement, describe algorithm design and software development for the calculation of some of these quantities in single-chromosome genomes, and report on the the results of applying these tools to a database of mitochondrial gene orders inferred from genomic sequences.

1 Role of Rearrangements in Evolution

Genes evolve largely through the processes of nucleotide substitution, insertion and deletion. Genomes, containing the entire genetic complement of an organism, evolve because the genes in them evolve. But there are other mechanisms of evolution at the genomic level as well. For example, entire genes, or segments of chromosomes containing several genes, may be deleted or inserted as a single event. Genes or chromosomal segments may migrate (i.e. be transposed) from one region of the genome to another. A segment of the chromosome may become inverted: *...xabcdy...* becomes *...xdcbay...*, where the inverted segment is transferred to the opposite strand of double stranded DNA so that it is still read in the original order.

2 Phylogenetic Inference

The mathematical study of evolution at the genomic level based on operations such as insertion, deletion, transposition and inversion of chromosomal segments containing one or more genes is necessarily different from traditional studies involving nucleotide

---

* Work supported by operating and infrastructure grants from the Natural Science and Engineering Research Council (Canada) and a team grant from the *Fonds pour la formation de chercheurs et l'aide à la recherche (Quebec).* The author is a Fellow of the Canadian Institute for Advanced Research. This paper was begun while a guest of Professor Claude Weber at the University of Geneva in 1991. I thank Natalie Antoine, Robert Cedergren, Franz Lang and Bruno Paquin who constructed the database from which the mitochondrial gene orders used here were drawn, and especially Guillaume Leduc for his perseverance in implementing DERANGE.

sequence comparison because some of these operations are non-local; inversion and transposition in particular may involve arbitrarily distant (at least in theory) terms in the gene order.

Nevertheless, since these processes are widespread and accumulate over time, it is interesting to ask whether the comparison of genomes, i.e. of their gene orders, can provide us a statistical basis for assessing the similarity or distance or evolutionary divergence between organisms, in analogy with nucleotide sequence comparison [1, 4, 5, 6, 8]. In this paper, we motivate and describe an algorithm capable of inferring the minimal set of inversions and transpositions necessary to convert one genome order into another. This, together with a measure of gene deletion, not only provides an estimate of the degree of evolutionary divergence, but characterizes it in terms of the predominant processes separating related organisms from their ancestors.

3   The Mitochondrial Data

The mitochondrion constitutes an ideal model for studying eukaryotic evolution through genome rearrangement. The organellar genome is small enough ($\approx 50,000$ bp) to be tractable by current sequencing technology, so that about 20 sequences are known completely or almost completely, with nearly all genes identified.

Table 1. Organisms providing mitochondrial genomes used in this study. References in [8]

Organism	Number of Genes
FUNGI	
Fission Yeast	
Schizosaccharomyces pombe	35
Budding Yeasts	
Torulopsis glabrata	34
Saccharomyces cerevisiae	39
Kluyveromyces lactis	31
Filamentous fungi	
Neurospora crassa	50
Aspergillus nidulans	44
Podospora anserina	45
MULTICELLULAR	
Human	37
Strongylocentrotus purpuratus (sea urchin)	37
Drosophila yakuba (fruitfly)	37
Ascaris suum (nematode)	36

These are widely dispersed among subgroups of the eukaryotes so that the phylogenetic range of our methodology may be evaluated.Most important, despite its minuscule size compared to nuclear genomes, the mitochondrion appears to undergo the same processes of genome rearrangement which interest us here as do nuclear chromosomes: mainly deletion, inversion and transposition of chromosomal segments.

We have sampled 11 mitochondrial genomes from the database of [8]. As listed in Table 1, this contains four multicellular organisms, extending from nematodes to humans, and seven fungi, of which three are budding yeasts, one is a fission yeast, and three are filamentous fungi. In our analyses, whenever we compare all pairs of genomes, we will distinguish comparisons within the two groups of fungi (grouping budding and fission yeasts), within the multicellular organisms, and among the three groups.

4   Genes in Common

Let us first consider only those differences between genomes due to gene deletion and insertion. The simplest distance measure between two genomes based on these processes is simply $D(x, y)$, the number of genes present in either one of the genomes but not the other. This is calculated as

$$D(x, y) = N(x) + N(y) - 2N(xy),$$

where xy represents the genes in common between x and y, and $N(s)$ is the number of genes in s. The intra- and intergroup averages of D are presented in Figure 1.

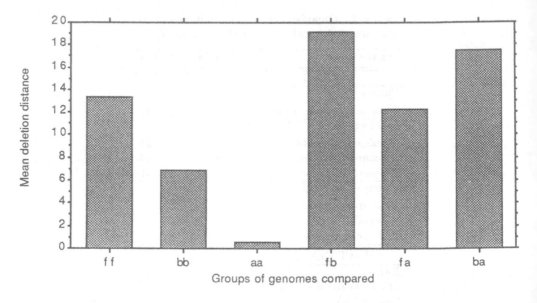

Fig. 1. Intra- and intergroup deletion distances $D(x, y)$, based on difference in gene content only. f=filamentous fungi, b=budding yeasts and *S. pombe*, and a=multicellular eukaryotes.

In some contexts $D(x, y)$ might be a perfectly reasonable measure; but it has at least three weaknesses. If genomes differ markedly in the total number of genes each contains, then a given distance between two small genomes may represent a greater amount of evolution than the same distance between two large genomes. This could possibly be compensated for by normalisation, such as by dividing the distance by half the sum of the two genome lengths. This, however, would implicitly involve the questionable assumption that deletion from already small mitochondrial genomes, even those affecting less "superfluous" genes, happens as readily as deletions from large mitochondrial genomes.

Second, if the deletion or insertion processes may involve a block of several contiguous genes, it may not seem appropriate to count each gene in such a block separately. We will discuss this problem further below.

Finally, it is clear that within the multicellular group of genomes (aa), deletion is too rare an event to be used as an indicator of phylogenetic relationships, though it does distinguish between these and the fungi (fa,ba), and within the fungal group, the yeasts (bb) show lower values of D among themselves than between them and the non-yeasts (fb).

5 Breakpoints

Suppose $ABCDEFGH$ and $ABCEFGDH$ are two genomes. Except for the adjacent pairs CD, DE and GH in the first genome and CE, GD and DH in the second, the gene orders are the same. We say that each order has three "breakpoints" with respect to the other. The number of such breakpoints is often considered an index of how much genome rearrangement has occurred (e.g. [2, 10]). Figure 2 displays the quantity $B(x, y)$ measuring the number of breakpoints between the genomes x and y when all genes absent from one or the other are excluded.

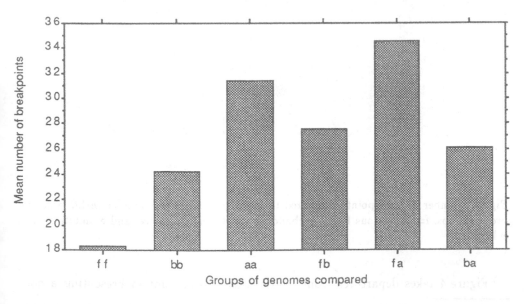

Fig. 2. Average number of breakpoints $B(x, y)$ in the comparisons between groups of mitochondrial genomes. f=filamentous fungi, b=budding yeasts and S. pombe, and a=multicellular eukaryotes.

While Figure 2 suggests that the number of breakpoints is indicative of phylogenetic relationships, especially within the fungi, we should take into account that $B(x, y)$ can be expected to be closely related to $N(xy)$, the number of genes in both genomes.

On the average, in fact, for two random genomes x and y having $n = N(xy)$ genes in common, the number of breakpoints is $n(n-2)/(n-1)$. This can be seen by considering the probability that in a random circular genome the right-hand neighbour of gene i is gene $i+1$ - otherwise there is a breakpoint between the two[2]. For $n \geq 2$, this probability is obviously $1/(n-1)$. Then the expected number of such genes i is $n/(n-1)$, and the expected number of breakpoints is then $n - n/(n-1) = n(n-2)/(n-1)$, which is very close to $n-1$ for the range of n that interests us.

Figure 3 presents the difference between $N(xy)$, the number of genes in common and $B(x, y)$, the number of breakpoints depicted in Figure 2. It can be seen that there is little if any non-randomness detectable in the comparisons between the fungal and the other mitochondria since the distribution is centered around the value 1, but within the fungi and within the multicellular organisms, there are clearly fewer breakpoints than between random genomes since the means of the distributions are much greater than 1.

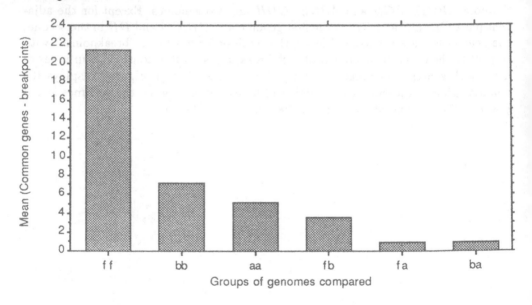

Fig. 3. Number of breakpoints compared to number of genes in common within and between groups. f=filamentous fungi, b=budding yeasts and *S. pombe*, and a=multicellular eukaryotes.

Figure 4 takes departures from randomness into account in presenting a nor-

---

[2] When, as a result of inversion, a genome contains genes of both left-to-right and right-to-left orientations, the latter is indicated by a minus sign. Thus if the j-th gene has right-to-left orientation, we write $-j$, and the condition for a breakpoint is the the right-hand neighbour not be $-(j-1)$

malised value, namely $100 \times B(x, y)/N(xy)$.

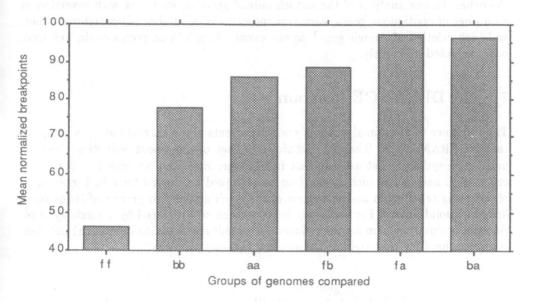

Fig. 4. $B(x, y)$ normalized, i.e. multiplied by $100/N(xy)$. f=filamentous fungi, b=budding yeasts and *S. pombe*, and a=multicellular eukaryotes.

The normalised values in Figure 4 are consistent with phylogenetic relationships within the non-fungal group, within the yeast group and within the filamentous fungi. In addition the comparison between the two groups of fungi is closer than between either of them and the multicellular group.

6 Edit Distance

In the example at the beginning of the previous section, *ABCDEFGH* versus *ABCEFGDH*, it is clear that the second genome could have been produced from the first by the simple operation of excising the gene *D* from between *C* and *E*, and transposing it to a position between *G* and *H*. Such a transposition creates three breakpoints. If we now transposed *B* in the second genome to the position between *F* and *G*, this would produce the order *ACEFBGDH*, which has six breakpoints compared to the first order, each transposition having created three. But if instead of transposing *B*, we had transposed *C* in the second genome to the position between *F* and *G*, this would have produced *ABEFCGDH*, which has only five breakpoints with respect to *ABCDEFGH*. And had we transposed *C* to the position after *D*, this would have created *ABEFGDCH*, which has only four breakpoints with respect to the original order (or three if we don't count *DC* as a breakpoint). Thus the number of breakpoints, while roughly reflecting the degree of rearrangement in comparing two genomes, is not necessarily an accurate measure of the number of rearrangement events which have occurred to produce one genome from the other. This motivates the definition of a rearrangement edit distance between two genomes, namely the minimal number of rearrangement events necessary to convert one to the other. In our analysis of the mitochondrial genome, we count each inversion of a number of contiguous genes, each transposition of a number of contiguous genes, and each deletion of a single gene[3], as one event, though these events could well have been weighted differently.

7 The DERANGE Program

The inference of a minimal series of rearrangements is the carried out by a program called DERANGE [9]. The key technique is that of alignment reduction. Two or more gene symbols that are adjacent in both genomes and are either of the same orientation and order, such as the four pairs linked by dotted lines in Figure 5, or of opposing orientation and in reverse order, such as the two groups of three pairs linked by solid lines in Figure 5, may be combined and replaced by a single symbol, since an optimal solution for the reduced (after this combination operation) problem is also optimal for the original (before combination) problem.

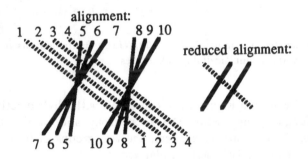

Fig. 5. Example of alignment reduction. Dotted lines represent homologous genes with same orientation in both genomes, solid lines indicate genes with opposite orientation.

Alignments may also be reduced by first applying inversion and transpositions so as to produce such combinable pairs of terms. This is illustrated with three inversions in Figure 6.

Inversions may reduce an alignment by up to two linkages, and transpositions by up three. The weight associated with each rearrangement operation (weight = 1 in the present study) contributes to the total distance between the genomes. Reducing an alignment through the combining operation of course costs nothing, since this is just a formalism for indicating that we have obtained a partial solution to the problem. The algorithm stops when the alignment has been reduced to exactly one linkage, representing the fact that one genome has been transformed so that all its terms are in the same order as in the other genome, with the same orientation. This algorithm is a generalization of a sub-optimal method of [10] for inversions only,

---

[3] This was calculated in Section 4 and, in the present formulation, is independent of the analysis of transposition and inversion.

in the single circular chromosome case without taking account of orientation; its current implementation [9] is based on a branch-and bound search to completely solve[4] the general case of an unknown number of overlapping inversions (e.g. data such as that of [3], requiring five superimposed inversions), and transpositions, with user-imposed weights on the different types of events.

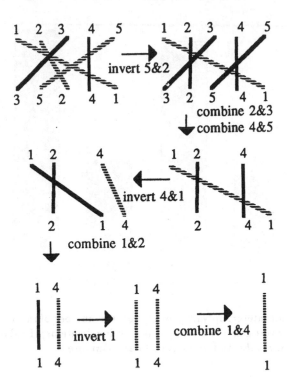

Fig. 6. Reducing an alignment using three inversions.

8 DERANGE Results

Figure 7[5] presents the $e(x, y)$, the output of DERANGE, once the genes absent from either member of the pair being compared are excluded, and combining adjacent genes not separated by a breakpoint.

In Figure 7, it appears as if $e(x, y)$ contains much phylogenetic information, in that for pairs (x, y) within the two fungal groups, or between them, the distance e tends to be small, while it is much greater between fungi and multicellular organisms. The multicellular organisms are relatively diverse as measured by e, but this was

---

[4] For even the small genomes discussed here, time and space requirements are prohibitive. Thus the search is limited in a number of ways, as discussed below.

[5] Figure 7 represents the bulk of the computation in this research. About 150 hours of computing time were distributed among a battery of various Macintosh models.

also visible in the analysis of B in Figure 4. Indeed, it must be remembered that for a pair where B is small, that the corresponding permutation is easily converted into another may simply be due to the fact that the reduced alignment is short. Thus, similar to the way we normalised B by dividing by N, we also normalise e by dividing by B. The results appear in Figure 8.

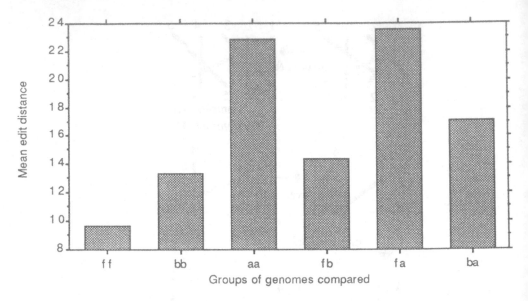

Fig. 7. Average intra- and intergroup values of $e(x, y)$, the number of inversions and transpositions necessary to convert genome x into y (discounting deletions). f=filamentous fungi, b=budding yeasts and *S. pombe*, and a=multicellular eukaryotes.

Normalizing the edit distances subtly changes the configuration seen in Figure 7. Within the fungi, the grouping of filamentous fungi versus the yeasts is lost. Nevertheless the fungal genomes appear consistently more closely related among themselves than they are to the non-fungal organisms.

We may conclude that e contains not only the phylogenetic information in B, but somewhat more.

9 Weights

As mentioned above, in our analysis of the mitochondrial genome, the two types of rearrangement event were assigned the same weight, and so was the deletion of each gene. This decision is of course somewhat arbitrary. Let us examine first the relative weights of transpositions versus inversions. Figure 9 shows some typical results of varying the relative weights on four comparisons. It can be seen that be when the effect of changing the weights is to increase the number of transpositions, the number of inversions decreases almost twice as fast.

We find that this relationship is quite general. Despite the fact that inversions can reduce an alignment by two lines, in practice few of these are found, and the

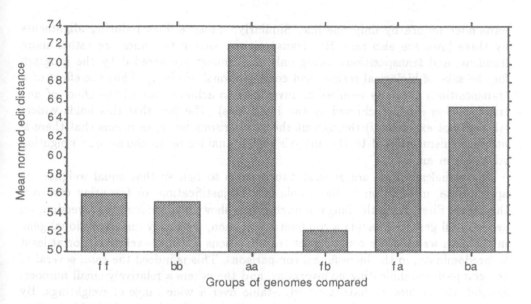

Fig. 8. Edit distances $e(x, y)$ normalized by number of breakpoints $B(x, y)$.

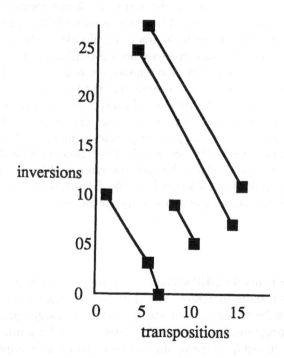

Fig. 9. Effect of changing inversion vs. transposition weights. Examples involve comparison of *Neurospora* with, from upper right to lower left, human, *Drosophila*, *S.pombe* and *K.lactis*.

remainder reduce by only one line. Similarly, transpositions reducing alignments by three lines are also rare. But transpositions saving two lines are rather more frequent, and transpositions saving only one linkage are avoided by the program for the sake of biological realism and computational efficiency. Thus the effect of x transpositions typically requires $2x$ inversions to achieve (indeed the effect of any transposition can be achieved by two inversions). The fact that this holds widely (though not exclusively) throughout the comparisons we make means that there is no mechanism internal to the analysis which enables us to choose one weighting scheme over another.

Nevertheless, there are at least two reasons to believe that equal weights are appropriate, in addition to the simple lack of justification for favouring one over the other. First, since the fungal mitochondria show little evidence of inversion, all genes in all genomes having a common orientation, with only one exceptional gene in the data we used, we could expect transpositions to occur exclusively, or at least to predominate, in the intra-fungal comparisons. This is indeed the case, several of the comparisons indicating no inversions, and the others a relatively small number. Second, the results we have are fairly stable over a wide range of weightings. By increasing the weight on inversions rather substantially, the number of inversions remains the same or decreases relatively slowly. It is somewhat easier to increase the number of of inversions, but only by weighting transpositions very heavily, and this leads to unlikely numbers of inversions in the intra-fungal comparisons.

As for gene deletion, its analysis could be integrated with that of inversion and transposition, but this does not seem biologically justifiable at the present time. If it were to be found that deletion often involves several contiguous genes at the same time, the cost of deletion and insertion of k contiguous genes should not be simply k times the cost for one gene, but a more slowly growing (concave) function of k. In addition, the insertion or deletion of a number of contiguous genes should be allowed to occur at any time during the transformation of one genome into another, rather than calculated at the outset as we do presently, since a number of genes deleted[6] together at one point in time might previously have been dispersed throughout the genome and then brought together through inversion and transposition. These changes would require integrating deletion into the DERANGE search algorithm, which would risk making computing requirements excessive.

10 Search depth

The computing requirements for DERANGE, both time and memory, grow rapidly as the number of breakpoints increases. When there are more than 12-15 breakpoints (there are up to 36 in the present study) , it is no longer practical, in the current version of the program, to carry out the entire search for a minimal series of rearrangement events. Instead, the program incorporates an user-imposed limit on

---

[6] It would be important to require of such events that they involve only the genes that are present in one genome and not the other; otherwise the minimizing sequence of genome changes might artificially turn out to involve the deletion of a large segment of one genome followed by the insertion of a long segment of the other, where both segments contain many of the same genes, biologically an implausible event.

the number of possible search paths to be explored, and eliminates those that seem
less likely. Figure 10 indicates the effect of varying this "search depth" parameter
on five typical genome comparisons where the "true" number of events ranges from
10 to 26. It can be seen that the number of events is badly overestimated when the
search depth is 50 or 100, but that it approaches the true value when the depth is
set at 500 and only diminishes slowly if at all with depths over 1000. In this study, a
depth of 4000 was used throughout. Though undoubtedly some of the distances are
overestimated, we can be sure that these are not large errors, and all the distances
are subject to proportionately the same risk of error, so that the phylogenetic results
should not be biased in any particular direction.

There are a number of ways we use to check whether a comparison gives artificially high results due to an insufficient search depth. One way is to reverse the
order in which the algorithm receives the two genomes to compare. If converting one
genome to the other is more costly than the conversion in the opposite direction,
the first of these two analyses is not optimal. Second, for selected comparisons only
we can afford to increase the search depth to 10 000 or more to see if a more economical solution is found. Third, our results should obey the triangular inequality:
if converting x to z costs more than the cost of converting x to y plus the cost of
converting y to z, then the x to z result is not optimal.

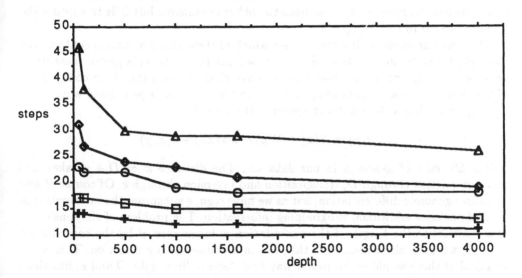

Fig. 10. The effect of limiting search depth on the number of events inferred. Comparison
of humans with, from top to bottom, *Ascaris, S. pombe, S. purpuratus, Drosophila* and
Allomyces macrogynus.

11 Nonrandomness

In Section 8, we found that normalizing e removes much of the apparent phylogenetic information it contains. We may go further and ask whether the edit distance

$e(x, y)$ between the various pairs (x, y) of mitochondrial genomes is significantly below the random noise level, in analogy to the discussion above of the departure from randomness of the breakpoint number $B(x, y)$ for closely related genomes. Because no analytical results are available for e, we generated a number of pairs of random permutations of various lengths, combined adjacent genes not separated by a breakpoint, and submitted them to the same analysis as the mitochondrial genomes. The results will be reported elsewhere, but it is clear that comparisons within and between the two fungal groups are clearly non-random and that the comparisons involving multicellular organisms are generally indistinguishable from random.

12 Genomic Distance

We have seen that there are three independent aspects of the differences among genomes, measured by D, B normalized by N, and e normalized by B. Each of these is pertinent to different subgroupings of the genomes. D clearly distinguishes the fungal mitochondria from the others and within the fungi, sets the yeasts apart from the others. B, when normalized by N, also distinguishes the yeasts from the other fungi, but shows much less striking homogeneity within the multicellular organisms. The only clear distinction made by e, when the effects of B are normalized away, distinguishes the fungi as a whole from the other organisms, but fails to capture the division within fungal group.

It is not our purpose, however, to see which of these three measures contains the most phylogenetic information. Rather we set out to construct a global measure of genome rearrangement and, now that we have characterised the three components of the distance between genomes, we may combine them in a principled way.

Figure 11 shows the results of applying the formula

$$d(x, y) = N(x) + N(y) - 2N(xy) + e(x, y)$$

to the 55 pairs of genomes in our data set. The distance $d = D + e$ takes into account deletion through D, transposition and inversion through e. Of course B also measures genomic differentiation, but as we have seen, e subsumes the information in B, and contains additional subgrouping information. The problem of the sensitivity of e (and B) to the length of the permutation is taken care of by the combination with D, which is also sensitive to this, but in a compensating direction. This is not to say that there would be no better way to differentially weight D and e, but there is as yet no justification for any particular weighting.

Figure 11 depicts more clearly than any we have seen in previous sections, the relatively low within-group distances for all three groups (ff, bb, and aa), and the somewhat closer relationship among the two fungal groups (fb) than between them and the multicellular group (fa, ba).

13 Conclusions

While gene sequences may in many cases serve as good indicators of evolutionary relationships, they do not in themselves bear on many phylogenetically important

134

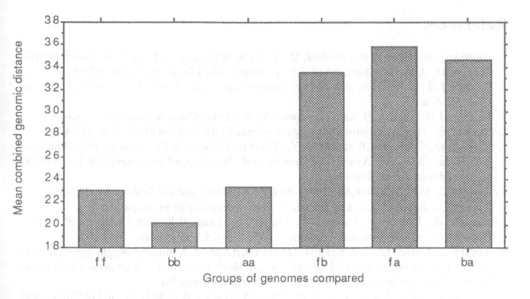

Fig. 11. Genomic rearrangement distance between pairs of genomes.

events. For example, gene-based phylogenies are constructed from the analysis of point mutations, with only minor insertions and/or deletions also being tabulated. In recent years, however, we have seen that evolutionarily relevant processes such as inversion, duplication, transposition, reciprocal translocation and deletion of entire segments of chromosomes, as well as recombination resulting in gene conversion generate much greater change, which is ignored (and in fact is effectively invisible) in gene sequence comparisons. In addition, in phylogenies near the radiative origin of many orders, phyla and kingdoms, current reconstructions of the branching order are too imprecise to provide clear distinctions among alternate tree topologies. Changes in gene sequence during the relatively short periods of time involved are often not extensive enough, and too difficult to infer precisely, to distinguish between opposing scenarios of evolution. On the other hand, major events at the level of genome arrangement may be able to discriminate decisively among competing phylogenetic accounts. In a broader sense, an understanding of the role of genome-level processes in evolutionary divergence and their consequences for the organization of life processes is of crucial importance to biology. Organellar DNAs are sufficiently small that one can realistically hope to have available the complete sequence of a great many of them in the near future, drawn from phylogenetically key organisms, thereby obtaining the type and amount of data required to evaluate these issues.

On the methodological level, progress is being made on the general problem of minimally generating elements of the group of permutations using inversions only, which is very closely related to the task of DERANGE. Better bounds for branch-and-bound algorithms, approximation algorithms and the statistics of solutions to this problem have been obtained by J. Kececioglu [7].

References

1. Cedergren,R., Abel,Y. and Sankoff, D. (1991) in Molecular Techniques in Taxonomy (G. M. Hewitt, A.W.B. Johnston and J.P.W. Young, eds.) Springer-Verlag. 87-99.
2. Nadeau,J.H. and Taylor, B.A. (1984) Proceedings of the National Academy of Sciences USA 81, 814-8.
3. Palmer, J.D., Osorio, B. and Thompson, W.F. (1988) Current Genetics 14, 65-74.
4. Sankoff, D. (1989) Bulletin of the International Statistical Institute 47.3, 461-475
5. Sankoff, D., Cedergren,R.and Abel, Y. (1990) in Methods in Enzymology 183. Molecular Evolution: Computer Analysis of Protein and Nucleic Acid Sequences, (R.F.Doolittle, ed.) Academic Press. 428-438.
6. Sankoff D. and Goldstein,M. (1988) Bulletin of Mathematical Biology 51, 117-124.
7. Sankoff,D., Kececioglu,J. and Leduc, G. (1992) manuscript in preparation.
8. Sankoff, D., Leduc, G., Antoine, N., Paquin, B., Lang, B.F., and Cedergren,R. (1992) Proceedings of the National Academy of Sciences USA to appear.
9. Sankoff,D, Leduc, G., and Rand,D. (1991) DERANGE. Minimum weight generation of oriented permutation by block inversions and block movements. Macintosh application, Centre de recherches mathématiques, Université de Montréal.
10. Watterson, G.A., Ewens, W.J., Hall, T.E. and Morgan, A. (1982) Journal of Theoretical Biology 99, 1-7.

3-D Substructure Matching in Protein Molecules

Daniel Fischer[1,2], Ruth Nussinov[2,3], Haim J. Wolfson[1,4]*

[1] Computer Science Department, Raymond and Beverly Sackler Faculty of Exact
Sciences, Tel Aviv University
[2] Sackler Inst. of Molecular Medicine, Faculty of Medicine, Tel Aviv University
[3] Lab of Math. Biology, PRI - Dynacor, NCI-FCRF, NIH
[4] Robotics Research Laboratory, Courant Inst. of Math. Sc., New York University

Abstract. Pattern recognition in proteins has become of central importance
in Molecular Biology. Proteins are macromolecules composed of an ordered
sequence of amino acids, referred to also as residues. The sequence of residues
in a protein is called its primary structure. The 3-D conformation of a protein
is referred to as its tertiary structure. During the last decades thousands of
protein sequences have been decoded. More recently the 3-D conformation of
several hundreds of proteins have been resolved using X-ray crystallographic
techniques.
Todate, most work on 3-D structural protein comparison has been limited
to the linear matching of the 3-D conformations of contiguous segments (allowing insertions and deletions) of the amino acid chains. Several techniques
originally developed for string matching have been modified to perform 3-D
structural comparison based on the sequential order of the structures.
We present an application of pattern recognition techniques (in particular
matching algorithms) to structural comparison of proteins. The problem we
are faced with is to devise efficient techniques for routine scanning of structural databases, searching for recurrences of inexact structural motifs not
necessarily composed of contiguous segments of the amino acid chain. The
method uses the Geometric Hashing technique which was originally developed for model-based object recognition problems in Computer Vision. Given
the three dimensional coordinate data of the structures to be compared, our
method automatically identifies every region of structural similarity between
the structures without prior knowledge of an initial alignment. Typical structure comparison problems are examined and the results of the new method
are compared with the published results from previous methods. Examples
of the application of the method to identify and search for non-linear 3-D
motifs are included.

1 Introduction

The functional properties of proteins depend on their three dimensional structures
(see [1]). Finding structural similarities in proteins is crucial for understanding key
biochemical processes such as catalysis, recombination, transcription and the initia-
tion steps regulating the genetic message itself. Immense practical potential exists in

---

* Work on this paper was supported by grant No. 89-00481 from the US-Israel Binational
Science Foundation (BSF), Jerusalem, Israel.

these solutions for medicine and biotechnology. Drug design, disease detection and protein design and engineering are just a few important applications.

A protein may be viewed as a sequence (linear chain) of amino acids which folds to generate a compact domain with a specific three-dimensional structure. The linearly ordered chain of amino acids is called the *primary structure* of the protein.

Studies of protein structure have indicated the presence of recurring motifs [5], conserved through evolution. Many of the specific folding motifs have been identified and classified. Moreover, it has become increasingly clear that proteins can be clustered into different structural families, built from a limited set of motifs. These structural motifs recur many times in different proteins. Finding structural motifs in proteins is crucial for understanding their role, and the mechanism by which they work. These can be inferred by analogy with other proteins containing the motif.

The most commonly used computerized macromolecule comparison approaches deal mainly with comparison of the primary structure of molecules. They are based on character string comparison algorithms. Most of them use variations of the dynamic programming technique (for a good survey see [20]). However, studies of protein structure have indicated that families of proteins retain a common underlying 3-D structure, even though their amino acid sequences may differ. This suggests that 3-D structure changes much more slowly than amino acid sequence during evolution. The above mentioned algorithms cannot deal with such 'truly 3-D' comparisons.

We propose a new method for protein comparison that uses 3-D information and is sequence independent. The method is computationally efficient and can be fully automated. It can find 3-D motifs contributed by different segments or isolated single amino acids. Moreover, one does not have to know in advance the required motif. Our method finds similar 'big enough' 3-D substructures in different proteins. In this sense it can be viewed as the geometric 3-D analog of the simpler (1-D) common subsequence finding algorithm. It should be noticed, however, that in our case we do not assume any linear order on the amino acids. On the contrary, we want an algorithm which performs independently of this order.

Protein comparison techniques are usually composed of three steps: 1) Find a relatively small initial alignment of the proteins; 2) Compute the superposition of the proteins that achieves the closest match of corresponding atoms in the initial alignment; 3) Transform one of the proteins according to this superposition, extend the initial alignment by choosing additional pairs of atoms that lie close enough under the given superposition.

Usually, step (1) above is the most time consuming. For proteins with considerable structural similarity eye inspection using computer graphic devices can be helpful in finding the initial match. For less similar proteins, quantitative methods are required, since an initial match cannot usually be unambiguously detected by eye inspection. As opposed to many instances of Pattern Recognition (e.g. object recognition in Computer Vision) no human eye can deal with the problem of protein comparison without the aid of computers. This is so because proteins are dense molecules containing thousands of atoms. [6]

---

[5] A motif can be viewed as a somewhat loosely defined pattern.

[6] Richardson cites the impressions of two workers on the first low resolution model of myoglobin back in 1958. Kendrew said: "Perhaps the most remarkable features of the

Quantitative methods to find the initial match usually perform a search of the large space of possible subsets (for a review see [9]). Remington & Matthews ([13, 14]) compare all possible pairs of linear structural fragments of a chosen length between two structures. A drawback of this method is its sensitivity to insertions and deletions between the sequences being compared. Rossmann & Argos ([16, 17, 18]) compare the structures in all possible orientations in order to find similar substructures.

Some of the current computerized methods for structural analysis concentrate on the comparison of the secondary structure of proteins. [7] Chothia & Lesk ([2]) propose a method where corresponding segments of secondary structures in the proteins being compared are individually identified and superimposed. In [10] it is suggested to build an undirected, labeled and fully connected graph whose nodes correspond to the linear representation of the secondary structure elements (i.e. helices and strands), and the edges correspond to the angles between these elements. A modification of the subgraph isomorphism algorithm ([24]) is used in the search procedure. Recently, [22] developed a method for defining topological equivalences in proteins. They build a graph whose nodes are the atoms of the protein and the edges are the inter-atomic distances. To find a set of equivalent atoms in both proteins they search for the maximal common subgraph between the two graphs representing the proteins. By applying several constraints they claim to reduce the exponential complexity of the general problem. One of the constraints used is the requirement that the sequential order of the two proteins be preserved in the equivalence.

All these methods are computationally expensive, exploit the linear order of the amino acid chain or require that the structures being compared be relatively similar.

Techniques for 3-D structural protein comparison based on the dynamic programming algorithm have been also applied (e.g. [23, 19]). These techniques again depend on the order imposed by the linear sequence of the proteins being compared.

Our approach is inspired by techniques which have been originally developed in Computer Vision for recognition of objects in cluttered scenes. In particular, we use the Geometric Hashing Paradigm for model-based object recognition due to Lamdan, Schwartz and Wolfson ([6, 5, 26]) and adapted for macromolecule comparison by Nussinov and Wolfson ([11]).

The major difference between previous work and ours is that whereas previous methods compare 3-D structures belonging to contiguous amino acids on the primary chain, ours is completely independent of the order of the amino acids in the chain. Our method compares proteins in a "real" 3-D approach. 3-D comparison conserving amino acid 'linearity' cannot focus on 3-D motifs that are strictly spatial. The methods, which exploit sequence linearity, are intrinsically one-dimensional, hence they are dealing with a much easier problem, and their performance is restricted.

Our method for protein comparison searches for non-predefined, sequence inde-

---

molecule are its complexity and its lack of symmetry. The arrangement seems to be almost totally lacking in the kind of regularities which one instinctively anticipates." Perutz commented: "Could the search for ultimate truth really have revealed so hideous and visceral-looking an object?". Richardson concludes by noting that appreciation of the aesthetics of protein structure has evolved during the last years.

[7] Secondary structure refers to the path that a local contiguous segment of amino acids follows in space. The most common regular secondary structures in proteins are the alpha-helices and beta-strands.

pendent matches, requires no prior knowledge of the motifs nor an initial alignment of the proteins. The method is highly efficient, fully automated, and is not constrained to linear motifs (i.e. the residues need not be contiguous on the chain). Our procedure is efficient for both non-predefined motifs, or when searching for occurrences of a known motif in new proteins.

In order to show the accuracy and efficiency of the method, we examine typical structure comparison problems and compare our results with published ones of previous methods. Some of the examples below compare homologous structures for which previous methods were used. In these examples, the matches produced by our method are equivalent to those of conventional alignment techniques. Thus, these examples appear to follow the 'sequence linearity' assumed by previous methods, because in these cases the best match coincides with the sequential alignment. It should be noted however, that no information about the order of the residues in the primary chain has been exploited by the algorithm. However, some additional structural, 3-D matches (i.e. not conserving the linearity) are also obtained by our method. In other examples, a motif is searched for within a protein. The motif can be a sequential motif or a real 3-D, non sequential motif. In these examples our method correctly detects the motif within the protein.

2   3-D Substructure Matching

In this section we give a short description of our algorithmic approach for matching rigid 3-D molecular substructures.

In the sequel we will frequently use a purely geometric language. The corresponding biological equivalents are as follows. An (rigid) object is analogous to a molecule. Such an object consists of a set of points, which correspond to atoms. Each point may have a label (= name of an atom). Additional relations between points (links) may be defined.

The structure comparison problem can be stated as follows. Given the 3-D coordinates of the atoms of two different molecules, find a rigid transformation (rotation and translation) in space, so that a 'sufficient' number of atoms of one molecule matches the atoms of the other molecule. Note that we do not intend to constrain ourselves to the linear order of the amino acid sequence. Hence the problem solved is inherently 3-D as opposed to the inherently 1-D problem of atom sequence matching.

The pairwise molecule comparison problem can be generalized to the comparison of a target molecule to the molecules of a given data-base. Namely, if one has a data-base of molecules and is faced with a new target molecule, one may be interested to find all the occurrences of "big enough" substructures of the data-base molecules in the target molecule. The pairwise molecule comparison problem is just a particular case, where the data-base consists of one model. To simplify the exposition we describe our method for the pairwise comparison only.

We assume that the structures to be compared are described by sets of *interest points* and their 3-D coordinates. For example, one may consider C_α atoms of the proteins as interest points. [8]

Next, we briefly outline the three major steps of our approach.

---

[8] Amino acids are composed of several atoms. Every amino acid has a C_α atom which

1) **Finding seed matches.** The first step of the algorithm searches through the structures to find (relatively small) candidate initial matches (i.e. possible initial alignments), which we call 'seed matches'. This step requires an extensive search on the structures. The Geometric Hashing Paradigm for model based object recognition (see [6, 5]) adopted from Computer Vision is applied to generate these seed matches. In order to restrict the search in this first step, each molecule is covered by a set of 'balls' of a pre-specified radius, and matching structures are discovered only within single 'balls'. This restriction to balls of a specified radius (external parameter of the algorithm) follows the practically tested assumption, that atoms of a matching substructure are in spatial proximity. The algorithm produces 'seed matches' which score relatively high in the process. A seed match is represented by a list of matching pairs of atoms and by a 3-D rigid transformation (rotation and translation). The number of the matching pairs should be above a threshold (minimal score) which is either a static or dynamic parameter of the algorithm. Each pair in the list specifies a correspondence between an atom from one structure and an atom from the other structure. The transformation represents the 3-D rotation and translation, which superimposes the atoms of the first structure onto the corresponding atoms of the second structure. Note that this step may produce several candidate seed matches which may have (almost) the same transformation, obtained from different pairs of 'balls'. Thus the next step clusters all these matching atoms together. The complexity of the first step is of the order of n^3, where n is the number of 'interest atoms'.

2) **Clustering of the seed matches.** In the second step those seed matches that represent almost identical transformations are clustered, and their correspondence lists are merged. Since a 3-D rigid motion can be described by six parameters, three for rotation and three for translation, we use these parameters to cluster the candidate seed matches.

The clustering algorithm iteratively joins transformations into groups according to the proximity of their parameters. The distance between two seed matches (transformations) is defined as a 6-dimensional distance between their parameter vectors.

At the end we are left with a relatively small number of significant clusters. Each cluster represents one transformation obtained from the individual transformations that were joined into the group. The seed match of a group is obtained by choosing matching pairs from the original seed matches that compose the group. To improve accuracy we choose only pairs that appear at least in a certain percentage of the seed matches. The complexity of this second step is $O(k^2)$, where k is the number of the initial seed matches.

3) **Extending the seed matches.** The relatively reliable correspondence lists of the seed matches obtained by Step 2 are extended to contain additional matching pairs. This is done by first transforming one of the structures according to the transformation specified by the seed match of the cluster. Then, pairs of atoms that lie "close" enough after the transformation are candidate additional

---

serves as a good "representative" of all the atoms of the amino acid. This atom is part of the main chain of the protein (backbone). Using this choice of interest points, the number of points per structure ranges in the low hundreds.

matches. Because proteins are usually quite dense in space, each atom from one structure may have several "close" neighbors in the other structure. To choose the appropriate pairs a heuristic iterative matching algorithm is applied which minimizes the sum of the distances between all the matched pairs.

At the end, the best extended matches are reported. The quality of the match is determined by the number of the matching pairs of atoms and by the least squares distance between these matching atoms.

2.1 The Geometric Hashing Paradigm

We give a short description of the first step of our algorithm, which is based on the *Geometric Hashing* paradigm for model based object recognition ([4, 6, 5]). This approach is especially suitable for detection of a partial match between a scene and objects belonging to a large data-base. The adaptation of the technique for 3-D molecule matching is presented in [11]. Although in Molecular Biology we use a particular version of *Geometric Hashing* for comparison of 3-D structures under rigid motion, we are in addition faced with specific problems because of the nature of our input. Especially, in Molecular Biology we are dealing with structures consisting of hundreds of atoms (see Appendices), while in Computer Vision a typical object has 10-20 'interest points'. This stresses the importance of developing a low complexity matching algorithm in the first stages of seed match finding.

In order to design an efficient matching algorithm one needs an appropriate representation of the 'interest point' sets.

Motion Invariant Representation. Let us assume for the moment that our 'interest points' have no special salient features, except their spatial location. If the points were colored, each by a different color, one could readily recognize a matching subset between the model and target molecules. Hence, we shall look for a 'geometric color' which describes the spatial location of our points. From Analytic Geometry we know that a natural representation of a point is by its coordinates in a given reference frame. Then, the coordinates are the required geometric color. There still remains the problem of choice of the reference frame which should address the following issues :

i) Rotational and translational invariance.
ii) Choice of the same reference frame both on the model and the target molecules.
iii) Handling of partial matches.

A 3-D reference frame is uniquely defined by three ordered non-collinear points. The three points define the $x - y$ plane, where the first point can serve as the origin, and the second defines the direction of the x-axis. The directions of the y-axis and z-axis are then defined uniquely.

Thus all 'interest points' of a molecule have unique coordinates in such a reference frame. These coordinates can be indices to a hash-table where the molecule and reference frame are stored. Such a representation enables fast detection of the existence of many similar coordinates on the target molecule for some reference frame (for details see [11]).

The above mentioned representation relies heavily on the reference frame points. In particular, a reference frame, which was chosen for the model molecule might not appear in the matching portion of the target molecule. To overcome this difficulty, one could represent the 'interest points' of a molecule in *all* relevant reference frames. In general, for an n-point set, $O(n^3)$ reference frames are available. However, exploitation of biological knowledge can help us in construction of fewer reference frames. In particular, we are currently experimenting with the representation of n-atom sets in just $O(n)$ reference frames, where each frame is based on a single amino acid. However, even without any biological information, one can represent the atoms in just $O(n^2)$ reference frames by compromising on the uniqueness of definition of a single point coordinate. This compromise can be resolved in later stages of the algorithm.

Let us sketch the technique that we have implemented for partial matching of 3-D molecules, which may undergo a rigid motion (rotation and translation).

Preprocessing. Each data base molecule is preprocessed separately. For a given molecule, one encodes the coordinates of the atoms, with respect to a reference set based on a pair of atoms. Specifically, pick an ordered pair of atoms (which is subsequently nicknamed 'basis'), and for each atom of the molecule compute the distance of this atom from each of the basis atoms. Use the lengths of the sides of the resulting triangle as an address to a hash-table and store in the appropriate entry the basis atoms and the third atom, and the name of the molecule to which they belong. This procedure is done for all the molecule atoms (per basis), and for all possible bases (pairs of atoms). In our implementation we have introduced minimal and maximal distance constraints on the lengths of the triangle sides, so not all triangles have been considered in the preprocessing stage. Note that the representation of an atom in a given basis is not unique, since the triangle sides define a circle in space, but this ambiguity can be easily removed in subsequent verification stages. The complexity for preprocessing a single molecule of n atoms is $O(n^3)$ (practically, it is less because of the distance constraints). The preprocessing is done off-line once for each new molecule joining the data base.

Recognition. In the on-line recognition stage we are faced with a target molecule, and want to discover substructures of this molecule which match substructures of the data base molecules modulo rotation and translation (see the 'seed matches' of the previous section). In this procedure we pick a pair of atoms on the target molecule (basis) and consider the triangles that it creates with the other target molecule atoms (given that the triangle complies with the side distance constraints). The lengths of the triangle sides define an address to the hash-table, and we cast a vote for the pairs *(molecule, basis)* which appear in that hash-table entry. (Due to the inaccuracy of the input data, one may have to retrieve the data also from the neighboring bins of the hash-table.) If such a pair scores a large number of votes, it implies a large match between a substructure of that molecule with our target molecule, when the appropriate bases are transformed to each other. One can take all the atoms that have participated in the voting for that match, eliminate all the inconsistent entries, due to the nonuniqueness of the third triangle point location in space, and compute a more accurate spatial transformation using a least squares procedure ([21]). The final score takes into account both the number of votes of a candidate match and its least squares distance. Note that due to the redundant and transformation invariant encoding of each atom into the hash-table we are able to detect partially matching structures regardless of their rotation and translation.

The above mentioned procedure is repeated for each pair of target molecule atoms. Assuming a well distributed hash-table the complexity is $O(n^3)$, where n is the number of atoms on the target molecule. Since the hash-table is known in advance from the preprocessing stage, one can avoid processing entries with high occupancy, since their processing is time consuming and does not contribute significant salient information. In general, a weighted voting scheme can be introduced.

2.2 Current Experimental Results

A version of the proposed algorithm has recently been programmed and applied ([3]) to proteins which have previously been compared using other methods. We have successfully run tens of experiments. Few of the results are cited below. Tables of actual matching examples are gives in Appendices A, B, C. In appendices A and B we give examples of a predefined motif search, while appendix C gives an example of structural comparison between two molecules which recovers a non-predefined motif.

(i) In searches for non-predefined, similar domains in bacterial ferredoxin from Peptococcus aerogenes, excellent fit of our results with those of Rossman and Argos ([17]) has been obtained.

(ii) Two members from the phospholipase A2 proteins were compared - Bovine pancreas and Crotalus atrox venom. These proteins have been previously compared by Renetseder et al. ([15]) using standard techniques, i.e. finding by eye a homologous core and then aligning using the least square procedure. Again, our alignment corresponds exactly to that reported by Renetseder et al.

(iii) The helix-turn-helix motif was located in several bacterial repressor proteins (tryptophan repressor, lambda cro and phage 434 cro) just as noted in the annotated protein databank.

(iv) Two proteins from the calmodium calcium binding protein group were compared - parvalbumin and intestinal calcium binding protein. Several matches were obtained. Two of these correspond to the alignment reported by Taylor and Orengo ([23]).

(v) Bovine liver rhodanese contains two motifs, which have been compared both by Taylor and Orengo and by Ploegman et al. ([12]), yielding similar results. Our matches are almost identical to the previously obtained ones.

(vi) Two lysozymes have been compared: hen egg-white and T4 phage. Our matches compare favorably with those of Rossman and Argos ([17]), Weaver et al ([25]) and Taylor and Orengo ([23]).

It should be noted that previously published matches are based either on linear structural comparisons, where contiguous amino acids are matched or on a search of a known predefined motif. Our three dimensional comparisons had no such prior assumptions whatsoever and have also unraveled some "real" 3-D, sequence-order independent matches. (When our algorithm detects a candidate match which does preserve linear structure it may serve as an additional indication of the 'non-accidentalness' of the detected solution.) We expect that intensive applications of the method to the crystallographic database will yield hitherto undiscovered, recurring spatial motifs.

It should also be noted that the running time of all the described experiments has been very short . For example in experiment (ii) above, where each of the molecules contained about 120 'interest atoms', the substructure search took 12.3 sec of CPU time on a Sparc Sun processor. The search of a 19 'interest atom' predefined motif (pattern) in four molecules of about 100 'interest atoms' (Appendix A) took about 3.7 sec CPU in each of the molecules.

3 Conclusions and Future Research

We have presented a method for structural comparisons. Our algorithms deal with the comparison of a-priori unknown rigid substructures modulo translation and rotation in space. The algorithm is relatively efficient and has a very natural and highly efficient parallel implementation potential. We have implemented a prototype of the *Geometric Hashing* algorithm (originally developed for Computer Vision applications) for molecular biology applications and the results seem to be superior to previously achieved ones.

Until now most of the collaboration between computer scientists and biologists in this area has been channeled to the design of efficient string editing algorithms. The approach that we present is novel, since it is not yet another more efficient string matching, but it addresses directly the inherent three-dimensional structure of molecules.

We view the presented algorithm only as a basic paradigm. Additional biological information can be incorporated into this basic framework. In particular, one can consider the chemical links between various atoms and groups of atoms. Any such additional information adds additional matching constraints an may speed up the algorithms. We plan to further develop our algorithm both from its computational and biological aspects, so it can be efficiently applied to large data bases.

Currently we examine similar algorithmic ideas for the problem of the docking ligands into the active sites of their receptors, an extremely important problem which has direct application to synthetic drug design. Our preliminary results show an improvement by at least an order of magnitude in the complexity compared with previously developed methods. Another potential area of applicability is to protein-protein interactions in the immune system.

References

1. C. Branden and J. Tooze. *Introduction to Protein Structure*. Garland Publishing, Inc., New York and London, 1991.
2. C. Chothia and A.M. Lesk. The relation between the divergence of sequence and structure in proteins. *EMBO Jour.*, 5(4):823–826, 1986.
3. D. Fischer, O. Bachar, R. Nussinov, and H.J. Wolfson. An Efficient Computer Vision based technique for detection of three dimensional structural motifs in Proteins. *J. Biomolec. Str. and Dyn.*, 1992. in press.

4. Y. Lamdan, J. T. Schwartz, and H. J. Wolfson. On Recognition of 3-D Objects from 2-D Images. In *Proceedings of IEEE Int. Conf. on Robotics and Automation*, pages 1407–1413, Philadelphia, Pa., April 1988.

5. Y. Lamdan, J. T. Schwartz, and H. J. Wolfson. Affine Invariant Model-Based Object Recognition. *IEEE Trans. on Robotics and Automation*, 6(5):578–589, 1990.

6. Y. Lamdan and H. J. Wolfson. Geometric Hashing: A General and Efficient Model-Based Recognition Scheme. In *Proceedings of the IEEE Int. Conf. on Computer Vision*, pages 238–249, Tampa, Florida, December 1988.

7. A.M. Lesk and C. Chothia. *J. Mol. Biol.*, 136:225–270, 1980.

8. A.M. Lesk and C. Chothia. *J. Mol. Biol.*, 160:325–342, 1982.

9. B.W. Matthews and M.G. Rossman. *Methods Enzymol.*, 115:397–420, 1985.

10. E.M. Mitchel, P.J. Artymiuk, D.W. Rice, and P. Willet. Use of Techniques Derived from Graph Theory to Compare Secondary Structure Motifs in Proteins. *J. Mol. Biol.*, 212:151–166, 1989.

11. R. Nussinov and H.J. Wolfson. Efficient detection of three-dimensional motifs in biological macromolecules by computer vision techniques. *Proc. Natl. Acad. Sci. USA*, 88:10495–10499, 1991.

12. J. H. Ploegman, G. Drent, K. H. Kalk, and W. G. Jol. *J. Mol. Biol.*, 123:557–594, 1987.

13. S. J. Remington and B. W. Matthews. *Proc. Natl. Acad. Sci. USA*, 75:2180–2184, 1978.

14. S. J. Remington and B. W. Matthews. *J. Mol. Biol.*, 140:397–420, 1980.

15. R. Renetseder, S. Brunie, B. W. Dijkstra, J. Drent, and P. B. Sigler. *J. Biol. Chem.*, 206:11627–11634, 1985.

16. M. G. Rossman and P. Argos. *J. Biol. Chem.*, 250:7525–7532, 1975.

17. M. G. Rossman and P. Argos. *J. Mol. Biol.*, 105:75–96, 1976.

18. M. G. Rossman and P. Argos. *J. Mol. Biol.*, 109:99–129, 1977.

19. T.L. Sali, A. Blundell. *J. Mol. Biol.*, 212:403–428, 1990.

20. D. Sankoff and J.B. Kruskal. *Time Warps, String Edits and Macromolecules: The Theory and Practice of Sequence Comparison*. Addison-Wesley, 1983.

21. J.T. Schwartz and M. Sharir. Identification of Partially Obscured Objects in Two Dimensions by Matching of Noisy 'Characteristic Curves'. *The Int. J. of Robotics Research*, 6(2):29–44, 1987.

22. N. Subbarao and I. Haneef. Defining Topological Equivalences in Macromolecules. *Protein Engineering*, 4(8):887–884, 1991.

23. W. R. Taylor and C. A. Orengo. Protein structure alignment. *J. Mol. Biol.*, 208:1–22, 1989.

24. J.R. Ullman. An algorithm for subgraph isomorphism. *J. ACM*, 23:31–42, 1976.

25. L. H. Weaver, M. G. Grutter, S. J. Remington, T. M. Gray, N. W. Issacs, and B. W. Matthews. *J. Mol. Evol.*, 21:97–111, 1985.

26. H. J. Wolfson. Model Based Object Recognition by 'Geometric Hashing'. In *Proceedings of the European Conf. on Computer Vision*, pages 526–536, Antibes, France, April 1990.

A   Appendix : Detection of the heme pocket in some globins.

A classical example for protein comparison is the globin family. These studies range from sequence comparisons to analysis of crystals of hemoglobin to structural comparison of the X-ray crystallographic data (for a review and an exhaustive analysis and comparison of the atomic structures of nine different globins see [7]).

The known globins have different amino acid sequences but remarkably similar secondary and tertiary structures. They are mainly composed of alpha helices which assemble in a common pattern, enclosing the heme group in pockets of similar geometry made up from homologous portions of the molecules.

We demonstrate our method by searching for a predefined 3-D motif within several globins. We use the heme pocket of the α subunit of hemoglobin (PDB code: 4HHB), as a predefined 3-D motif. This motif is composed of 19 non-sequential residues that are involved in binding the heme ([7]).

We search for the recurrence of this motif within myoglobin (PDB code: 4MBN), the β subunit of hemoglobin (denoted as 4HBBB), horse hemoglobin (PDB code: 2DHB) and sea lamprey hemoglobin (PDB code: 2LHB). The residues from the target molecule that were matched to the motif define the recurring motif within the target molecule. In all matches, our results are identical to those by Lesk & Chothia([7]).

| votes: 19 | | votes: 19 | | votes: 19 | | votes: 18 | |
| rms : 0.93 | | rms : 0.60 | | rms : 1.03 | | rms 94 | |
MODEL heme	|TARG.| |4HHBB|	MODEL heme	|TARG.| | 2DHB|	MODEL heme	|TARG.| | 4MBN|	MODEL heme	|TARG.| | 2LHB|
	|.....|		|.....|		|.....|		| |
136-L	|141-L|	136-L	|136-L|	136-L	|142-I|	136-L	|145-L|
	|.....|		|.....|		|.....|		|.....|
132-V	|137-V|	132-V	|132-V|	132-V	|138-F|		| |
	|.....|		|.....|		|.....|		|.....|
101-L	|106-L|	101-L	|101-L|	101-L	|107-I|	101-L	|119-L|
	|.....|		|.....|		|.....|		|.....|
98-F	|103-F|	98-F	| 98-F|	98-F	|104-L|	98-F	|116-F|
97-N	|102-N|	97-N	| 97-N|	97-N	|103-Y|	97-N	|115-Y|
	|.....|		|.....|		|.....|		|.....|
93-V	| 98-V|	93-V	| 93-V|	93-V	| 99-I|	93-V	|111-V|
	| 97-H|		| 92-R|		| 98-K|		|110-Q|
91-L	| 96-L|	91-L	| 91-L|	91-L	| 97-H|	91-L	|109-F|
	|.....|		|.....|		|.....|		|.....|
87-H	| 92-H|	87-H	| 87-H|	87-H	| 93-H|	87-H	|105-H|
86-L	| 91-L|	86-L	| 86-L|	86-L	| 92-S|	86-L	|104-K|
	|.....|		|.....|		|.....|		|.....|
83-L	| 88-L|	83-L	| 83-L|	83-L	| 89-L|	83-L	|101-L|
	|.....|		|.....|		|.....|		|.....|
66-L	| 71-F|	66-L	| 66-L|	66-L	| 72-L|	66-L	|81-V |
65-A	| 70-A|	65-A	| 65-G|	65-A	| 71-A|	65-A	|80-A |
	|.....|		|.....|		|.....|		|.....|
62-V	| 67-V|	62-V	| 62-V|	62-V	| 68-V|	62-V	|77-I |
61-K	| 66-K|	61-K	| 61-K|	61-K	| 67-T|	61-K	|76-R |
	|.....|		|.....|		|.....|		|.....|
58-H	| 63-H|	58-H	| 58-H|	58-H	| 64-H|	58-H	|73-H |
	|.....|		|.....|		|.....|		|.....|
46-F	| 45-F|	46-F	| 46-F|	46-F	| 46-F|	46-F	|55-F |
45-H	| 44-S|	45-H	| 45-H|	45-H	| 45-R|	45-H	|53-P |
	| 43-E|		| 44-E|		| 44-D|		|.....|

```
43-F   | 42-F|    43-F   | 43-F|    43-F   | 43-F|   43-F  |52-F |
42-Y   | 41-F|    42-Y   | 42-Y|    42-Y   | 42-K|   42-Y  |51-F |
       |.....|            |.....|            |.....|          |.....|
TRAN: (-0.3,-1.1,2.7)                   TRAN:(4.6, 0.5,15.3)
ROT: (-0.01,0.05,-3.13)                 ROT: (-0.36,-0.51, 1.37)
               TRAN: (-0.3,  -1.5, -0.1)                 TRAN:(-25.2,-0.8,
               ROT: (-0.01, 0.02, -0.01)                 ROT:(1.35, 0.39,
```

Heme pocket search from the α-subunit of hemoglobin (heme) within the β-subunit of hemoglobin (4HHBB), horse hemoglobin (2DHB), myoglobin (4MBN) and sea lamprey hemoglobin (2LHB).

In the above table the C_α atoms of each residue are shown with their sequence position and the residue to which they belong. Under each match the 3 translational (TRAN) parameters and the 3 rotation (ROT) parameters of the rigid body transformation applied to obtain the match are shown.

B Appendix : Detection of a basic immunoglobulin fold.

Immunoglobulin molecules consist of six domains that have similar secondary and tertiary structures and low amino-acid sequence homology. Each domain contains two stacked β-sheets pinned together by a disulphide bridge. One β-sheet contains four strands and the other three. Lesk & Chothia ([8]) extensively studied the cores of several immunoglobulin domains. They observed that a group of 35 atoms is present in all domains which they compared and called it the β-sheet core. Within the core, they defined a group of 9 atoms that occurs at the center of the region between the two β-sheets and called it the pin.

The antigen-binding fragment of FAB NEW (PDB code: 3FAB) contains four of these domains designated VH, VL, CL and CH1. The VH and VL domains are highly homologous to each other. The CL and CH1 domains are also highly homologous to each other, and less homologous to VH and VL.

We defined a 3-D motif as the $C\alpha$ atoms from the VL domain pin, plus some other residues from the core. This motif can represent the basic fold of an immunoglobulin domain. The residues used to define the motif are: 20-I, 21-S, 22-C, 24-G, 32-V, 33-K, 34-W, 69-S, 70-A, 71-T, 72-L, 73-A, 85-Y, 86-Y and 87-C. It includes the disulphide bridge, some of its neighbor residues, a conserved tryptophan as well as additional ones.

We searched for this motif in the VH, CL and CH1 domains. The residues from each domain that matched the motif residues define the motif occurring within the domain. The results of our method show the same matches as those defined by Lesk & Chothia.

```
votes:15                 votes:15                 votes:15
rms  :0.69               rms  : 0.56              rms  : 0.58
   MODEL   |TARG.|          MODEL   |TARG.|          MODEL   |TARG.|
   VLpin   | ch1|           VLpin   | cl |           VLpin   | vh |
------------|    |        ------------|    |        ----------|    |
          |.....|                  |.....|                  |.....|
```

87-C	|200-C|	87-C	|195-C|	87-C	| 95-C|
86-Y	|199-I|	86-Y	|194-S|	86-Y	| 94-Y|
85-Y	|198-Y|	85-Y	|193-Y|	85-Y	| 93-Y|
	|.....|		|.....|		|.....|
73-A	|187-T|	73-A	|181-S|	73-A	| 81-R|
72-L	|186-V|	72-L	|180-L|	72-L	| 80-L|
71-T	|185-V|	71-t	|179-Y|	71-T	| 79-S|
70-A	|184-S|	70-A	|178-S|	70-A	| 78-F|
69-S	|183-S|	69-s	|177-S|	69-S	| 77-Q|
	|.....|		|.....|		|.....|
34-W	|158-W|	34-W	|150-W|	34-W	| 36-W|
33-K	|157-S|	33-K	|149-A|	33-K	| 35-T|
32-V	|156-V|	32-V	|148-V|	32-V	| 34-S|
	|.....|		|.....|		|.....|
24-G	|146-V|	24-G	|138-I|	24-G	| 24-V|
	|145-L|		|137-L|		| 23-T|
22-C	|144-C|	22-C	|136-C|	22-C	| 22-C|
21-S	|143-G|	21-S	|135-V|	21-S	| 21-T|
20-I	|142-L|	20-I	|134-L|	20-I	| 20-L|
	|.....|		|.....|		|.....|

TRAN:-27.3, 22.2,-40.0   TRAN:-54.7, 25.4,   8.2   TRAN:-23.8,   5.8, 44.3
ROT: 0.42, 0.12,-0.66   ROT: -1.52, -1.12, -2.09   ROT: -1.82, -0.25, -2.33

An immunoglobulin motif was defined as the $C\alpha$ residues from the pin of domain VL from 3FAB. This pin motif was detected in the CH1, CL and VH domains of 3FAB. In the above table the C_α atoms of each residue are shown with their sequence position and the residue to which they belong. Under each match the 3 translational (TRAN) parameters and the 3 rotation (ROT) parameters of the rigid body transformation applied to obtain the match are shown.

C   Appendix: Detection of the HELIX-TURN-HELIX motif.

The HELIX-TURN-HELIX (HTH) is a sequential motif found in some DNA-binding proteins. The HTH motif is involved in DNA-binding in some repressor proteins. Repressor proteins play an integral role in the control of gene transcription . The motif contains two alpha helices connected by a variable turn.

We compare three transcriptional regulatory proteins known to contain the HTH motif: tryptophan repressor (PDB code: 2WRP), lambda CRO (PDB code: 1CRO) and phage 434 CRO (PDB code: 2CRO).

The sequence positions where the HTH motifs appear are:

Prot	Pos	Sequence
2wrp:	66-88	MS QRELKNELGA GIATITRGSNS
1cro:	14-36	FG QTKTAKDLGV YQSAINKAIHA
2cro:	15-37	MT QTELATKAGV KQQSIQLIEAG

The motif position is not given to the program; the full proteins are compared to each other. In this case, the major structural similarity between every pair of proteins is the HTH motif itself.

In the three pairwise comparisons below our method succeeds in matching the HTH motif from one protein to the HTH motif from the other. Very few other pairs are matched, showing that the only equivalent substructure between the proteins is the HTH motif itself. The pairs outside the HTH motif are 3-D nonlinear matches.

We do not try to qualify the matches obtained by our method, or asses any biological significance. Since our method is a purely geometric one, no biologically true result or false positive can be verified. The fact that our method, unbiased by the order of the amino sequences, produces equivalent results to those previously reported (which exploit specifically the sequence order) indicates that the best structural superposition between the structures in these cases is indeed one that conserves the sequence order. Such a coincidence of results is definitely not random. Nevertheless, sporadic non linear matches which have not been discovered in earlier work, appear in our results. These matches reflect the fact that under the given transformation, the atoms in the matched pairs came sufficiently close together so as to be matched. The biological significance of such pairs is out of the scope of this paper.

votes: 27		votes: 29		votes: 32	
rms: 0.90		rms: 1.29		rms: 0.97	
MODEL	|SCENE|	MODEL	|SCENE|	MODEL	|SCENE|
2cro	| 2wrp|	2wrp	|1croB|	1croB	| 2cro|
-------------	| |	-------------	| |	-------------	| |
	| |		| |	55-V	| 60-Q|
	| |		| |		|.....|
	|.....|		|.....|	44-I	| 53-N|
	| |	55-V	| 40-I|		|.....|
	|.....|		|.....|	51-Y	| 50-M|
	| |		| |	52-A	| 49-A|
	|.....|	89-L	| 37-G|		|.....|
37-G	| 88-S|	88-S	| 36-A|	36-A	| 37-G|
36-A	| 87-N|	87-N	| 35-H|	35-H	| 36-A|
35-E	| 86-S|	86-S	| 34-I|	34-I	| 35-E|
34-I	| 85-G|	85-G	| 33-A|	33-A	| 34-I|
33-L	| 84-R|	84-R	| 32-K|	32-K	| 33-L|
32-Q	| 83-T|	83-T	| 31-N|	31-N	| 32-Q|
31-I	| 82-I|	82-I	| 30-I|	30-I	| 31-I|
30-S	| 81-T|	81-T	| 29-A|	29-A	| 30-S|
29-Q	| 80-A|	80-A	| 28-S|	28-S	| 29-Q|
28-Q	| 79-I|	79-I	| 27-Q|	27-Q	| 28-Q|
27-K	| 78-G|	78-G	| 26-Y|	26-Y	| 27-K|
26-V	| 77-A|	77-A	| 25-V|	25-V	| 26-V|
25-G	| 76-G|	76-G	| 24-G|	24-G	| 25-G|
24-A	| 75-L|	75-L	| 23-L|	23-L	| 24-A|
23-K	| 74-E|	74-E	| 22-D|	22-D	| 23-K|
22-T	| 73-N|	73-N	| 21-K|	21-K	| 22-T|
21-A	| 72-K|	72-K	| 20-A|	20-A	| 21-A|
20-L	| 71-L|	71-L	| 19-T|	19-T	| 20-L|
19-E	| 70-E|	70-E	| 18-K|	18-K	| 19-E|
18-T	| 69-R|	69-R	| 17-T|	17-T	| 18-T|
17-Q	| 68-Q|	68-Q	| 16-Q|	16-Q	| 17-Q|
16-T	| 67-S|	67-S	| 15-G|	15-G	| 16-T|

```
            | 66-M|      66-M    | 14-F|      14-F     | 15-M|
13-L        | 65-E|      65-E    | 13-R|      13-R     | 14-K|
11-I        | 64-G|               |.....|              |.....|
9-R         | 63-R|      63-R    | 9-D|      10-Y     | 10-R|
            |.....|      61-L    | 8-K|               | 9-R|
            |     |               | 7-L|               | 8-K|
            |     |      59-E    | 6-T|      8-K      | 7-K|
            |     |               |.....|      7-L      | 6-L|
            |.....|                                    |.....|
44-F        | 53-T|               39-K     | 2-L|
            |.....|                                    |.....|
43-R        | 50-A|                                    |     |

TRAN:16.6, -7.7, -2.4    TRAN:-16.6, -49.6, -20.1    TRAN:-23.4, -45.2, -19.7
ROT: -0.11, 0.29,-2.53   ROT:  2.34,  0.64, -0.90    ROT:  1.91, -0.77, -2.59
```

Comparison of three DNA-binding proteins: tryptophan represor (2wrp), lambda CRO (1croB) and lambda 434 (2cro). The HTH motif in each protein is correctly matched.

Fast Serial and Parallel Algorithms for Approximate Tree Matching with VLDC's* (Extended Abstract)

Kaizhong Zhang[1], Dennis Shasha[2], Jason Tsong-Li Wang[3]

[1] Department of Computer Science, The University of Western Ontario,
London, Ontario, Canada N6A 5B7 (kzhang@csd.uwo.ca)
[2] Courant Institute of Mathematical Sciences, New York University,
251 Mercer Street, New York, NY 10012, USA (shasha@cs.nyu.edu)
[3] Department of CIS, New Jersey Institute of Technology,
University Heights, NJ 07102, USA (jason@vienna.njit.edu)

1 Introduction

Problem. Ordered, labeled trees are trees in which each node has a label and the left-to-right order of its children (if it has any) is fixed. Suppose we define the distance between two ordered trees to be the weighted number (the user chooses the weighting) of edit operations (insert, delete, and relabel) to transform one tree to the other. This paper presents algorithms to perform approximate matching for such trees with variable-length don't cares (VLDC's). As far as we know, these are the first such algorithms ever to be presented.

Let P be a pattern tree and T a text tree. We allow P to contain two kinds of VLDC's: path-VLDC's (denoted |) and umbrella-VLDC's (denoted $\wedge$). When matching P with T,

1. | may substitute for part of a path from the root to a leaf of T (Fig. 1(a)).
2. $\wedge$ may substitute for part of such a path and all the subtrees emanating from the nodes of that path, except possibly at the lowest node of that path. At the lowest node, the VLDC can substitute for a set of leftmost subtrees and a set of rightmost subtrees. Formally, let the lowest node be n and let the children of n be $c_1, \ldots, c_k$ in left-to-right order. Let i, j be such that $0 \leq i < j \leq k+1$. An umbrella can substitute for the subtrees rooted at $c_1, \ldots, c_i$ and $c_j, \ldots c_k$ in addition to the node n, ancestors along a path starting at n, and the subtrees of those proper ancestors of n (Fig. 1(b)).

Given a pattern P that contains both kinds of VLDC's, and a text tree T, we design serial and parallel algorithms to answer the following questions:

---

* Work supported in part by NSF grants IRI-8901699 and CCR-9103953, by ONR contracts N00014-90-J-1110 and N00014-91-J-1472, by the Natural Sciences and Engineering Research Council of Canada under grant OGP0046373, and by an SBR grant from NJIT

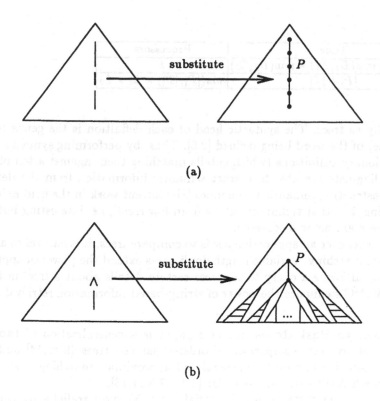

(a)

(b)

Fig. 1. Substitution of VLDC's. (a) | substitutes for the nodes on P; (b) ∧ substitutes for the nodes on P and shaded subtrees.

1. What is the minimum distance between P and T after performing the substitutions of the VLDC's?
2. Let cutting at node n from tree T mean removing the subtree rooted at n. What is the minimum distance between P and T after performing the substitutions of VLDC's, allowing zero or more cuttings at nodes from T? (We refer to the case as "matching with cut".)

Main Results. Let trees P and T have depths D_1 and D_2 and leaves L_1 and L_2 respectively. The asymptotic running times (assuming a concurrent-read concurrent-write parallel random access machine) for both of the above questions are shown in Table 1.

Significance of the Work. Ordered trees represent data in image processing, RNA secondary structures, language parsing, and generalized markup languages. For the sake of illustration, let us consider the application of our algorithms (which have been implemented in a toolkit) to natural language processing [12]. Linguists often store dictionary definitions in a lexical database. The definitions are represented

Table 1.

Time	Processors
$\|P\| \times \|T\| \times \min(D_1, L_1) \times \min(D_2, L_2)$	1
$\|P\| + \|T\|$	$\min(\|P\|, \|T\|) \times L_1 \times L_2$

syntactically as trees. The syntactic head of each definition is the genus term (superordinate) of the word being defined [2,6]. Thus, by performing syntactic analysis of the dictionary definitions (which entails matching them against a template with cuttings), linguists are able to extract semantic information from the definitions, thereby constructing semantic taxonomies [1]. Current work in the field relies on exact matching, but that technique suffers from low recall, i.e. interesting but slightly different trees are never discovered.

In fact, whenever an application needs to compare trees, it is natural to ask about approximate matching. Variable length don't cares extend the power of approximate matching by allowing a query to suppress certain details about a tree, in the same way that VLDC's enhance the power of string-based information retrieval systems.

Comparison to Past Research. This paper is a generalization of two lines of work: (1) approximate comparison of ordered labeled trees [9,10,14] and (2) approximate matching in strings in general and approximate matching in strings with variable-length don't cares in particular [3,4,5,7,8,11,13].

In this extended abstract, we give serial (Sect. 3) and parallel algorithms (Sect. 4) for matching with cut only. The entire set of algorithms is in the full paper and is available from the authors. We have chosen to present this relatively simple result, because we believe it is the most useful of our results to date concerning approximate tree matching with VLDC's.

2 Preliminaries

We first review several definitions for trees without VLDC's, and then define the VLDC substitutions.

2.1 Edit Operations

Our distance metric for trees is a generalization of the editing distance between sequences. The edit operations are relabel, delete, and insert. We represent these operations as $u \to v$, where each of u and v is either a node or the null node (Λ). We call $u \to v$ a relabeling operation if $u \neq \Lambda$ and $v \neq \Lambda$; a delete operation if $u \neq \Lambda = v$; and an insert operation if $u = \Lambda \neq v$. Let T_2 be the tree that results from the application of an edit operation $u \to v$ to tree T_1; this is written $T_1 \Rightarrow T_2$ via $u \to v$. Figure 2 illustrates the edit operations.

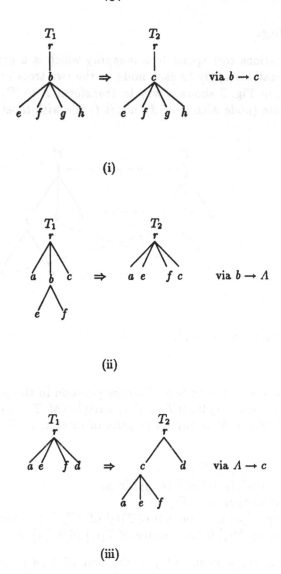

Fig. 2. (i) Changing node label b to c. (ii) Deleting node b. (iii) Inserting node c. (A consecutive sequence of children of r become the children of c.)

Let S be a sequence $s_1, s_2, \ldots, s_k$ of edit operations. S transforms tree T to tree T' if there is a sequence of trees $T_0, T_1, \ldots, T_k$ such that $T = T_0, T' = T_k$ and $T_{i-1} \Rightarrow T_i$ via s_i for $1 \leq i \leq k$.

Let γ be a cost function that assigns to each edit operation $u \rightarrow v$ a nonnegative real number $\gamma(u \rightarrow v)$. By extension, the cost of a sequence is simply the sum of the costs of constituent edit operations. The *distance* between T and T', denoted $\delta(T, T')$, is simply the minimum cost of all sequences of edit operations taking T to T'.

2.2 Mappings

The edit operations correspond to a *mapping* which is a graphical specification of what edit operations apply to each node in the two trees (or two ordered forests). The mapping in Fig. 3 shows a way to transform T to T'. It corresponds to the sequence (delete (node with label d), insert (node with label d)).

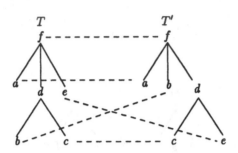

Fig. 3. A mapping from T to T'.

Let $T[i]$ represent the node of T whose position in the postorder for nodes of T is i. Formally, a mapping from T to T' is a triple (M, T, T') (or simply M if there is no confusion), where M is any set of pairs of integers (i, j) satisfying the following conditions:

1. $1 \leq i \leq |T|$, $1 \leq j \leq |T'|$;
2. For any pair of (i_1, j_1) and (i_2, j_2) in M,
   (a) (one-to-one) $i_1 = i_2$ iff $j_1 = j_2$;
   (b) (sibling) $T[i_1]$ is to the left of $T[i_2]$ iff $T'[j_1]$ is to the left of $T'[j_2]$;
   (c) (ancestor) $T[i_1]$ is an ancestor of $T[i_2]$ iff $T'[j_1]$ is an ancestor of $T'[j_2]$.

The cost of M, denoted $\gamma(M)$, is the cost of deleting nodes of T not touched by a mapping line plus the cost of inserting nodes of T' not touched by a mapping line plus the cost of relabeling nodes in those pairs related by mapping lines with differing labels.

Lemma 1. *Given S, a sequence $s_1, s_2, \ldots, s_k$ of edit operations from T to T', there exists a mapping M from T to T' such that $\gamma(M) \leq \gamma(S)$. Conversely, for any mapping M, there exists a sequence of edit operations S such that $\gamma(S) = \gamma(M)$.*

Hence, $\delta(T, T') = \min\{\gamma(M) \mid M \text{ is a mapping from } T \text{ to } T'\}$.

2.3 Substitution of VLDC's

Let P be a pattern tree that contains both umbrella-VLDC's and path-VLDC's and let T be a text tree. A VLDC-substitution s on P replaces each path-VLDC in P by a path of nodes in T and each umbrella-VLDC in P by an umbrella pattern of nodes in T (cf. Fig. 1). We require that any mapping from the resulting (VLDC-free) pattern $\bar{P}$ to T map the substituting nodes to themselves. (Thus, no cost is induced by VLDC substitutions.) Define the distance between P and T w.r.t. s, denoted $\delta(P, T, s)$, as the cost of the best mapping from $\bar{P}$ to T. Then, $\delta(P, T) = \min_{s \in S}\{\delta(P, T, s)\}$ where S is the set of all possible VLDC-substitutions.

2.4 Cut Operations

Let *cutting at node* $T[i]$ mean removing the subtree rooted at $T[i]$. We define a subtree set *Subtrees*(T) as a set of numbers satisfying

1. $i \in Subtrees(T)$ implies that $1 \le i \le |T|$;
2. $i, j \in Subtrees(T)$ implies that neither is an ancestor of the other.

Let $Cut(T, Subtrees(T))$ represent the text tree T with subtree removals at all nodes in $Subtrees(T)$. Formally, the term "matching with cut" is defined as calculating

$$\hat{\delta}(P, T) = \min_{Subtrees(T)}\{\delta(P, Cut(T, Subtrees(T)))\}.$$

Our problem is to find $\hat{\delta}(P, T)$. (Figure 4 shows an example.)

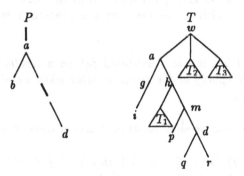

Fig. 4. Assume T_1, T_2, T_3 do not contain a, b or d. In matching P with T, the two |'s would substitute for w and h, m respectively. All leaves of T would be removed as would the subtrees T_1, T_2, T_3. The distance would be 1 (representing the cost of relabeling b by g).

2.5 Notation

Let $l(i)$ denote the number of the leftmost leaf descendant of the subtree rooted at $T[i]$. When $T[i]$ is a leaf, $l(i) = i$. The parent of $T[i]$ is $par(i)$; the number of children of t is $deg(t)$.

$T[i..j]$ is the ordered subforest of T induced by the nodes numbered i to j inclusive (Fig. 5). If $i > j$, then $T[i..j] = \emptyset$. $T[l(i)..i]$ will be referred to as $tree(i)$.

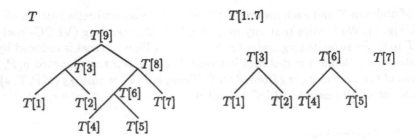

Fig. 5.

The distance (with cuttings) between $P[i'..i]$ and $T[j'..j]$ is $fd(P[i'..i], T[j'..j])$ or $fd(i'..i, j'..j)$. The distance between the $tree(i)$ and $tree(j)$ is $td(i, j)$.

3 The Serial Algorithm

The following lemma shows that path and umbrella VLDC's are the same in the presence of free cuts (which is the case we are concerned with in this extended abstract).

Lemma 2. *A path-VLDC can be substituted for an umbrella-VLDC or vice versa without changing the mapping or the distance value when we allow subtrees to be cut freely from the text tree.*

Because of this lemma, we shall focus on pattern trees containing path-VLDC's only.[4]

We compute $td(i, j)$ for $1 \leq i \leq |P|$ and $1 \leq j \leq |T|$. In the intermediate steps, we need to calculate $fd(l(i)..s, l(j)..t)$ for $l(i) \leq s \leq i$ and $l(j) \leq t \leq j$. The algorithm considers the following two cases separately: (1) $P[l(i)..s]$ or $T[l(j)..t]$ is a forest; (2) both are trees. The overall strategy is to try to find a best substitution for the VLDC's in $P[l(i)..s]$, and ask whether or not $tree(t)$ is cut. (Note that in the algorithm, $\gamma(P[s] \to \Lambda) = 0$ and $\gamma(P[s] \to T[t]) = 0$ when $P[s] = |.)$

3.1 When $P[l(i)..s]$ or $T[l(j)..t]$ is a Forest

If $tree(t)$ is cut, then $fd(l(i)..s, l(j)..t) = fd(l(i)..s, l(j)..l(t)-1)$. Otherwise, consider a minimum-cost mapping M between $P[l(i)..s]$ and $T[l(j)..t]$ after performing an optimal removal of subtrees of $T[l(j)..t]$. The distance is the minimum of the following three cases.

---

[4] The case for matching without cuttings is much more involved. In that case, we have to consider the two kinds of VLDC's separately and need an auxiliary suffix forest distance measure when dealing with umbrella-VLDC's.

(1) $P[s]$ is not touched by a line in M. (This includes the case where $P[s] = |$ is replaced by an empty tree.) So, $fd(l(i)..s, l(j)..t) = fd(l(i)..s - 1, l(j)..t) + \gamma(P[s] \to \Lambda)$.

(2) $T[t]$ is not touched by a line in M. So, $fd(l(i)..s, l(j)..t) = fd(l(i)..s, l(j)..t-1) + \gamma(\Lambda \to T[t])$.

(3) $P[s]$ and $T[t]$ are both touched by lines in M (Fig. 6). (This includes the case where $P[s] = |$ is replaced by a path of nodes in T.) By the ancestor and sibling conditions on mappings, (s, t) must be in M. By the ancestor condition on mapping, any node in $tree(s)$ can be touched only by a node in $tree(t)$. Hence, $fd(l(i)..s, l(j)..t) = fd(l(i)..l(s) - 1, l(j)..l(t) - 1) + fd(l(s)..s - 1, l(t)..t - 1) + \gamma(P[s] \to T[t])$.

Note that $fd(l(i)..s, l(j)..t) \leq fd(l(i)..l(s) - 1, l(j)..l(t) - 1) + td(s, t)$, since the distance is the cost of a minimum cost mapping and the latter expression represents a particular (and therefore possibly suboptimal) mapping from $P[l(i)..s]$ to $T[l(j)..t]$. For the same reason, $td(s, t) \leq fd(l(s)..s - 1, l(t)..t - 1) + \gamma(P[s] \to T[t])$. Thus we can use the equation $fd(l(i)..s, l(j)..t) = fd(l(i)..l(s) - 1, l(j)..l(t) - 1) + td(s, t)$.

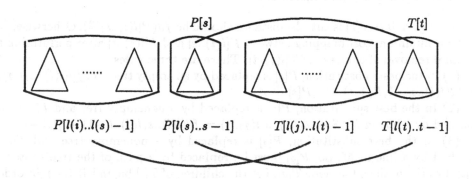

Fig. 6. Case 3 holds when (s, t) is in mapping.

3.2 When $P[l(i)..s]$ and $T[l(j)..t]$ are Trees

Lemma 3. *If $P[s] \neq |$ or $t = l(j)$, then*

$$fd(l(i)..s, l(j)..t) = \min \begin{cases} fd(P[l(i)..s], \emptyset), \\ fd(l(i)..s - 1, l(j)..t) + \gamma(P[s] \to \Lambda), \\ fd(l(i)..s, l(j)..t - 1) + \gamma(\Lambda \to T[t]), \\ fd(l(i)..s - 1, l(j)..t - 1) + \gamma(P[s] \to T[t]) \end{cases}$$

Proof. If $tree(t)$ is cut, then the distance should be $fd(P[l(i)..s], \emptyset)$. Otherwise, consider a minimum-cost mapping M between $P[l(i)..s]$ and $T[l(j)..t]$ after performing an optimal removal of subtrees of $T[l(j)..t]$. There are two cases.

(1) $P[s] \neq |$. Depending on whether $P[s]$ or $T[t]$ is touched by a line in M, we argue similarly as in Sect. 3.1.

(2) $P[s] = |$ and $t = l(j)$. Then, in the best substitution, either $|$ is replaced by an empty tree, in which case $fd(l(i)..s, l(j)..t) = fd(l(i)..s - 1, l(j)..t) + \gamma(P[s] \to \Lambda)$, or $|$ is replaced by $T[t]$, in which case $fd(l(i)..s, l(j)..t) = fd(l(i)..s - 1, l(j)..t - 1) + \gamma(P[s] \to T[t])$. The distance is the minimum of these two cases.

Since $t = l(j)$; $fd(l(i)..s, l(j)..t - 1) + \gamma(\Lambda \to T[t]) = fd(P[l(i)..s], \emptyset) + \gamma(\Lambda \to T[t]) \geq fd(P[l(i)..s], \emptyset) = fd(P[l(i)..s - 1], \emptyset) = fd(l(i)..s - 1, l(j)..t - 1) + \gamma(P[s] \to T[t])$. Thus, we can add an additional item $fd(l(i)..s, l(j)..t - 1) + \gamma(\Lambda \to T[t])$ to the minimum expression, obtaining the formula asserted by the lemma. $\square$

Lemma 4. *If* $P[s] = |$ *and* $t \neq l(j)$, *then*

$$fd(l(i)..s, l(j)..t) = \min \begin{cases} fd(P[l(i)..s], \emptyset), \\ fd(l(i)..s - 1, l(j)..t) + \gamma(P[s] \to \Lambda), \\ fd(l(i)..s, l(j)..t - 1) + \gamma(\Lambda \to T[t]), \\ fd(l(i)..s - 1, l(j)..t - 1) + \gamma(P[s] \to T[t]), \\ \min_{t_k}\{td(s, t_k)|1 \leq k \leq n_t\} \end{cases}$$

where t_k, $1 \leq k \leq n_t$, *are children of* t.

Proof. Again, if $tree(t)$ is cut, the distance should be $fd(P[l(i)..s], \emptyset)$. Otherwise, Let M be a minimum-cost mapping between $P[l(i)..s]$ and $T[l(j)..t]$ after performing an optimal removal of subtrees of $T[l(j)..t]$. There are three cases.

(1) In the best substitution, $P[s]$ is replaced by an empty tree. So, $fd(l(i)..s, l(j)..t) = fd(l(i)..s - 1, l(j)..t) + \gamma(P[s] \to \Lambda)$.

(2) In the best substitution, $P[s]$ is replaced by a nonempty tree and $T[t]$ is not touched by a line in M. So, $fd(l(i)..s, l(j)..t) = fd(l(i)..s, l(j)..t - 1) + \gamma(\Lambda \to T[t])$.

(3) In the best substitution, $P[s]$ is replaced by a nonempty tree and $T[t]$ is touched by a line in M. So, $P[s]$ must be replaced by a path of the tree rooted at $T[t]$. Let the path end at node $T[d]$. Let the children of $T[t]$ be, in left-to-right order, $T[t_1], T[t_2], \ldots, T[t_{n_t}]$. There are two subcases.

(a) $d = t$. Thus, $|$ is replaced by $T[t]$. So $fd(l(i)..s, l(j)..t) = fd(l(i)..s - 1, l(j)..t - 1) + \gamma(P[s] \to T[t])$.

(b) $d \neq t$. Let $T[t_k]$ be the child of $T[t]$ on the path from $T[t]$ to $T[d]$ (Fig. 7). We can cut all subtrees on the two sides of the path. So, $fd(l(i)..s, l(j)..t) = td(s, t_k)$. The value of k ranges from 1 to n_t. Therefore, the distance is the minimum of the corresponding costs. $\square$

We want to calculate $fd(l(i)..s, l(j)..t)$ in time $O(1)$, rather than $O(deg(t))$. To accomplish this, we compute the minimum of subtree-to-subtree distances $td(s, t_k)$ progressively. Define $mintree(s, t_i) = \min_{1 \leq j \leq i}\{td(s, t_j)\}$. Inductively, $mintree(s, t_i) = \min\{\min_{1 \leq j \leq i-1}\{td(s, t_j)\}, td(s, t_i)\} = \min\{mintree(s, t_{i-1}), td(s, t_i)\} = \min\{mintree(s, l(t_i) - 1), td(s, t_i)\}$.

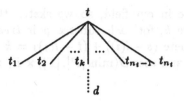

Fig. 7.

Lemma 5. *Suppose $P[l(i)..s]$ is a tree (i.e., $l(s) = l(i)$).*
(i) If $T[l(j)..t]$ is a forest (i.e., $l(t) \neq l(j)$) and $l(par(t)) = l(j)$, then

$$mintree(s,t) = \min \begin{cases} mintree(s, l(t) - 1), \\ td(s,t) \end{cases}$$

(ii) If $T[l(j)..t]$ is a tree, then $mintree(s,t) = td(s,t)$.

Thus, we can replace the term $\min_{t_k}\{td(s,t_k)|1 \leq k \leq n_t\}$ in Lemma 4 by $mintree(s, t_{n_t}) = mintree(s, t - 1)$. The following summarizes the algorithm.

Procedure treedist(i,j)
1. for $s := l(i)$ to i do
2.   for $t := l(j)$ to j do
3.   begin
4.   if $l(s) \neq l(i)$ or $l(t) \neq l(j)$ then
5.   begin
6.   Calculate fd as in Sect. 3.1;
7.   if $l(s) = l(i)$ and $l(par(t)) = l(j)$ then
8.   Calculate $mintree$ as in Lemma 5(i);
9.   end
10.   else begin   /* $l(s) = l(i)$ and $l(t) = l(j)$ */
11.   if $(P[s] \neq |$ or $t = l(j))$ then
12.   Calculate fd as in Lemma 3;
13.   if $(P[s] = |$ and $t \neq l(j))$ then
14.   Calculate fd as in Lemma 4;
15. $td(s,t) := fd(l(i)..s, l(j)..t);$
16. $mintree(s,t) := td(s,t);$
17.   end
18.   end;

4 The Parallel Algorithm

Since each formula above is calculated in constant time, we can use the framework presented in [14] to parallelize the algorithm. That is, we compute in "waves" along counter-diagonals for all subtree pairs $tree(i)$ and $tree(j)$ simultaneously.

This is nearly a cliche in our field, so we sketch the details. The algorithm starts at wave 0. At wave k, for each subtree pair $tree(i)$ and $tree(j)$, we compute $fd(l(i)..s, l(j)..t)$, where $(s - l(i)) + (t - l(j)) = k$. As in [14], the algorithm runs in time $O(|P| + |T|)$ and uses $O(\min(|P|, |T|) \times L_1 \times L_2)$ processors.

5 Conclusion

Using this simple, efficient algorithm, a user can submit a pattern tree containing variable length don't cares and obtain those portions of data trees that approximately match that pattern tree. To our knowledge, this is the first algorithm ever presented to solve this problem.

References

1. Chodorow, M. S., Klavans, J. L.: Locating syntactic patterns in text corpora. Tech. Rep. IBM T. J. Watson Research Center, New York (1990)
2. Chodorow, M. S., Byrd, R. J., Heidorn, G. E.: Extracting semantic hierarchies from a large on-line dictionary. Proc. Annual Meetings of the Association for Computational Linguistics (1985) 299–304
3. Galil, Z., Park, K.: An improved algorithm for approximate string matching. SIAM J. Comput. 19, 6 (1990) 989–999
4. Kosaraju, S. R.: Efficient tree pattern matching. Proc. 30th Annual Symp. on Foundations of Computer Science (1989) 178–183
5. Landau, G. M., Vishkin, U.: Fast parallel and serial approximate string matching. J. Algorithms 10 (1989) 157–169
6. Markowitz, J., Ahlswede, T., Evans, M.: Semantically significant patterns in dictionary definitions. Proc. Annual Meetings of the Association for Computational Linguistics (1986) 112–119
7. Myers, E. W., Miller, W.: Approximate matching of regular expressions. Bulletin of Mathematical Biology 51, 1 (1989) 5–37
8. Pinter, R. Y.: Efficient string matching with don't care patterns. Combinatorial Algorithms on Words, Apostolico, A. and Galil, Z. editors, Springer-Verlag (1985) 11–29
9. Shasha, D., Zhang, K.: Fast algorithms for the unit cost editing distance between trees. J. Algorithms 11, 4 (1990) 581–621
10. Tai, K.-C.: The tree-to-tree correction problem. J. ACM 26, 3 (1979) 422–433
11. Ukkonen, E.: Finding approximate pattern in strings. J. Algorithms 6 (1985) 132–137
12. Wang, J. T. L., Zhang, K., Jeong, K., Shasha, D.: A tool for tree pattern matching. Proc. 3rd IEEE International Conf. on Tools for AI (1991) 436–444
13. Wu, S., Manber, U.: Fast text searching with errors. Tech. Rep. TR 91-11, Dept. of Computer Science, University of Arizona (1991)
14. Zhang, K., Shasha, D.: Simple fast algorithms for the editing distance between trees and related problems. SIAM J. Computing 18, 6 (1989) 1245–1262

Grammatical Tree Matching*

Pekka Kilpeläinen and Heikki Mannila

University of Helsinki, Department of Computer Science
Teollisuuskatu 23, SF-00510 Helsinki, Finland
e-mail: kilpelai@cs.helsinki.fi, mannila@cs.helsinki.fi

Abstract. In structured text databases documents are represented as parse trees, and different tree matching notions can be used as primitives for query languages. Two useful notions of tree matching, *tree inclusion* and *tree pattern matching* both seem to require superlinear time. In this paper we give a general sufficient condition for a tree matching problem to be solvable in linear time, and apply it to tree pattern matching and tree inclusion. The application is based on the notion of a *nonperiodic* parse tree. We argue that most text documents can be modeled in a natural way using grammars yielding nonperiodic parse trees. We show how the knowledge that the target tree is nonperiodic can be used to obtain linear time algorithms for the tree matching problems. We also discuss the preprocessing of patterns for grammatical tree matching.

1 Introduction

Context-free grammars are often used for describing the structure of text documents [7, 2, 3, 6, 19, 11]. When the structure of the text is described using a grammar, the text itself is represented as a parse tree.

To extract information from a parse tree one needs query language primitives that are oriented towards tree-structured data. *Tree pattern matching* is a widely studied problem with many applications. (See, e.g., [9] and its references.) Consider two ordered and labeled trees P (the *pattern*) and T (the *target*). The trees that can be obtained by attaching new subtrees to the leaves of P are *instances* of P. The task in tree pattern matching is to locate subtrees of the target that are instances of the pattern. (See Fig. 1.)

Denote the number of nodes in pattern P by $|P| = m$ and the number of nodes in target T by $|T| = n$. The tree pattern matching problem can be solved by a trivial algorithm that traverses the target and compares the pattern against each subtree in turn. The $O(mn)$ worst case complexity of the trivial algorithm has only recently been improved by Kosaraju [15] and Dubiner, Galil, and Magen [4]; the algorithm of [4] requires time $O(n\sqrt{m}\,polylog(m))$.

To locate a subtree of the target using tree pattern matching one has to know the exact structure of the relevant parts of the target. For example, in order to locate a subtree where a node v is a descendant of a node u, the pattern must describe the exact path between u and v.

---

* Work supported by the Academy of Finland.

163

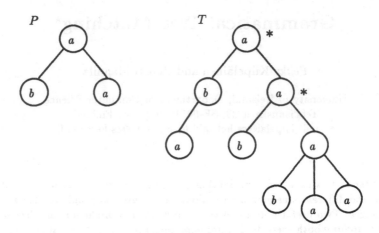

Fig. 1. The subtrees of T that are instances of P are denoted by stars.

In text databases the exact structure of the tree representing the text is often unknown. Thus one needs a more relaxed way of defining the occurrences of the target. Such a way is obtained by choosing the pattern P to stand for trees that *include* P. Intuitively, a tree includes P if a subset of its nodes agrees with the nodes of P with regard to labeling, ancestorship and left-to-right ordering. The precise definition of tree inclusion is given later; for an example see Fig. 2. Note that

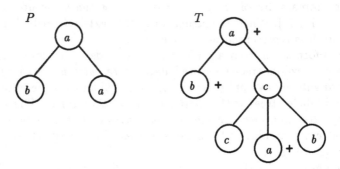

Fig. 2. A pattern P and an including tree T. The nodes of T corresponding to the pattern nodes are denoted by plus signs.

many details of the including trees can be omitted by describing them using a single pattern; hence the use of tree inclusion as a query language primitive provides some degree of data independence.

A tree U is a *minimal including tree of P*, if U includes P but no subtree of U does. The problem of locating subtrees of the target that are minimal including trees of the pattern is called the *tree inclusion problem* [13].

Gonnet and Tompa [7] have described an elegant grammatical model and query language for text databases. Following this work, Mannila and Räihä [17] proposed tree inclusion as a primitive for expressing the operations of the query language of Gonnet and Tompa. The tree inclusion problem has been shown solvable in $O(mn)$ time and space by Kilpeläinen and Mannila [13]. A query language and its implementation based on the concept of tree inclusion is described in [14].

Tree pattern matching and tree inclusion both seem to require superlinear time in the general case. In this paper we consider situations where the target tree represents a structured text, i.e., the target trees are parse trees over the grammar G defining the structure of the text.

If G is an arbitrary grammar, then the target tree can have an arbitrary structure. However, for structured text databases the grammars usually have a specific structure. Namely, iteration is expressed by using regular expressions in the right-hand-sides of productions, and no other means for recursion or iteration is used. We call such grammars *nonperiodic*. Note that nonperiodic grammars define exactly the regular languages, and that the height of a nonperiodic parse tree is bounded by the number of nonterminals in the grammar.

In this paper we show that tree pattern matching and tree inclusion can be solved in linear time for target trees that are parse trees over a nonperiodic grammar. For tree pattern matching the result is quite obvious, as even the trivial algorithm works in linear time for trees with nonperiodic underlaying grammars. For tree inclusion, the result is not so easy. We give these results using a general sufficient condition for linear solvability of tree matching problems; the condition can also be used to derive a couple of recent linearity results [16, 5, 8].

The rest of the paper is organized as follows. In Section 2 we define *general tree matching* and its special cases tree pattern matching and tree inclusion. We give a sufficient condition for general tree matching problems to be solvable in linear time. Then in Section 3 we define *grammatical* and *nonperiodic* tree matching problems and show that nonperiodic tree pattern matching is solvable in linear time. In Section 4 we show that nonperiodic tree inclusion problem is also solvable in time $O(n)$. Section 5 discusses possibilities to preprocess the pattern when the grammar of the target is known.

2 General Tree Matching

The *general tree matching* problem is to locate subtrees of the target that are *instances* of the pattern. A specific matching problem is defined by fixing the *instance relation* that specifies which trees are instances of which patterns. We say that the pattern *matches* at the root of the subtrees of the target that are instances of the pattern.

The instance relation of tree pattern matching is the relation between trees and their pruned counterparts: Tree P is a *pruned tree* of a tree U, if P can be obtained from U by deleting all descendants of zero or more nodes.

Problem 1 (Tree pattern matching). Given a pattern tree P and a target tree T, locate the subtrees U of T for which P is a pruned tree of U. □

The tree inclusion problem is defined via embeddings between trees. An injective function f from the nodes of P into the nodes of T is an *embedding* of P into T, if it preserves labels, ancestorship, and left-to-right-order, that is for all nodes u and v of pattern P we have

1. $label(u) = label(f(u))$,
2. u is an ancestor of v in P if and only if $f(u)$ is an ancestor of $f(v)$ in T, and
3. u precedes v in the postorder of P if and only if $f(u)$ precedes $f(v)$ in the postorder of T.

An embedding f of P into T is *root preserving* if $f(root(P)) = root(T)$.

Tree T *includes* P, or P is an *included tree* of T, if there is an embedding of P into T. A tree U is a *minimal including tree of P*, if there is an embedding of P into U and every embedding of P into U is root preserving.

Problem 2 (Tree inclusion problem). Given a pattern tree P and a target tree T, locate the subtrees of T that are minimal including trees of P. □

An instance relation is *linearly solvable*, if there is a constant c such that the question "Is U an instance of P?" can be answered in time bounded by $c|U|$ for all trees P and U. The relation "P and U are isomorphic *unordered* trees" is an example of a nontrivial linearly solvable instance relation [1, p. 84–86].

If P matches at a node v of T, we say that v is an *occurrence* of P. A set of nodes of T is a *candidate set of occurrences* of P, if it is superset of the set of occurrences of P. A set of nodes N is k-*thin*, if any node $n \in N$ has at most $k-1$ proper ancestors in N. A 1-thin set of nodes is *flat*. That is, a flat set of nodes does not contain two nodes one of which is an ancestor of the other. Note that in tree pattern matching there need not be a flat set of occurrences, since the pattern may match both at a node and at some of its descendants. The following lemma is almost immediate.

Lemma 3. Assume that for a tree matching problem a k-thin candidate set of occurrences can be computed in time $O(kn)$, and that the instance relation is linearly solvable. Then the tree matching problem is solvable in time $O(kn)$.

Proof. First compute a k-thin candidate set C in time $O(kn)$. For each node u in C, test whether P matches at u. Since the instance relation is linearly solvable, this requires time at most $c \sum_{u \in C} |tree(u)|$ for some constant c. Because C is k-thin, each node of T can belong to at most k trees rooted by nodes in C, and therefore

$$c \sum_{u \in C} |tree(u)| \leq ckn = O(kn) .$$

□

Lemma 3 gives us an easy way to show the linearity of some tree matching problems. For example, the following result is immediate.

Corollary 4. Locating subtrees of the target that are isomorphic to the pattern as unordered trees can be done in time $O(n)$.

Proof. A flat candidate set of occurrences can be formed in time $O(n)$ by traversing the target bottom-up and counting the sizes of its subtrees. The roots of the subtrees of size m form a flat candidate set of occurrences. As stated above, the instance relation is linearly solvable. The result follows from Lemma 3. □

The idea of Lemma 3 was used by Grossi in [8] for showing that locating subtrees of the target that are identical to the pattern or differ only with regard to don't care labels or up to m mismatching labels can be done in $O(n)$ sequential time.

3 Grammatical Tree Matching

Next we consider describing the structure of text databases by context-free grammars. We define a *grammar* to be a quadruple $G = (V, T, P, S)$, where V is the set of *nonterminals*, T is the set of *terminals*, P is the set of *productions* and S is the *start symbol*.

It is useful to allow regular expressions on the right-hand-sides of productions. This leads to fewer nonterminals and seems to be a form easily comprehensible also to nonspecialists. Therefore, we define the productions to be of the form $A \rightarrow \alpha$, where α is a regular expression over $V \cup T$. We say that a production $A \rightarrow w$ is an *instance* of $A \rightarrow \alpha$, if w belongs to the regular language defined by α.

As an example we show a grammar for describing the structure of a list of bibliographic references stored in a text database system.

publications	$\rightarrow$	publication*
publication	$\rightarrow$	authors title journal volume year pages
authors	$\rightarrow$	author*
author	$\rightarrow$	initials name
initials	$\rightarrow$	text
name	$\rightarrow$	text
title	$\rightarrow$	text
journal	$\rightarrow$	text
volume	$\rightarrow$	number
year	$\rightarrow$	number
pages	$\rightarrow$	start end
start	$\rightarrow$	number
end	$\rightarrow$	number
text	$\rightarrow$	character*
number	$\rightarrow$	digit*

The obvious productions for nonterminals *character* and *digit* have been excluded. The grammar is allowed to be ambiguous, since we do not use it for parsing. Producing string representations out of a database and parsing strings into database instances can be performed by using versions of the grammar that are *annotated* with extra terminal symbols. The methodology is explained in [11] and in [18].

To define *parse trees* over a grammar G, we define sets $T(G, a)$ for terminals $a \in T$ and sets $T(G, A)$ for nonterminals $A \in V$.

$$T(G, a) = \{a\} \text{ for terminals } a \ ;$$

$$T(G, A) = \{A(t_1, \ldots, t_n) \mid A \to B_1, \ldots, B_n$$

is an instance of a production in P

and $t_i \in T(G, B_i)$ for each $i = 1, \ldots, n\} \ ;$

Here $A(t_1, \ldots, t_n)$ stands for a tree whose root is labeled by the nonterminal A and whose ith immediate subtree is t_i for all $i = 1, \ldots, n$. That is, elements of $T(G, A)$ represent derivations of terminal strings from the nonterminal A according to G. Finally, the trees that represent derivations from the start symbol of G, i.e., the trees in $T(G, S)$, are the parse trees over G.

A tree matching problem is *G-grammatical*, if the target is a parse tree over a grammar G. Grammatical matching problems are in general no easier than the unrestricted ones, since a grammatical problem for the grammar

$$S \to a_1 \mid a_2 \mid \ldots \mid a_l$$
$$a_1 \to (a_1 \mid a_2 \mid \ldots \mid a_l)^*$$
$$\vdots$$
$$a_l \to (a_1 \mid a_2 \mid \ldots \mid a_l)^* \ ,$$

where $\{a_1, \ldots, a_l\}$ is the set of labels, is the same as the problem for arbitrary trees.

A grammar is *nonperiodic*[2] if it has no nonterminal A that can derive a string of the form $\alpha A \beta$. A tree T is *k-periodic* if any nonterminal appears at most k times on a single root-to-leaf path in T. A 1-periodic tree is *nonperiodic*, or equivalently, a tree T is nonperiodic if and only if T is a parse tree over some nonperiodic grammar. A matching problem (P, T) is *nonperiodic*, if it is G-grammatical for a nonperiodic grammar G (i.e., the target T is nonperiodic).

Although nonperiodic grammars are too weak for modeling programming languages, we argue that they are powerful enough to model the structure of most text databases. Recall that we allow regular expressions in the productions of the grammar. Therefore, a language defined by a nonperiodic grammar could be defined by giving one regular expression. For example, we could describe the previous list of bibliographic references also by a single production of the following form.

publications $\longrightarrow$ ((character* character*)*
character* character*
digit* digit* digit* digit*)*

The subexpressions *character* and *digit* that simply describe the set of recognized characters and the digits '0'–'9' have been left unspecified. It should be clear from this example that a single regular expression is not a convenient description for the

---

[2] Nonperiodic grammars are usually called *nonrecursive*. We use the term nonperiodic to avoid confusing nonperiodic matching problems with nonrecursive, i.e., undecidable problems.

logical structure of a text database. In practice, nonperiodic grammars with regular expressions in productions support modeling long lists of, say, dictionary articles, but unlimited nesting of structures is of course not possible.

In many applications the grammar G describing the structure of the text is fairly small. However, this is not always the case. For example, in a dictionary for Finnish the first third of the dictionary (about 17 000 words) has 1300 different structures for the entries, and there does not seem to be any simple description of this structure using a small grammar. Still, the parse tree is 2-periodic, and it can be simply transformed to a nonperiodic form.

Note that if T is a parse tree over a nonperiodic grammar G, then the height of T is at most $|\mathcal{V}| + 1$. It is known that restricting the height of the pattern improves the running time of the trivial algorithm for tree pattern matching:

Lemma 5. [4] If the height of the pattern is h, then the trivial algorithm for tree pattern matching takes time $O(nh)$. □

In a matching problem the height of P is at most the height of T. Thus nonperiodic tree pattern matching can be solved using the trivial algorithm in time $O(|\mathcal{V}|\, n)$. Next we show how Lemma 3 makes it possible to solve the tree pattern matching problem in time $O(kn)$ for a k-periodic target, and hence in time $O(n)$ for an arbitrary nonperiodic grammar G.

Lemma 6. A k-thin candidate set of occurrences for tree pattern matching and for tree inclusion can be computed in a k-periodic target in time $O(n)$.

Proof. The nodes of the target which have the same label as the root of the pattern form a candidate set of occurrences; they can be located by a simple traversal of the target. The set is k-flat, because the target is k-periodic. □

Lemma 7. The instance relation "P is a pruned tree of U" is linearly solvable.

Proof. The relation can be tested simply by comparing the corresponding nodes of the trees against each other; at most $\min\{|P|, |U|\}$ nodes of U are examined. □

Theorem 8. Tree pattern matching with a k-periodic target can be solved in time $O(kn)$, and the nonperiodic tree pattern matching can be solved in time $O(n)$.

Proof. Follows directly from Lemmas 3, 6, and 7. □

Theorem 8 shows that nonperiodic tree pattern matching is solvable in linear time. To obtain the same result for nonperiodic tree inclusion, we show in the next section that the instance relation "U is a minimal including tree of P" is solvable in linear time, when U is nonperiodic.

4 Solving Nonperiodic Tree Inclusion

Testing for the existence of a root preserving embedding of P into U amounts to searching appropriately related embeddings of the subtrees of P into the subtrees of

T. There may be exponentially many ways to embed the subtrees of P in T, but it was shown in [13] that this complexity can be avoided by concentrating on so called left embeddings.

A *forest* is an ordered sequence of trees $\langle T_1, \ldots, T_k \rangle$, $k \geq 0$. An embedding of a forest into another one is defined in the same way as an embedding between trees. Let P be a tree with immediate subtrees $\langle P_1, \ldots, P_k \rangle$ and U a tree with immediate subtrees $\langle U_1, \ldots, U_l \rangle$. It is obvious that there is a root preserving embedding of P into U if and only if $label(root(P)) = label(root(U))$ and there is an embedding of $\langle P_1, \ldots, P_k \rangle$ into $\langle U_1, \ldots, U_l \rangle$.

We denote the preorder and postorder numbers of a node n by $pre(n)$ and $post(n)$. In order to discuss the order of images of nodes in embeddings we define the *right relatives* of a node. Let F be a forest, N the set of its nodes, and u a node in F. The set of right relatives of u is defined by

$$rr(u) = \{x \in N \mid pre(u) < pre(x) \ \wedge \ post(u) < post(x)\} \ ,$$

i.e., the right relatives of u are those nodes that follow u both in preorder and in postorder. (See Fig. 3.)

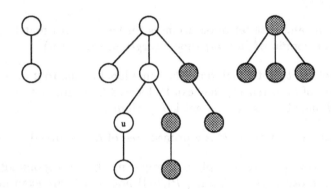

Fig. 3. The right relatives of the node u.

The following lemma tells us how to construct an embedding of a forest out of the embeddings of its trees.

Lemma 9. [13] Let $F = \langle T_1, \ldots, T_k \rangle$ and G be forests. There is an embedding of F into G if and only if for every $i = 1, \ldots, k$ there is an embedding f_i of T_i into G such that $f_{i+1}(root(T_{i+1}))$ is a right relative of $f_i(root(T_i))$, whenever $1 \leq i < k$. □

Let $F = \langle T_1, \ldots, T_n \rangle$ and G be forests. An embedding f of F into G is a *left embedding* of F into G if for every embedding g of F into G

$$post(f(root(T_i))) \leq post(g(root(T_i))), \ i = 1, \ldots, n \ .$$

Theorem 10. [13] Let F and G be forests. There is an embedding of F into G if and only if there is a left embedding of F into G. □

Next we give an algorithm for testing the instance relation of nonperiodic tree inclusion between a pattern P and a tree U. Denote the set of proper descendants of a node v by $desc(v)$. The nonperiodicity of U means that if v is a node in U then there are no nodes labeled by $label(v)$ in $desc(v)$. This implies two further facts utilized in the algorithm. First, tree U is a minimal including tree of P if and only if there is a root preserving embedding of P into U. Second, let N be a nonempty set of nodes of U all of which have the same label. Then the first node of N in the preorder of U and the first node of N in the postorder of U are the same node.

Let $u_1, \ldots, u_k$ be the children of the root of P, and $P_1, \ldots, P_k$ the corresponding immediate subtrees of P. The algorithm uses a pointer p for traversing the descendants of the root of U. First our algorithm searches for the image $f(u_1)$ under a left embedding f of $\langle P_1 \rangle$ into the forest of immediate subtrees of U, if one exists. After finding a left embedding f for the forest $\langle P_1, \ldots, P_i \rangle$, the pointer p points at the node $f(u_i)$. In order to extend f to a left embedding of $\langle P_1, \ldots, P_{i+1} \rangle$ we search the closest right relative x of p in U, such that there is a root preserving embedding of P_{i+1} into $tree(x)$.

Algorithm 11. Testing the instance relation of nonperiodic tree inclusion.

Input: Trees P and U, where U is nonperiodic, and nodes u and v,
   where $u = root(P)$ and $v = root(U)$.
Output: **true** if and only if U is a minimal including tree of P.
Method: If $label(u) = label(v)$, call emb(u, v); otherwise return **false**

1. **function** emb(u, v);
2.   **if** u is a leaf **then return true**;
3.   **else** Let $u_1, \ldots, u_k$ be the children of u;
4.   Let p be the first descendant of v in preorder
         satisfying $label(p) = label(u_1)$;
         if there is no such node p **then return false**; **fi**;
5. $i := 1$;
6.   **while** $i \leq k$ **do**
7.   **if** emb(u_i, p) **then** $i := i + 1$; **fi**;
8.   **if** $i \leq k$ **then**
9.   Let p be the first node in $rr(p) \cap desc(v)$ in
                 preorder of U satisfying $label(p) = label(u_i)$;
                 if there is no such p **then return false**; **fi**;
10.   **fi**;
11.   **od**;
12.   **return true**;
13.   **fi**;
14. **end**;

Lemma 12. Algorithm 11 tests the relation "U is a minimal including tree of P" correctly for all trees P and all nonperiodic trees U.

Proof. If P consists of a single node u, the claim is obvious. Then assume that the height of P is $h > 0$ and and that the algorithm works correctly on all patterns of height less than h. Let the immediate subtrees of P rooted by $u_1, \ldots, u_k$ be $P_1, \ldots, P_k$. Now the following two invariants can be shown to hold for the loop on lines 6–11. First, before each execution of the loop, $post(f(u_i)) \geq post(p)$ for all embeddings f of $\langle P_1, \ldots, P_i \rangle$ into the forest of immediate subtrees of U. Second, after each execution of the loop, the forest $\langle P_1, \ldots, P_{i-1} \rangle$ has a left embedding into the forest of immediate subtrees of U. The correctness of the algorithm follows from these invariants. □

Lemma 13. The instance relation "U is a minimal including tree of P" is linearly solvable for nonperiodic trees U.

Proof. Algorithm 11 tests the relation for nonperiodic trees U correctly by Lemma 12. We show that there is a constant c such that the algorithm works in time bounded by $c|U|$ for all trees P and U.

Denote by $t(n)$ the maximum time needed to compare the root labels of P and U and perform the function call $emb(root(P), root(U))$, when P and U are trees and $|U| = n$. First, consider testing a single node. Obviously there is a constant c' such that $t(1) \leq c'$. Then assume that $|U| = n > 1$ and $t(m) \leq c'm$ for all targets of size $m < n$. Let n'' be the number of nodes of U that are examined during the traversal on lines 4 and 9 of the algorithm, excluding the roots of the subtrees of U that are examined in the recursive calls on line 7. Let n' be the total size of the subtrees of U that are examined in the recursive calls. There is a constant c'' such that the traversal can be performed in time $c''n''$. Therefore for $n > 1$ we get $t(n) \leq c'n' + c''n''$. Since the search proceeds in preorder, and the subtrees examined in the recursive calls do not overlap, $n' + n'' \leq n$. Hence for all $n > 0$ we have $t(n) \leq cn$ by selecting $c = \max\{c', c''\}$. □

Theorem 14. Nonperiodic tree inclusion problem can be solved in $O(n)$ time.

Proof. By Lemma 6 a flat candidate set of occurrences can be computed in $O(n)$ time, and by Lemma 13 the instance relation is linearly solvable. Therefore, by Lemma 3 the problem is solvable in linear total time. □

Periodic targets seem to be more difficult in the case of tree inclusion. Trying to apply the approach of Algorithm 11 to testing the instance relation with k-periodic targets seems to lead to $\Omega(2^k n)$ worst case running times. This means that the specialized approach of this paper would be for k-periodic targets more efficient than the general $O(mn)$ tree inclusion algorithm of [13] if $k < \log m$.

5 Preprocessing Grammatical Patterns

In the nonperiodic tree matching problems considered above obviously only nonperiodic patterns can have occurrences in the target. More generally, in a G-grammatical matching problem one can check, before performing the actual matching, whether P can have an instance in a parse tree over G. For example, in a text database application patterns not passing this test could result in informative diagnostics about the impossibility of locating data in the database using such patterns. Such preprocessing is probably useful only when the grammar of the database is essentially smaller than the parse tree of the database.

In what follows, we can assume that the grammar of the target $G = (V, T, P, S)$ contains only useful nonterminals. That is, every nonterminal in V appears in a derivation of a string of terminals from the start symbol S of G. (See [10, p. 88–90].)

For tree pattern matching one checks that each node of pattern P is labeled by a symbol of grammar G and that the children of each internal node of P correspond to an appropriate production in G. Let u be an internal node of P with label A and with children $u_1, \ldots, u_k$. Checking u is performed by applying to the string of labels in $u_1, \ldots, u_k$ an automaton that recognizes the instances of the right-hand sides of the productions for nonterminal A. These finite automata need to be constructed only once for a grammar. If the automata are deterministic, checking pattern P takes only time $O(|P|)$. If the worst case $O(2^{|G|})$ size of the deterministic automata is prohibitive, it is also possible to construct nondeterministic finite automata for the same task; this can be done in time $O(|G|)$. The NFAs can be simulated for checking the children of each pattern node yielding total time $O(|G||P|)$. (See [1].)

For the tree inclusion problem the check is slightly more complicated. For each node of P labeled by A and having children labeled by $a_1, \ldots, a_k$ the following should hold in G:

$$A \overset{*}{\Rightarrow} \beta_0 a_1 \beta_1 \ldots \beta_{k-1} a_k \beta_k \ ,$$

where $\beta_i \in (V \cup T)^*$. This can be checked in the following manner. For each nonterminal $B \in V$ let B' be a unique terminal not belonging to T, and for a set of nonterminals N let $N' = \{A' \mid A \in N\}$. For each terminal $t \in T$ let t' be a unique new nonterminal not belonging to V, and for a set of terminals C let $C' = \{t' \mid t \in C\}$. For a production p denote by p' the production obtained by replacing each terminal t in p by t'. For a set of productions Q denote by Q' the set $\{p' \mid p \in Q\}$. Finally, for grammar $G = (V, T, P, S)$ and a nonterminal $A \in V$ let $G'_A = (V'', T'', P'', S'')$ be the grammar with

$$V'' = V \cup T'$$
$$T'' = T \cup V'$$
$$P'' = P' \cup$$
$$\{A \rightarrow (A' \mid \epsilon) \mid A \in V\} \cup$$
$$\{t' \rightarrow (t \mid \epsilon) \mid t \in T\}$$
$$S'' = A \ .$$

The idea is that grammar G'_A generates the subsequences of the sentential forms that are derivable from nonterminal A in grammar G. (Note that G'_A and G'_B may differ only with regard to the start symbol.) Now checking a pattern node labeled by A and having children labeled by $a_1, \ldots, a_k$ is done by first substituting B' for each nonterminal B in the sequence $a_1, \ldots, a_k$ and then parsing this sequence using G'_A.

We have above outlined possibilities to check the patterns against the grammar before performing a grammatical tree matching. Another promising direction for preprocessing patterns with regard to the grammar, analogical to query optimization in databases, is trying to transform the given matching problem to an easier one that still yields the same set of occurrences as the original problem.

For some patterns P and grammars G we may be able to compute a unique sequence of labels on any path between two nodes labeled by a and b in any parse tree over G, when a and b are labels of a node u and its descendant v in pattern P. Such knowledge allows us to complete the tree inclusion pattern by adding nodes labeled by the corresponding sequence of labels between every node-descendant pair u, v in P, and to solve the problem, possibly more efficiently, as a variant of tree pattern matching. Another problem that is feasible for this transformation is *unordered* tree inclusion. In this variation of the problem the order of the subtrees is not significant. It would sometimes be convenient to express queries on a grammatical database using tree inclusion but ignoring the left-to-right order of subtrees. Unfortunately, unordered tree inclusion is an NP-complete problem [12]. If the unordered tree inclusion pattern P and grammar G allow transforming P as above, the problem reduces to a subgraph isomorphism problem between trees. The latter problem is solvable in polynomial time [20].

6 Conclusion

We have shown that the restriction to nonperiodic targets makes two tree matching problems solvable in linear time. While the nonperiodicity assumption makes tree pattern matching and tree inclusion easier, it does not always help. For example, unordered tree inclusion is NP-complete, and remains NP-complete even for nonperiodic targets [13]. Similarly, adding logical variables to tree inclusion makes the problem NP-complete, and the restriction to nonperiodic targets does not help [14].

The restriction to nonperiodic targets does not seem severe for text database applications, but can be problematic in other applications, for example in code generation. We have shown how one can obtain tree pattern matching algorithms whose running times are proportional to the amount of periodicity in the target.

We have also outlined possibilities to check and preprocess patterns with regard to the grammar in grammatical tree matching problems.

In practical applications queries accessing structured texts contain in most cases a string search component. An interesting topic for further research is how one could integrate efficient string search mechanisms and tree inclusion algorithms. A simple starting point would be to use string searches to locate a set of candidate occurrences, and then find the real occurrences using a tree inclusion algorithm.

Acknowledgements

We wish to thank Jyrki Kivinen for his comments on the previous version of this paper.

References

1. A. V. Aho, J. E. Hopcroft, and J. D. Ullman. *The Design and Analysis of Computer Algorithms.* Addison-Wesley, 1974.

2. F. Bancilhon and P. Richard. Managing texts and facts in a mixed data base environment. In G. Gardarin and E. Gelenbe, editors, *New Applications of Data Bases*. Academic Press, 1984.

3. G. Coray, R. Ingold, and C. Vanoirbeek. Formatting structured documents: Batch versus interactive. In J.C. van Vliet, editor, *Text Processing and Document Manipulation*. Cambridge University Press, 1986.

4. M. Dubiner, Z. Galil, and E. Magen. Faster tree pattern matching. In *Proc. of the Symposium on Foundations of Computer Science (FOCS'90)*, pages 145–150, 1990.

5. P. Dublish. Some comments on the subtree isomorphism problem for ordered trees. *Information Processing Letters*, 36:273–275, 1990.

6. R. Furuta, V. Quint, and J. André. Interactively editing structured documents. *Electronic Publishing*, 1(1):19–44, 1988.

7. G. H. Gonnet and F. Wm. Tompa. Mind your grammar - a new approach to text databases. In *Proc. of the Conference on Very Large Data Bases (VLDB'87)*, pages 339–346, 1987.

8. R. Grossi. A note on the subtree isomorphism for ordered trees and related problems. *Information Processing Letters*, 39:81–84, 1991.

9. C. M. Hoffman and M. J. O'Donnell. Pattern matching in trees. *Journal of the ACM*, 29(1):68–95, January 1982.

10. J. E. Hopcroft and J. D. Ullman. *Introduction to Automata Theory, Languages, and Computation*. Addison-Wesley, 1979.

11. P. Kilpeläinen, G. Lindén, H. Mannila, and E. Nikunen. A structured document database system. In Richard Furuta, editor, *EP90 – Proceedings of the International Conference on Electronic Publishing, Document Manipulation & Typography*, The Cambridge Series on Electronic Publishing. Cambridge University Press, 1990.

12. P. Kilpeläinen and H. Mannila. Ordered and unordered tree inclusion. Report A-1991-4, University of Helsinki, Dept. of Comp. Science, August 1991.

13. P. Kilpeläinen and H. Mannila. The tree inclusion problem. In Samson Abramsky and T.S.E. Maibaum, editors, *TAPSOFT'91, Proc. of the International Joint Conference on the Theory and Practice of Software Development, Vol. 1: Colloqium on Trees in Algebra and Programming (CAAP'91)*, pages 202–214. Springer-Verlag, 1991.

14. P. Kilpeläinen and H. Mannila. A query language for structured text databases. Manuscript in preparation, February 1992.

15. S. R. Kosaraju. Efficient tree pattern matching. In *Proc. of the Symposium on Foundations of Computer Science (FOCS'89)*, pages 178–183, 1989.

16. E. Mäkinen. On the subtree isomorphism problem for ordered trees. *Information Processing Letters*, 32:271–273, September 1989.

17. H. Mannila and K.-J. Räihä. On query languages for the p-string data model. In H. Kangassalo, S. Ohsuga, and H. Jaakkola, editors, *Information Modelling and Knowledge Bases*, pages 469–482. IOS Press, 1990.

18. E. Nikunen. Views in structured text databases. Phil.lic. thesis, University of Helsinki, Department of Computer Science, December 1990.

19. V. Quint and I. Vatton. GRIF: An interactive system for structured document manipulation. In J.C. van Vliet, editor, *Proceedings of the International Conference on Text Processing and Document Manipulation*. Cambridge University Press, 1986.

20. S. W. Reyner. An analysis of a good algorithm for the subtree problem. *SIAM Journal of Computing*, 6(4):730–732, December 1977.

Theoretical and Empirical Comparisons of Approximate String Matching Algorithms [*]

William I. Chang[1] and Jordan Lampe[2]

[1] Cold Spring Harbor Laboratory, Cold Spring Harbor, NY 11724
[2] University of Washington, Seattle, WA 98195

Abstract. We study in depth a model of non-exact pattern matching based on *edit distance*, which is the minimum number of substitutions, insertions, and deletions needed to transform one string of symbols to another. More precisely, the *k differences approximate string matching problem* specifies a text string of length n, a pattern string of length m, the number k of differences (substitutions, insertions, deletions) allowed in a *match*, and asks for all locations in the text where a match occurs. We have carefully implemented and analyzed various $O(kn)$ algorithms based on dynamic programming (DP), paying particular attention to dependence on b the alphabet size. An empirical observation on the average values of the DP tabulation makes apparent each algorithm's dependence on b. A new algorithm is presented that computes much fewer entries of the DP table. In practice, its speedup over the previous fastest algorithm is 2.5X for binary alphabet; 4X for four-letter alphabet; 10X for twenty-letter alphabet. We give a probabilistic analysis of the DP table in order to prove that the expected running time of our algorithm (as well as an earlier "cut-off" algorithm due to Ukkonen) is $O(kn)$ for random text. Furthermore, we give a heuristic argument that our algorithm is $O(kn/(\sqrt{b}-1))$ on the average, when alphabet size is taken into consideration.

1 Introduction and Summary of Results

Beginning in the 1980s, genetics and DNA sequence analysis research provided the impetus for advances in non-exact string matching. The *k differences approximate string matching problem* specifies, in addition to text string T and pattern string P, the parameter k of *differences* (insertions, deletions, substitutions) allowed in a *match*. The problem is to find all locations in the text where a match *ends*. (So the output is of linear size. This problem formulation is due to Sellers [30], and is equivalent to finding where matches begin, by reversing the strings.) In this paper, the text is assumed to be given on-line and to be scanned sequentially; the space requirement should (preferably) be linear in the length of the pattern. While this is a simple model of non-exact matching, we note that it has a rich history ([30, 33, 19, 20, 8, 9, 4, 16, 3] chronologically) and interesting combinatorics, and is a natural starting point before more complex, parametric cost functions are to be considered (e.g. Gusfield, et al. [11, 12]).

[*] This research was conducted at the University of California, Berkeley, and was supported in part by Department of Energy grant DE-FG03-90ER60999.

Notation. Text $T[1, \ldots, n]$ and pattern $P[1, \ldots, m]$ over fixed, finite alphabet of size b. The *edit distance* (or Levenshtein distance [22]) ed(u, v) of two strings u, v is the minimum number of substitutions, insertions, deletions (edit operations) needed to transform one string into the other.

Seller's dynamic programming (DP) algorithm [30] (**mn.dp**) computes (column by column) an $m + 1$ by $n + 1$ table whose entry $D(j, i)$ is the minimum number of edit operations needed to transform the length j *prefix* of the pattern into *some* text fragment ending at the i-th letter. (Boundary conditions are $D(j, 0) = j$ and $D(0, i) = 0$. There is a match ending at text position i if and only if entry $D(m, i)$ is at most k.) There is a simple recursive formula giving each entry in terms of the three adjacent entries above and to the left:

$$D(j, i) = \min \{\, 1 + D(j - 1, i),\ 1 + D(j, i - 1),\ I_{ji} + D(j - 1, i - 1)\, \}$$

where $I_{ji} = 0$ if $P[j] = T[i]$; $I_{ji} = 1$ if $P[j] \neq T[i]$. The three expressions in the min correspond respectively to deleting $P[j]$ from the pattern; inserting $T[i]$ into the pattern; and substituting $T[i]$ for $P[j]$.

Remark. The classical dynamic programming algorithm for computing the edit distance of two strings u, v differs from the above only in the boundary condition $D(0, i) = i$. Speedups to $O(\text{ed}(u, v) \cdot \text{length of shorter string})$ are due to Ukkonen [32] and Myers [25]. A different model of approximate matching based on *longest common subsequence* (each $I_{ji} = 2$; equivalent to not allowing substitutions) has received a great deal of attention from Myers [26] and Manber, Wu [24].

It can be seen from the recurrence that (∗) adjacent entries along rows and columns differ by at most one; and (∗∗) forward diagonals ($\searrow$) are non-decreasing and adjacent entries differ by at most one. More recent methods by Ukkonen, et al. [33, 35, 16]; Landau, Vishkin [19, 20] (survey and refinements by [8]); and Galil, Park [9] take advantage of these geometric properties in order to compute $O(kn)$ instead of mn entries.

The simplest of these, Ukkonen's "cut-off" algorithm (**kn.uk**) [33] never computes the bottom portion of a column if those entries can be inferred to be greater than k. Despite statement in [33] that "It should be quite obvious," no rigorous analysis was done that shows kn.uk has $O(kn)$ expected running time [34]. We give the first proof of this fact in Section 2. The locations of the first $k + 1$ *transitions* (x to $x + 1$) along each forward diagonal are sufficient to characterize the solution, by the (non-decreasing) diagonal monotonicity property (∗∗). Landau, Vishkin (**kn.lv**) [19, 20] computes each transition in constant time. Several practical improvements (see [4]) have made kn.lv the best among $O(kn)$ worst case algorithms. It turns out, however, that by replacing the $O(1)$ time diagonal transition subroutine with a brute-force method, the resulting algorithm (**kn.dt**) has $O(kn)$ *expected* running time and is *faster* in practice [26]. See Section 3 for a succinct description.

We have done careful theoretical and empirical comparisons of these methods; apart from questions of overhead, they do not have the same dependence on b the alphabet size. We discovered a speedup of the dynamic programming method whose running time depends on the *row averages* of table D (the higher the averages, the *faster* our algorithm). Our method works by *partitioning* each column into *runs of consecutive integers* (e.g. 012 23 234), and is many times faster than previous algorithms based on dynamic programming. Its expected running time is $O(kn)$ because it is always faster than kn.uk. In addition, it has given us special insights into the *statistics* of sequence matching. With these insights we are able to formulate empirical running times of various algorithms as functions of n, m, k, and b (see Table 1). Variations of column partitioning are given in Section 4.

Recall that the minimum number of substitutions, insertions and deletions needed to transform one string into another is called *edit distance*. It is a surprising fact that relatively little is known about the average case behavior of edit distance. Early qualitative results of Chvátal, Sankoff [6] and Deken [7] on *longest common subsequence* (*LCS*) can be carried over to edit distance: the expected edit distance between two uniformly random strings of size m (as $m \to \infty$) is $C_b m$ for some constant C_b that depends only on alphabet size. But exact bounds are not easily converted, and there has been no "formula" given in the literature for C_b. (Sankoff, Mainville [28] conjectured $\lim_{b\to\infty} C_b'\sqrt{b} = 2$ where C_b' are the corresponding constants for length of LCS.) Since any match where the text differs from the pattern by much fewer than C_b-fraction differences can be considered "significant," a basic understanding of these constants is of paramount importance. The primary difficulties are (1) the proof of convergence by the Subadditive Ergodic Theorem is non-constructive; and (2) the algorithmic formulation of edit distance (like approximate matching) is highly recursive, and leads to exponentially many states in the natural Markov model.

We have made the following empirical observation: columns of the dynamic programming table D for approximate matching consist of runs of consecutive integers, of average length very close to $\sqrt{b}$. This observation leads to Conjecture 1, row m of table D has average value $(1 - 1/\sqrt{b} + o(1/\sqrt{b})) \cdot m$ as $m \to \infty$; and Conjecture 2, $C_b = 1 - 1/\sqrt{b} + o(1/\sqrt{b})$. A probabilistic analysis of partitions of columns, subject to several simplifying assumptions, has yielded heuristic arguments (but no proof) in favor of our conjectures (see Section 5). In addition, we state Conjecture 3, the expected *minimum value* of row m (i.e. best match) is $(1 - 1/\sqrt{b} + o(1/\sqrt{b})) \cdot (m - \Theta(\log_b n))$ as $m, n \to \infty, n < b^m$. This conjecture implies *linear expected time, polynomial space* (in length of pattern), *constant-fraction differences* approximate string matching (see Chang [3]).

We have carefully optimized the simpler algorithms based on dynamic programming: mn.dp (Sellers [30]); kn.uk (Ukkonen [33]); kn.dt (see Galil, Park [9]); mn.clp and kn.clp (described below). Generally, our code for these is faster than others we have seen. Several more complicated algorithms have also been implemented but are unoptimized: kn.gp (verbatim from Galil, Park [9]); kn.lv (Landau, Vishkin [20]); let.cl and set.cl (Chang, Lawler [4]). Speedup by a small factor can be expected from careful optimization. As presently implemented the *linear expected time* let.cl and *sublinear expected time* set.cl [4] are not competitive with kn.clp, primarily because the hash coded suffix tree [23] is not the fastest implementation possible. An algorithm similar to let.cl (discovered independently, but without analysis of threshold) was implemented by Jokinen, Tarhio, and Ukkonen [16], and was the fastest for small k among algorithms they tested. In addition, [16] includes extensive tables of running times for mn.dp; kn.uk; kn.gp; and a new algorithm [31].

Our programs are allowed only $O(m)$ space, except kn.gp which is $O(m^2)$. Text is read *on-line*. While text buffering on-the-fly is slightly slower compared to reading the entire input into real or virtual memory, we feel it is more realistic not to make a copy of the input, for the types of applications we envision: at least n in the millions; m in the hundreds; and k in the tens. Our results are summarized in Table 1.

2   Cut-Off Algorithm and Its Average Case Analysis

Recall that dynamic programming table D has two important geometric properties:
  (∗) Adjacent entries along horizontal and vertical directions differ by 0 or 1
  (∗∗) Forward diagonals are non-decreasing and adjacent entries differ by 0 or 1.

Ukkonen Cut-Off Algorithm (kn.uk). Let $l_i = \max j$ s.t. $D(j,i) \leq k$ $(l_0 = k)$. Given l_{i-1}, compute $D(j,i)$ for j up to $l_{i-1}+1$, and set l_i to largest $j \leq l_{i-1}+1$ such that $D(j,i) \leq k$. Correctness follows immediately from property (∗∗).

Despite statement in [33] that "It should be quite obvious," it was not previously proved that kn.uk has $O(kn)$ expected running time. We will prove this fact. Assume the text is uniformly random over a size b alphabet. Given two strings u, v, let lcs(u, v) denote the length of the longest common subsequence between u, v.

Proposition 1. $D(j,i) \geq (1/2) \cdot \mathrm{ed}(P[1,\ldots,j], T[i-j+1,\ldots,i])$.

Proof. Consider the edit distance y between the length j strings $P[1,\ldots,j]$ and $T[i-j+1,\ldots,i]$. The value $D(j,i) = x$ must come from matching $P[1,\ldots,j]$ against a text fragment ending at i whose length differs from j by at least $y-x$ (edit distance satisfies the triangle inequality). Then $y-x$ is a lower bound on x just by consideration of length. So $x \geq y/2$. Q.E.D.

Proposition 2. *There exist constants $c < 1$, $\beta < 1$ and α s.t. Pr[two random, length j strings have a common subsequence of length cj] $< (1/j) \cdot \alpha \beta^j$.*

Proof. For convenience assume cj is an integer. By Stirling's formula $j!/(cj)!(j-cj)! = (1+o(1)) \cdot (\sqrt{2\pi c(1-c)j} \cdot c^{cj}(1-c)^{(1-c)j})^{-1}$. Let $p =$ Pr[length cj common subsequence]. Then $p \leq \sum b^{-cj}$ where the summation is over all size cj bipartite matchings of positions. Hence

$$p \leq \binom{j}{cj}^2 \cdot b^{-cj} = \frac{1+o(1)}{2\pi c(1-c)} \cdot \frac{1}{j} \cdot (c^c(1-c)^{1-c})^{-2j} \cdot b^{-cj}.$$

This last expression decreases exponentially in j if $\beta = (c^c(1-c)^{1-c})^{-2} \cdot b^{-c} < 1$. This condition is satisfied for all $b \geq 2$ by the choice $c = 7/8$. As $b \to \infty$, it suffices to choose $c > e/\sqrt{b}$. Choose $\alpha > (2\pi c(1-c))^{-1}$, sufficiently large to overcome the error term in Stirling's formula. Q.E.D.

Theorem. *The expected running time of algorithm kn.uk is $O(kn)$.*

Proof. It suffices to prove E$[l_i] = O(k)$ since l_i bounds the work in column $i+1$. Let $l = 2k/(1-c)$, so $l-2k = cl$ and $j-2k \geq cj$ for all $j \geq l$. We have E$[l_i] < l-1+\sum_{j \geq l} j \cdot$ Pr$[D(j,i) \leq k]$. By Proposition 1, $D(j,i) \leq k$ implies ed$(u,v) \leq 2k$ where $u = P[1,\ldots,j]$ and $v = T[i-j+1,\ldots,i]$ are length j strings. Since $j-$lcs(u,v) is clearly a lower bound on ed(u,v), this implies lcs$(u,v) \geq j-2k \geq cj (j \geq l)$. By Proposition

2, for $j \geq l$, $\Pr[D(j,i) \leq k] \leq \Pr[\mathrm{lcs}(u,v) \geq cj] < (1/j) \cdot \alpha\beta^j$ for some constant α and constant $\beta < 1$. Hence $\mathrm{E}[l_i] < l - 1 + \sum_{j \geq l} j \cdot (1/j) \cdot \alpha\beta^j = l - 1 + O(1) = O(k)$. $Q.E.D.$

3 Diagonal Transition Algorithms

Diagonal monotonicity $(**)$ implies the locations of the first $k+1$ transitions *along each diagonal* are sufficient to characterize D for the solution to the k differences problem [20]. A key ingredient is the "jump" $J(j,i)$ = length of the longest exact match $P[j,\ldots] = T[i,\ldots]$. The following algorithms differ only in how jumps are computed.

Landau & Vishkin Algorithm (kn.lv). Call cell $D(j,i)$ an entry of diagonal $i-j$. But instead of D, compute column by column a $(k+1) \times (n+1)$ table L where $L(x,y) = \max j$ s.t. $D(j,j+y-x) \leq x$ ($0 \leq x \leq k$; $0 \leq y \leq n$). That is, $L(x,y)$ is the row number of the last x along diagonal $y-x$. Let us first look at D the original table. Since $D(j,0) = j$, every cell of diagonal $-j$ is at least j. We can define $L(x,-1) = -\infty$ because there is no j s.t. $D(j,j-1-x) \leq x$. Likewise it is convenient to define $L(x,-2) = -\infty$. It is easy to see $L(0,y) = J(1,1+y)$. Entry $L(x,y)$ can be computed using jumps and the three cells above and to the left: $\alpha = L(x-1,y-2), \beta = L(x-1,y-1), \gamma = L(x-1,y)$, which are respectively the row numbers of the last $x-1$'s in diagonals $y-x-1, y-x, y-x+1$. More precisely, it can be inferred that $D(\alpha, \alpha+y-x) \leq 1 + D(\alpha, \alpha+y-x-1) = x$ (by an insertion into P of $T[\alpha+y-x]$). Similarly, $D(\beta+1,\beta+1+y-x) \leq x$ (by substitution) and $D(\gamma+1,\gamma+1+y-x) \leq x$ (by deletion of $P[\gamma+1]$). So along diagonal $y-x$, three cells at rows $\alpha, \beta+1, \gamma+1$ are known to be at most x. Let $j = \max(\alpha, \beta+1, \gamma+1)$. Then it is easy to see that for $j' > j$, $D(j', j'+y-x) = x$ iff $J(j+1, j+1+y-x) \geq j' - j$. To summarize, $L(x,y) = j + J(j+1, j+1+y-x)$.

Jumps are computed according to Chang, Lawler [4]. Two key ingredients are *matching statistics* (a summary of all exact matches between the text and pattern) and *lowest common ancestor* (LCA). In our implementation [2] of the Schieber, Vishkin LCA algorithm [29], only simple machine instructions are used (such as add, decrement, and complement, but not *bit-shift*). Logarithm in [29] is replaced by bit magic, using a table of *reversals* of binary representation of numbers. Fewer than sixty machine instructions suffice to compute an LCA. The worst case running time of $O(kn)$ for kn.lv is modulo hashing in $O(m)$ space, or deterministic in $O(bm)$ space. The $O(m)$ space hash coded implementation [23] is slower in practice.

Diagonal Transition Algorithm (kn.dt). Compute jumps by brute force. This is algorithm *MN2* in [9]; it is a variation of an edit distance algorithm given in [32]. Expected running time can be shown to be $O(kn)$, first stated by Myers [26]. Briefly, whether or not a jump at (j,i) is needed is determined solely by P and $T[1,\ldots,i-1]$, so $\mathrm{E}[J(j,i)|\text{jump is needed}] = 1/(b-1)$; an extra comparison is needed to find the mismatch that ends a jump. Our optimized code for kn.dt is faster than kn.uk.

Galil & Park Algorithm (kn.gp). See [9]. Strictly $O(kn)$, but requires $O(m^2)$ space for a table of lengths of exact matches $P[j,\ldots] = P[j',\ldots]$.

4 Column Partition Algorithms

Column Partition Algorithm (mn.clp). Each column of table D can be partitioned into runs of consecutive integers: entry $D(j,i)$ belongs to *run* δ of column i iff $j - D(j,i) = \delta$ (note $j - D(j,i)$ is non-decreasing in j). For $\delta \geq 0$, we say run δ of column i ends at j if j is smallest possible such that $D(j+1,i)$ belongs to run $\delta' > \delta$. A run may be of zero length (whenever $D(j+1,i) < D(j,i)$), but (∗) implies no two consecutive runs δ, $\delta + 1$ may both be of zero length. The goal is to compute where each run ends in *constant time*; the algorithm would then perform $O(m - D(m,i))$ work on column i, for a total of $O((m - \mu)n)$ where μ is the average of row m of D.

Proposition 3. *If run δ of column i ends at j and is of zero length, then run δ of column $i + 1$ ends at $j + 1$.*

Proof. The condition means $D(j+1,i) < D(j,i)$ and $\delta = j - D(j,i) + 1$; (∗∗) implies $D(j+1, i+1) = D(j,i)$ but $D(j+2, i+1) \leq D(j,i)$. Q.E.D.

Proposition 4. *If run δ of column i ends at j and is of length $l \geq 1$, and $j' \in [j - l + 2, j + 1]$ is smallest possible such that $P[j'] = T[i+1]$, then run δ of column $i + 1$ ends at $j' - 1$. If no such j' exists and run $\delta + 1$ of column i is not of zero length, then run δ of column $i + 1$ ends at $j + 1$; otherwise it ends at j.*

Proof. We know $D(j - l, i) \geq D(j - l + 1, i)$, so $D(j - l + 1, i+1) \geq D(j - l + 1, i)$ by (∗∗). Also, $D(j+2, i+1) \leq D(j+1,i) + 1 \leq D(j,i) + 1$. Run δ of column $i + 1$ must therefore end within the range $[j - l + 1, j + 1]$. The proposition then follows easily from the recurrence. Q.E.D.

Implementation. Pre-compute and tabulate the partial function $\mathbf{loc}(j, x) = \min j'$ s.t. $P[j'] = x$ and $j' \geq j$ (this requires $O(bm)$ space). Keep track of only the column partitions, not the actual entries of D. An alternative, $O(m)$ space implementation, using linked lists $\mathbf{loc}_x$ consisting of those j s.t. $P[j] = x$, is $O(mn)$ worst case but has the same running time in practice.

Remark. This can be viewed as a sparse matrix computation, cf. [10].

k **Differences Column Partition Algorithm (kn.clp).** In a manner similar to kn.uk, mn.clp can be "cut off" at k. The expected running time is $O(kn)$ because it is always faster than kn.uk. Empirically, it is much faster than previous algorithms based on dynamic programming (2.5X for binary alphabet; 4X for four-letter alphabet; 10X for twenty-letter alphabet compared to kn.dt).

Sparser *k* Differences Column Partition Algorithm (kn'.clp). Using sophisticated data structures it is possible to reduce the work on column i to $O(\log \log m)$ for each j s.t. $P[j] = T[i]$ and $O(1)$ for each run of length zero. The locations of ends of runs are stored by their *diagonal* number modulo m in a data structure that allows $O(\log \log m)$ insertion, deletion, and *nearest neighbor* lookup. When $P[j] = T[i]$ the run that needs to be modified according to Proposition 4 (i.e. would have contained cell $D(j,i)$ had $P[j] \neq T[i]$) can be looked up in the table, as a nearest neighbor of $i - j \bmod m$. Run ends that need to be modified because of runs of zero length can be handled separately, by keeping a sublist of runs of zero length. Finally, the remaining run ends stay on the same diagonal so are automatically taken care of.

The expected running time of this algorithm is $O(b^{-1}kn \log\log m)$. Unfortunately the overhead appears to be very high.

Remark. A similar result, for the *longest common subsequence metric* (equivalent to not allowing substitutions), is described in Manber, Wu [24].

5 Heuristic Analysis of Column Partitions

We showed in section 2 that $E[D(j,i)] = \Theta(j)$; the bound we obtained is not tight, and does not fully characterize the running times of "cut-off" algorithms kn.uk and kn.clp. In this section we give a sketch of a heuristic argument that $E[D(j,i)] \approx (1 - 1/\sqrt{b}) \cdot j$, which agrees very well with simulation results.

The first simplification we make is to throw away the strings and consider instead an abstract dynamic programming model given by the same recurrence and boundary conditions as $D(j,i)$, but with one exception: I_{ji} are replaced by random variables $I'_{ji} = 1$ w.p. $1 - 1/b$; 0 w.p. $1/b$. That is,

$$D'(j,i) = \min \{ 1 + D'(j-1,i), \ 1 + D'(j,i-1), \ I'_{ji} + D'(j-1,i-1) \}$$

and $D'(j,0) = j$; $D'(0,i) = 0$. Furthermore, we let $m, n \to \infty$.

Let us call a run of length l an l-run. Let $\phi =$ probability that a run is of zero length (all runs equally likely to be chosen). Assume $\phi = O(1/b)$. Next, focus on a column. Let x denote a cell chosen uniformly at random. Let $S_l = \Pr[x$ belongs to an l-run] ($l \geq 1$); so $\sum_{l \geq 1} S_l = 1$. A given, longer run is more likely to be hit than a given, shorter one: $S_l = l \cdot \#l$-runs/area. Then $S_l/l = \Pr[x$ is the end of an l-run], and is also the *odds* that a run of positive-length is of length l. Furthermore, the average length of a positive-length run is given by $1/\sum_{l \geq 1} S_l/l$ (*call this* λ). (We also have: $\Pr[x$ is the end of a run] $= \sum_{l \geq 1} S_l/l$; $\Pr[x$ is the k-th cell of a run] $= \sum_{l \geq k} S_l/l$; $E[\text{length of run containing } x] = \sum_{l \geq 1} S_l \cdot l$.)

The assumption that run-lengths of positive-length runs are *geometrically distributed* by length (this fits simulation data) is equivalent to the assumption that for a random cell x the events (1) it is the k-th cell of a run; (2) it is the end of a run, are *independent*. If we make this simplifying assumption, then it follows by a calculation that $S_l = (1/\lambda^2) \cdot l \cdot (1 - 1/\lambda)^{l-1}$. Also $E[\text{length of run containing } x] = 2\lambda - 1$.

Next, calculate $E[x - y]$ where y is the cell adjacent and to the left of x. This expectation approaches 0 as $n \to \infty$. A case analysis of the column partitioning process yields the following after some calculation: (1) $\Pr[x - y = 1] = \Pr[y$ is first cell of a run, and there is no "match" for the entire run above $y] = (1+o(1)) \cdot (\lambda^{-1} + b^{-1})$; (2) $\Pr[x - y = -1] = \Pr[y$ is *not* first cell of a run, and there is some "match" above y in the run] $= (1 + o(1)) \cdot (\lambda/b)$. Hence $\lambda \approx \sqrt{b}$ (highest order term in b), and also average run length $\approx \sqrt{b}$ (highest order term in b).

Open Problems. Show $\phi = O(1/b)$. Remove the independence assumption.

Conjecture 1. $E[D(j,i)] = (1 - 1/\sqrt{b} + o(1/\sqrt{b})) \cdot j$ as $j \to \infty$.

Conjecture 2. $E[\text{edit distance between two strings of length } l] = (1 - 1/\sqrt{b} + o(1/\sqrt{b})) \cdot l$ as $l \to \infty$.

Conjecture 3. $E[\min D(m, i), 1 \le i \le n] = (1 - 1/\sqrt{b} + o(1/\sqrt{b})) \cdot (m - \Theta(\log_b n))$ as $m, n \to \infty, n < b^m$.

Conclusions

What has been achieved is an empirical understanding of approximate string match-ing given by edit distance, for the case of uniformly random strings. A benefit of this understanding is a new algorithm that runs several times faster in practice. The stated conjectures, if true, will have profound consequences on the future develop-ment of approximate string matching algorithms.

Table 1. Summary of Theoretical and Empirical Running Times (based on runs with $n = 100,000$; $m = 100$; $k = 10, 20, 30, \ldots, C_b m$; $b = 2, 3, 4, 8, 16, 32, 64$ on a VAX 8600 using the Unix program gprof; empirical running times for random text are formulated as functions of n, m, k, b and given in *microseconds*)

Alg'm	Worst case	Empirical ($k < C_b m$)	Notes	Attribution
mn.dp	$O(mn)$	$3.5\,mn$		Sellers [30]
kn.uk	$O(mn)$	$4.1\sqrt{b}/(\sqrt{b}-1) \cdot kn$	a	Ukkonen [33]
kn.dt	$O(mn)$	$4.2\,b/(b-1) \cdot kn$	b	"diagonal transition" [9, 26]
kn.gp	$O(kn)$	$50\,kn$	c	Galil, Park [9]
kn.lv	$O(kn)$	$40\,kn$	d	Landau, Vishkin [20]
mn.clp	$O((m-\mu)n)$	$1.4/\sqrt{b} \cdot mn$	e	Chang, Lampe
kn.clp	$O((m-\mu)n)$	$1.4/(\sqrt{b}-1) \cdot kn$	e	Chang, Lampe
let.cl		$80\,n$	f	Chang, Lawler [4]
set.cl		$160\,(k\log_b m)(n/m)$	g	Chang, Lawler [4]

Notes.

a. We showed in section 2 that kn.uk is $O(kn)$ on the average.

b. Myers [26] was first to state it is $O(kn)$ on the average (proof is simple).

c. (Unoptimized.) Requires $O(m^2)$ space.

d. (Unoptimized.) This is Landau, Vishkin [20] using McCreight *suffix tree* [23]; Chang, Lawler *matching statistics* [4]; and Schieber, Vishkin *lowest common ancestor* [29] with logarithm replaced by bit magic [2]. The worst case running time of $O(kn)$ is modulo hashing in $O(m)$ space, or deterministic in $O(bm)$ space.

e. Running time depends on d.p. table row averages; μ = average of last row. To guarantee the worst case running time of $O((m - \mu)n)$, $O(bm)$ space is needed. An alternative, $O(m)$ space implementation has the same running time in practice. Expected running time of kn.clp is $O(kn)$ because it is always faster than kn.uk.

f. (Unoptimized.) Linear expected time when error tolerance k is less than the thresh-old $k^* = m/(\log_b m + c_1) - c_2$ (for suitable constants c_i). In practice, for m in the hundreds the error thresholds k^* in terms of percentage of m are 35 ($b = 64$); 25 ($b = 16$); 15 ($b = 4$); and 7 ($b = 2$) percent. Worst case performance is same as dynamic programming based subroutine.

g. (Unoptimized.) Sublinear expected time when $k < k^*/2 - 3$ (in the sense that not all letters of the text are examined). The expected running time is $o(n)$ when k, treated as some fraction of m and not as a constant, is $o(m/\log_b m)$.

183

Acknowledgment

We would like to thank Gene Lawler for guidance and support. David Aldous, K. Balasubramanian, Maxime Crochemore, Dan Gusfield, Dick Karp, Gad Landau, Dalit Naor, Frank Olken, Kunsoo Park, Esko Ukkonen, Sun Wu, and the referees provided helpful comments and encouragement.

References

1. R. Arratia and M.S. Waterman, Critical Phenomena in Sequence Matching, *The Annals of Probability* 13:4(1985), pp. 1236–1249.
2. W.I. Chang, Fast Implementation of the Schieber-Vishkin Lowest Common Ancestor Algorithm, computer program, 1990.
3. W.I. Chang, *Approximate Pattern Matching and Biological Applications*, Ph.D. thesis, U.C. Berkeley, August 1991.
4. W.I. Chang and E.L. Lawler, Approximate String Matching in Sublinear Expected Time, *Proc. 31st Annual IEEE Symposium on Foundations of Computer Science*, St. Louis, MO, October 1990, pp. 116–124.
5. W.I. Chang and E.L. Lawler, Approximate String Matching and Biological Sequence Analysis (poster), abstract in *Human Genome II Official Program and Abstracts*, San Diego, CA, Oct. 22–24, 1990, p. 24.
6. V. Chvátal and D. Sankoff, Longest Common Subsequences of Two Random Sequences, *Technical Report STAN-CS-75-477*, Stanford University, Computer Science Department, 1975.
7. J. Deken, Some Limit Results for Longest Common Subsequences, *Discrete Mathematics* 26(1979), pp. 17–31. *J. Applied Prob.* 12(1975), pp. 306–315.
8. Z. Galil and R. Giancarlo, Data Structures and Algorithms for Approximate String Matching, *Journal of Complexity* 4(1988), pp. 33–72.
9. Z. Galil and K. Park, An Improved Algorithm for Approximate String Matching, *SIAM J. Comput.* 19:6(1990), pp. 989–999.
10. Z. Galil and K. Park, Dynamic Programming with Convexity, Concavity, and Sparsity, manuscript, October 1990.
11. D. Gusfield, K. Balasubramanian, J. Bronder, D. Mayfield, D. Naor, PARAL: A Method and Computer Package for Optimal String Alignment using Variable Weights, in preparation.
12. D. Gusfield, K. Balasubramanian and D. Naor, Parametric Optimization of Sequence Alignment, submitted.
13. P.A.V. Hall and G.R. Dowling, Approximate String Matching, *Computing Surveys* 12:4(1980), pp. 381–402.
14. D. Harel and R.E. Tarjan, Fast Algorithms for Finding Nearest Common Ancestors, *SIAM J. Comput.* 13(1984), pp. 338–355.
15. N.I. Johnson and S. Kotz, *Distributions in Statistics: Discrete Distributions*, Houghton Mifflin Company (1969).
16. P. Jokinen, J. Tarhio, and E. Ukkonen, A Comparison of Approximate String Matching Algorithms, manuscript, October 1990.
17. S. Karlin, F. Ost, and B.E. Blaisdell, Patterns in DNA and Amino Acid Sequences and Their Statistical Significance, in M.S. Waterman, ed., *Mathematical Methods for DNA Sequences*, CRC Press (1989), pp. 133–157.
18. R.M. Karp, *Probabilistic Analysis of Algorithms*, lecture notes, U.C. Berkeley (Spring 1988; Fall 1989).

19. G.M. Landau and U. Vishkin, Fast String Matching with k Differences, *J. Comp. Sys. Sci.* 37(1988), pp. 63–78.
20. G.M. Landau and U. Vishkin, Fast Parallel and Serial Approximate String Matching, *J. Algorithms* 10(1989), pp. 157–169.
21. G.M. Landau, U. Vishkin, and R. Nussinov, Locating alignments with k differences for nucleotide and amino acid sequences, *CABIOS* 4:1(1988), pp. 19–24.
22. V. Levenshtein, Binary Codes Capable of Correcting Deletions, Insertions and Reversals, *Soviet Phys. Dokl.* 6(1966), pp. 126–136.
23. E.M. McCreight, A Space-Economical Suffix Tree Construction Algorithm, *J. ACM* 23:2 (1976), pp. 262–272.
24. U. Manber and S. Wu, Approximate String Matching with Arbitrary Costs for Text and Hypertext, manuscript, February 1990.
25. E.W. Myers, An O(ND) Difference Algorithm and Its Variations, *Algorithmica* 1(1986), pp. 252–266.
26. E.W. Myers, Incremental Alignment Algorithms and Their Applications, *SIAM J. Comput.*, accepted for publication.
27. D. Sankoff and J.B. Kruskal, eds., *Time Warps, String Edits, and Macromolecules: The Theory and Practice of Sequence Comparison*, Addison-Wesley (1983).
28. D. Sankoff and S. Mainville, Common Subsequences and Monotone Subsequences, in D. Sankoff and J.B. Kruskal, eds., *Time Warps, String Edits, and Macromolecules: The Theory and Practice of Sequence Comparison*, Addison-Wesley (1983), pp. 363–365.
29. B. Schieber and U. Vishkin, On Finding Lowest Common Ancestors: Simplification and Parallelization, *SIAM J. Comput.* 17:6(1988), pp. 1253–1262.
30. P.H. Sellers, The Theory and Computation of Evolutionary Distances: Pattern Recognition, *J. Algorithms* 1(1980), pp. 359–373.
31. J. Tarhio and E. Ukkonen, Approximate Boyer-Moore String Matching, Report A-1990-3, Dept. of Computer Science, University of Helsinki, March 1990.
32. E. Ukkonen, Algorithms for Approximate String Matching, *Inf. Contr.* 64(1985), pp. 100–118.
33. E. Ukkonen, Finding Approximate Patterns in Strings, *J. Algorithms* 6(1985), pp. 132–137.
34. E. Ukkonen, personal communications.
35. E. Ukkonen and D. Wood, Approximate String Matching with Suffix Automata, Report A-1990-4, Dept. of Computer Science, University of Helsinki, April 1990.
36. M.S. Waterman, Sequence Alignments, in M.S. Waterman, ed., *Mathematical Methods for DNA Sequences*, CRC Press (1989), pp. 53–92.
37. M.S. Waterman, L. Gordon, and R. Arratia, Phase transitions in sequence matches and nucleic acid structure, *Proc. Natl. Acad. Sci. USA* 84(1987), pp. 1239–1243.

Fast and Practical Approximate String Matching

Ricardo A. Baeza-Yates and Chris H. Perleberg

Depto. de Ciencias de la Computación
Universidad de Chile
Casilla 2777, Santiago, Chile
{rbaeza,pchris}@dcc.uchile.cl *

Abstract. We present new algorithms for approximate string matching based in simple, but efficient, ideas. First, we present an algorithm for string matching with mismatches based in arithmetical operations that runs in linear worst case time for most practical cases. This is a new approach to string searching. Second, we present an algorithm for string matching with errors based on partitioning the pattern that requires linear expected time for typical inputs.

1 Introduction

Approximate string matching is one of the main problems in combinatorial pattern matching. Recently, several new approaches emphasizing the expected search time and practicality have appeared [3, 4, 20, 22, 23, 13], in contrast to older results, most of them only of theoretical interest. Here, we continue this trend, by presenting two new simple and efficient algorithms for approximate string matching.

First, we present an algorithm for string matching with k mismatches. This problem consists of finding all instances of a pattern string $P = p_1p_2p_3....p_m$ in a text string $T = t_1t_2t_3....t_n$ such that there are at most k mismatches (characters that are not the same) for each instance of P in T. When $k = 0$ (no mismatches) we have the simple string matching problem, solvable in $O(n)$ time. When $k = m$, every substring of T of length m qualifies as a match, since every character of P can be mismatched.

Various algorithms have been developed to solve the problem of string matching with k mismatches. Running times have ranged from $O(mn)$ for the brute force algorithm to $O(kn)$ [15, 9] or $O(n \log m)$ [4, 12]. In this paper we present a simple algorithm (one page of C code) that runs in $O(n)$ time, worst case, if all the characters p_i in P are distinct (i.e. none of the characters in P are identical) and in $O(n + R)$ worst case time if there are identical characters in P, where R is the total number of ordered pairs of positions at which P and T match. Note that $0 \leq R \leq f_{max}n$, where f_{max} is the number of times the most commonly occurring character of P appears in P. The space used is $O(2m + |\Sigma|)$ where Σ is the underlying alphabet (typically ASCII symbols). The only algorithm which is similar to ours is a $O(n + R \log \log m)$ result for insertions and deletions, reported in [17]. Related ideas can be found in [8, 1].

---

* This work was partially supported by Grant C-11001 from Fundación Andes and Grant DTI I-3084-9222 of the University of Chile.

String matching with errors consists of finding all substrings of the text that have at most k errors with the pattern, where the errors are counted as the minimal number of insertions, deletions, or substitutions of characters needed to convert one string to another. The best worst case running time is $O(kn)$ [16, 10]. Recently, practical algorithms have been proposed [20, 22, 23], the later of them called "agrep" which is several times faster that "grep" (the well known Unix file searching utility). We present a new algorithm based on partitioning the pattern, which achieves a $O(n)$ expected search time for $k \leq O(m/\log m)$ using $O(m^2)$ extra space. This result generalizes to any algorithm based in partitioning the pattern, as in [22], and is simpler than Chang and Lawler's algorithm [6].

For a complete set of references on these problems we refer the reader to [11].

2 String matching with mismatches

First let us describe the string matching with k mismatches algorithm using the best (and practical) case of only distinct characters in P. In this case, each character in P has a unique offset from the beginning of P, and every time a character of P is encountered in T, the position of the beginning of this instance of P is uniquely determined. An array of the same size as the alphabet of characters can be used to indicate the offset of each character from the beginning of P. If a character is not in P, a special flag value can replace the offset.

Suppose there is a counter c_i for every t_i. The counters are initialized to zero, and as each character of T is read and examined, its offset (if it is a character in P) is used to find the counter at the beginning of this instance of P, and this counter is incremented. If there is an instance of P with $k = 0$, an exact match, the counter at the position of this instance of P will be incremented m times. The number of mismatches is equal to m minus the value of the counter, and if it is less than or equal to k, a match is reported. Only m counters are necessary at any one time, so the counters can be implemented with an array of size m that is traversed in a circular manner. This idea is shown in Figure 1.

Pattern = t h a n

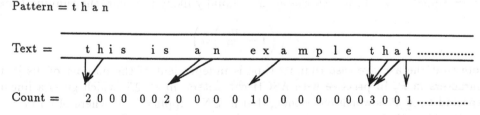

Fig. 1. Searching example for the counting approach.

The case of non-distinct characters in P can be handled by converting the array of offsets into an array of linked lists of offsets. If a character p_i occurs two times in P, it has two distinct offsets from the beginning of P, and these two offsets can be obtained by indexing into the array using the character p_i and traversing the linked list which contains the two offsets in two nodes. The two offsets can then be used to increment two distinct counters.

The C code for the algorithm is given in Figure 2. Several optimizations are included that deviate from the simple description of the algorithm given above. The array **count** that contains the m counters has 256 entries so it can be accessed in a circular manner using the fast AND (&) operator and not the slow MOD (%) operator. Counters are maintained at the position of the last character (not at the position of first character) of each possible instance of P. Each counter is initialized with the value m, and the counters are decremented with each occurrence of a match, so the final value of each counter is the number of mismatches. To handle the special case of the first $m - 1$ characters in T, the first $m - 1$ counters are initialized to high values so they can never reach a valid value of mismatches. Finally, it is worth noting that if all the characters of the pattern are distinct, the innermost "for" loop can be removed from the subroutine **search**.

Now we discuss the running time. Consider first the simple case of only distinct characters in P. In this case, for each character in T, the array of offsets is accessed, and a counter is possibly updated. This case has a running time of $O(n + R)$, where R is the number of ordered pairs of positions where characters of P and T match. Note that $0 \leq R \leq n$ for this case, having $O(n)$ worst case. Space used is $O(m + |\Sigma|)$ which includes the array of offsets which has an entry for every possible character of the alphabet Σ, and the array of m counters. Running time for preprocessing is $O(2m + |\Sigma|)$ as each entry of the array of offsets (size $|\Sigma|$) is initialized, m entries for the m characters of P are written into the array, and the m counters are initialized.

The space used for the case of non-distinct characters in P is $O(2m + |\Sigma|)$ (we use up to $m - 1$ more linked list nodes than in the simple case), and the running time for preprocessing is of the same order as described above. Worst case running time is $O(n + R)$. For this case, we have $0 \leq R \leq f_{max} n$ where f_{max} is the number of times the most commonly occurring character of P appears in P. Every time this character appears in T, a linked list of size f_{max} is traversed, and f_{max} counters are updated. In the absolute worst situation, P consists of m occurrences of a single distinct character, and T consists of n instances of this character, giving a running time of $O(mn)$. However, this is neither a common nor useful situation. In the average case, for an alphabet in which each character is equally likely to occur, the running time is

$$O\left(\left(1 + \frac{m}{|\Sigma|}\right) n\right)$$

Note that the average case running time is independent of the number of distinct characters in P. In practice with ASCII characters, $m < |\Sigma|$, which gives a linear expected time. It is worth mentioning that in DNA applications, where often $m >> |\Sigma|$, the algorithm approaches its worst case running time of $O(mn)$.

Because the algorithm is independent of k, it can be easily adapted to find the "best match" (smallest error). Many times this is desirable when a bound on k is not known *a priori*.

```
#define SIZE     256         /* size of alpha index and count array */
#define MOD256  0xff         /* for the mod operation */
typedef struct idxnode {     /* structure for index of alphabet */
    int offset;              /* distance of char from start of pattern */
    struct idxnode *next;    /* pointer to next idxnode if it exists */
} anode;
anode alpha[SIZE];  /* offset for each alphabetic character */
int   count[SIZE];  /* count of the characters that don't match */


int search(t,n,m,k,alpha,count)      /* string searching with mismatches */
char *t;           /* text */
int   n,m;         /* number of characters in text and pattern */
int   k;           /* the Hamming Distance */
anode alpha[];     /* index of alphabet */
int   count[];     /* circular buffer of count of mismatched characters */
{
    int i, off1;    anode *aptr;

    for (i=0; i<n; i++) {
        if ((off1=(aptr=&alpha[*t++])->offset) >= 0) {
            count[(i+off1)&MOD256]--;
            for (aptr=aptr->next; aptr!=NULL; aptr=aptr->next)
                count[(i+aptr->offset)&MOD256]--;
        }
        if (count[i&MOD256] <= k)
            printf("Match in position %d with %d mismatches\n"
                                   ,i-m+1,count[i&MOD256]);
        count[i&MOD256] = m;
    }
}
int preprocess(p,m,alpha,count)               /* preprocessing routine */
char *p;           /* pointer to pattern */
int   m;           /* number of characters in pattern */
anode alpha[];     /* alphabetical index giving offsets */
int   count[];     /* circular buffer for counts of mismatches */
{
    int    i,j;    anode *aptr;

    for (i=0; i<SIZE; i++) {
        alpha[i].offset = -1;  alpha[i].next  = NULL;  count[i] = m;
    }
    for (i=0,j=128; i<m; i++,p++) {
        count[i] = SIZE;
        if (alpha[*p].offset == -1) alpha[*p].offset = m-i-1;
        else {
            aptr = alpha[*p].next;              alpha[*p].next = &alpha[j++];
            alpha[*p].next->offset = m-i-1;  alpha[*p].next->next = aptr;
        }
    }
}
```

Fig. 2. C code for string matching with mismatches

3 String Matching with Errors

In this section we describe a new algorithm which is based in two classical algorithms and the partition approach mentioned in [22]. We show that this very simple algorithm has linear expected running time for most values of k.

The partition approach is based on the following fact: an occurrence with at most k errors of a pattern of length m implies that at least one substring of length r in the pattern matches a substring of the text occurrence exactly, where

$$r = \left\lfloor \frac{m}{k+1} \right\rfloor .$$

There are many ways to use this idea. Perhaps the simplest one, used in [22], is to search for the first $k+1$ blocks of size r of the pattern P. If any of the blocks is an exact match, we try to extend the match, checking if there are at most k errors. This idea was used in conjunction with the extension of the shift-or algorithm [3] to string matching with errors. Here we combine the idea with traditional multiple string searching algorithms.

The simplest algorithm is to build an Aho-Corasick machine [2] (the extension of the Knuth-Morris-Pratt algorithm [14] to search for multiple patterns) for the $k+1$ blocks of length r (less blocks if some of them are equal). For every match found, we extend the match, checking if there are at most k errors, by using the standard dynamic programming algorithm to check the edit distance between two strings. We can decompose the running time in the search phase and the checking phase. The search phase requires linear worst case time for finite alphabets. The checking phase will depend on the number of potential matches found. Assuming that both the pattern and the text are random, the expected number of occurrences of any of $k+1$ patterns of length r in a text of length n is

$$\frac{(k+1)n}{|\Sigma|^r}$$

Because we may have the block in more than one position of the pattern (the case of repeated blocks), instead of storing all the possible positions and checking each case, it is much easier to check the whole pattern in the neighborhood of the block occurrence. Thus, we search for the pattern between positions $i - (m - r) - k$ and $i + m + k$ if the block was found in the i-th position of the text. Then, in the worst case, for every match checked we need $m(2(m+k) - r)$ operations by using dynamic programming. Although in average less operations are needed, this worst case bound is sufficient to prove expected time linearity. Thus, to achieve expected linear time, we should have

$$\frac{m(2(m+k) - r)(k+1)n}{|\Sigma|^r} \leq cn$$

for some fixed constant c. To solve for r in this transcendental equation, we use the following weaker inequality (replacing k by $m - 1$):

$$\frac{(4m - 3)m^2 n}{|\Sigma|^r} \leq cn$$

Solving for r, and then using $k \leq m/r - 1$, we have

$$k \leq \frac{m}{\log_{|\Sigma|}(m^2(4m-3)/c)} - 1$$

which shows that the maximum value for k is $O(m/\log m)$. Note that this bound is pessimistic, as we are using the worst case number of comparisons when checking. Also note that c can be used as parameter to tune the maximum value of k. This bound will be less restrictive if we use better algorithms in the checking phase. That is, this expected linearity result is valid for other algorithms, by example for [22]. Just as an example, for $c = 1.5$, $m = 10$ and $|\Sigma| = 32$, we have $k \leq 4$.

We can improve the searching phase by using multiple string searching algorithms based on the Boyer-Moore algorithm [7, 19, 5] or the shift-or algorithm [3, 23]. This improvement is significant when the number of blocks found is small (or in other words, when the alphabet is large).

To improve the check phase, we need to decrease the number of potential matches. Two possible solutions are:

- Select the blocks to be searched (the possible number of distinct sets of blocks depends on $m - (k+1)r$) according to character frequency in the text. Another possibility is to use the frequency of small sequences of characters [21].
- Search more than $k + 1$ blocks, say $2(k + 1)$ blocks. This forces to have at least two exact matches in a text window of size $m + k$, which decreases the number of potential matches. On the other hand, it may increase the search time phase, but the optimal number of blocks will depend on the input.

Also, the same algorithm can be used for string matching with mismatches, or other error measures that obey the partition scheme used. The partition idea may be also useful to solve this problem allowing preprocessing (that is, when some kind of index of the text is available).

4   Final Remarks

Our first algorithm, based on counting, achieves $O(n)$ worst case time independent of k and without restrictions on m (which [3, 22] have), when all or almost all the symbols in the pattern are different. This seems to show a difference between the bit-parallelism of the shift-or algorithm [3] which is bounded by the word size, and counting, which seems to be bounded by symbol frequency.

This algorithm, in the vein of [3, 22, 23], shows the potential of logical and arithmetical operations over text comparisons. This indicates that there might be new ways of improving algorithms for similar problems by changing the computation model. In fact, undergoing research is exploiting these ideas for simple and fast algorithms to find the longest common substring between a text and a pattern, and variations of string matching with errors.

The second algorithm achieves linear expected time in most cases, and the bound on k is similar to the algorithms of Chang et al. [6]. One way to improve it would be to have a good algorithm to search patterns with at most 1 or 2 errors. This could be used to search for blocks of length $2r$ or $3r$, decreasing the number of potential matches. This algorithm shows that reducing a text searching problem to simpler problems to find potential answers may lead to simpler and fast algorithms, if the number of potential matches to check is small. This approach has also been used for two dimensional text searching [5]. This problem reduction should satisfy:

- The reduction itself and solving the new problem is faster than the original problem, and if possible, with running time independent of the number of potential matches.
- The number of potential matches is small, where "small" may mean on average or related to the real number of matches.

Finally, as in many text searching problem, there is no algorithm that is the best for all cases. That will depend on the application or the input. This suggests the use of hybrid algorithms as in "agrep" or adaptive algorithms as in [18] for string searching.

References

1. K. Abrahamson. Generalized string matching. *SIAM J on Computing*, 16:1039–1051, 1987.
2. A.V. Aho and M. Corasick. Efficient string matching: An aid to bibliographic search. *C.ACM*, 18(6):333–340, June 1975.
3. R. Baeza-Yates and G.H. Gonnet. A new approach to text searching. In *Proc. of 12th ACM SIGIR*, pages 168–175, Cambridge, Mass., June 1989. (Addendum in ACM SIGIR Forum, V. 23, Numbers 3, 4, 1989, page 7.). To appear in *Communications of CACM*.
4. R. Baeza-Yates and G.H. Gonnet. Fast string matching with mismatches. *Information and Computation*, 1992. (to appear). Also as Tech. Report CS-88-36, Dept. of Computer Science, University of Waterloo, 1988.
5. R. Baeza-Yates and M. Régnier. Fast algorithms for two dimensional and multiple pattern matching. In R. Karlsson and J. Gilbert, editors, *2nd Scandinavian Workshop in Algorithmic Theory, SWAT'90*, Lecture Notes in Computer Science 447, pages 332–347, Bergen, Norway, July 1990. Springer-Verlag.
6. W. Chang and E. Lawler. Approximated string matching in sublinear expected time. In *Proc. 31st FOCS*, pages 116–124, St. Louis, MO, Oct 1990. IEEE.
7. B. Commentz-Walter. A string matching algorithm fast on the average. In *ICALP*, volume 6 of *Lecture Notes in Computer Science*, pages 118–132. Springer-Verlag, 1979.
8. M. Fischer and M. Paterson. String matching and other products. In R. Karp, editor, *Complexity of Computation (SIAM-AMS Proceedings 7)*, volume 7, pages 113–125. American Mathematical Society, Providence, RI, 1974.
9. Z. Galil and R. Giancarlo. Improved string matching with k mismatches. *SIGACT News*, 17:52–54, 1986.
10. Z. Galil and K. Park. An improved algorithm for approximate string matching. In *ICALP'89*, pages 394–404, Stressa, Italy, 1989.
11. G.H. Gonnet and R. Baeza-Yates. *Handbook of Algorithms and Data Structures - In Pascal and C*. Addison-Wesley, Wokingham, UK, 1991. (second edition).
12. R. Grossi and F. Luccio. Simple and efficient string matching with k mismatches. *Inf. Proc. Letters*, 33(3):113–120, July 1989.
13. A. Hume and D.M. Sunday. Fast string searching. *Software - Practice and Experience*, 21(11):1221–1248, Nov 1991.

14. D.E. Knuth, J. Morris, and V. Pratt. Fast pattern matching in strings. *SIAM J on Computing*, 6:323–350, 1977.
15. G. Landau and U. Vishkin. Efficient string matching with k mismatches. *Theoretical Computer Science*, 43:239–249, 1986.
16. G. Landau and U. Vishkin. Fast string matching with k differences. *JCSS*, 37:63–78, 1988.
17. U. Manber and S. Wu. An algorithm for approximate string matching with non uniform costs. Technical Report TR-89-19, Department of Computer Science, University of Arizona, Tucson, Arizona, Sept 1989.
18. P.D. Smith. Experiments with a very fast substring search algorithm. *Software - Practice and Experience*, 21(10):1065–1074, Oct 1991.
19. M.A. Sridhar. Efficient algorithms for multiple pattern matching. Technical Report Computer Sciences 661, University of Wisconsin-Madison, 1986.
20. J. Tarhio and E. Ukkonen. Boyer-moore approach to approximate string matching. In J.R. Gilbert and R.G. Karlsson, editors, *2nd Scandinavian Workshop in Algorithmic Theory, SWAT'90*, Lecture Notes in Computer Science 447, pages 348–359, Bergen, Norway, July 1990. Springer-Verlag.
21. S. Wu. personal communication. 1992.
22. S. Wu and U. Manber. Fast text searching with errors. Technical Report TR-91-11, Department of Computer Science, University of Arizona, Tucson, Arizona, June 1991.
23. S. Wu and U. Manber. Agrep - a fast approximate pattern-matching tool. In *Proceedings of USENIX Winter 1992 Technical Conference*, pages 153–162, San Francisco, CA, Jan 1992.

DZ
A text compression algorithm
For natural languages

Dominique REVUZ
revuz@litp.ibp.fr

Marc ZIPSTEIN
zipstein@litp.ibp.fr

INSTITUT GASPARD MONGE°

Abstract:
«Texts written in a natural language are essentially made of words of this language».
We use this obvious fact, together with an extensive lexicon to define a good model of
the statistical behavior of letters in texts. This model is used with the arithmetic
coding scheme to build an efficient universal data compression method. Initially our
method was specialized in the compression of French texts. However it can be easily
adapted to other languages. Tests show that the compression ratio obtained by our
method is on the average 30% on French texts. On the same texts Ziv & Lempel's
method yields an average ratio of 40%. On other kinds of test files (English text,
executable files, sources) the use of an order 1 Markov chain leads to results of the
same order as Ziv & Lempel's. We present a new approach to dynamic dictionary
construction for natural language compression. The fact well known to linguists that
the number of different words is small, makes a dynamic construction possible.

Introduction

The aim of text compression is to reduce the number of symbols necessary to represent a text. It is
generally used in transmission and to enhance storage efficiency. The performances of hardware are
growing, so are the needs of the users. This growth justifies the search for more efficient methods.
We call a probability model (model for short) a mechanism that produces estimates of the
probability of appearance of the letters.
Finding the perfect model for natural language is an interesting subject. See [2] for bibliography,
and the nice result given by Brown & all [3]. Our model is simple and has the advantage of being
usable in a compression mechanism. The model is based on two simple assumptions, first a text
is made of words i.e. strings of letter separated by punctuation marks, second the great majority of
those words are in the own lexicon of the language.
The model presented is a variable length context model.
The model described is used with a statistical encoding method (arithmetic coding). Statistical
methods splits the input into blocks of fixed length (generally the letters), and each block is
encoded in accordance with its probability of appearance, a frequent letter (i.e. a letter appearing
frequently) having a shorter code than a less frequent one. The oldest statistical method is that of
Huffman [6] which uses a prefix code of minimal average length. Arithmetic coding, a more

---

° I.G.M. Université de Marne La Vallée   2, allée Jean-Renoir   93160 NOISY LE GRAND
    FRANCE

efficient statistical method, has been presented by Rissanen [8], Rissanen & Langdon [9], Guazzo [5].

The efficiency of statistical methods depends on the adequateness of the estimate of the probability of appearance of letters and the real behavior of letters in the text. If the probabilities given by the model describe exactly the behavior of the letters of the text, the compression is optimal. In natural languages this behavior is highly context dependent. For example, in French after the letter "q" the letter "u" is much more frequent than any other letter. The originality of our method is in the choice of the probability model: an automaton constructed over a 655,000 word lexicon of French words. This automaton produces the probabilities which are used by the arithmetic coding method. The choice of arithmetic coding is motivated by the clear separation that exists between the probability model and the encoding itself. The version of arithmetic coding used is the version presented by Witten, Neal et Cleary [11].

Our work at Institut Gaspard Monge* gives us access to the LELAF. (the LADL‡'s electronic lexicon of inflected forms). This extensive lexicon is the base of our probability model: the prefixes of the words contained in the lexicon are used as left context to predict the following letter. In the case of new words we use a formal definition of word to add dynamically new words into the lexicon, this definition is derived from linguistic remarks.

The use of a French lexicon makes the method specially well-suited to text written in French. But the method is universal and compresses any kind of files with better results than Ziv & Lempel on English texts, and with results of the same order as Ziv & Lempel's method on other files. Its behavior is satisfactory in all cases where the entropy is not too low.

A presentation of arithmetic coding is given in appendix. A good survey and introductory material can be found in [2]. The only information needed by the encoding / decoding mechanism is the *set of probabilities associated with the letters at each step in the process*. The set of probabilities can be totally changed after the treatment of each symbol and this without increasing the complexity of the encoding mechanism. This possibility has already been used to create an adaptive arithmetic coding using the frequencies in the text already encoded [11]. The [11] model does not use context information. Abrahamson [1] has enhanced the model with an adaptive Markov chain of order 1. The probability of a letter is related to the preceding letter and this more accurate estimate increases the compression ratio. The lexicon used in our model is presented in section 1 at the same time as interesting properties of a DAWG representation for a huge lexicon (space saving, access speed). In section 2 is also shown how to transform the DAWG into a transducer recognizing any text and producing the frequencies of appearance of letters. The use of both the transducer and arithmetic coding is detailed in section 3 with the adaptive mechanism. The evolutive mechanism, the word Markov model and the dynamic dictionary approach is presented in section 4. The results of different versions of the algorithm is given in section 5. Those results and the experiments presented in [2] show that it is getting harder to improve the model.

1. The Lexicon and the DAWG

The L.E.L.A.F. is, to our knowledge, the most exhaustive lexicon of French. It contains most of the words of the language in their different inflected forms, but no proper names nor geographical names. It has 650,000 word entries. The lexicon was originaly stored as a 6.5Mega bytes file containing one word per line with and average length of ten ASCII characters.

---

* A young French laboratory in computer science and applied linguistics.

‡ LADL Laboratoire d'Automatique Documentaire et Linguistique, Jussieu University.(Paris 7).

One of our first results in data compression is in the compression of the L.E.LA.F. itself. Our task was in fact to represent in a space efficient way the L.E.L.A.F and keep access time reasonable.

The representation by a DAWG compresses it to 450K bytes (a very good compression result, less than a byte per word [7]). This representation is not only efficient in size, it also speeds up the access time to a given word. Our home-made spelling checker achieves a speed of 200,000 words per minute on a 80386 at 25MHz computer. The home-made spelling advisor produces words at edit distance (number of substitutions, insertions, deletions between two words) less or equal to 5 at a rate of 10,000 words per minute.

For the DZ algorithm the representation of a state of the DAWG is realized by a pair: the number of transitions followed by a compacted array of triple (letter; pointer to a state; frequency of use), and all states are stored in an array. In practice the automaton and the frequencies occupy 1M of central memory.

The present* version of the automaton has 49000 states, 128000 edges, 6200 terminal states and 23000 states with a unique output edge. The average number of output edges per state is 1.6, this is very important as the size of the local alphabet is one of the reasons of the efficiency of DZ. This local alphabet yields an accurate estimation of the probabilities of appearance of the letters after a given prefix.

2 .   The automaton and the frequencies

The DAWG is transformed into a transducer where every edge is labelled by a letter and the frequency of use of this letter after the prefix that leads to the state.

Example 1

We consider the alphabet $\mathcal{A} = \{a,d,e,l,n,s,u\}$ and a language reduced to $\mathcal{L}=\{de,du,des,le,la,les,un,une\}$.

This yields the following automaton of figure 1. (the frequencies shown are the ones computed after reading "delelesla").

The edges leaving from a state of the automaton describe the probability of appearance of the letters after the prefix that leads to the state.

In state-2 of the automaton the sum of the frequencies of the edges starting at state-2 is 3. The frequency of the edge labelled by the symbol "e" is 2 therefore the probability of the symbol "e" is estimated to 2/3 when the current state is state-2.

In state-3 the probability of the symbol "e" is estimated to 3/5.

This example shows the varying length context. In state-6 the context for the symbol "e" is of length 2 and for the same symbol in states 2 and 3 the context is of length 1. ◊

---

* We use the version 4 of the DELAF in the Experiments of section five.

Figure 1.

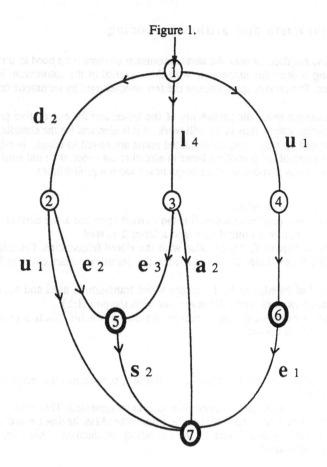

2.1. Default edges

To make the model universal the automaton must recognize any given text on the alphabet $\mathcal{A}$.

We make the following additions :

On every state except the initial state a default edge is added. Theses edges are all labelled by the empty word and and are all directed towards the initial state.

The initial state is completed so has to have an edge labelled by every symbol in the alphabet (0..255 in practice)

With those two transformations we have the property that every text in ASCII is recognized by a unique path in the automaton, so as the encoder and the decoder work in lock step they will follow the same path.

3 . Using automata and arithmetic coding

Both the encoder and the decoder uses the same automaton so there is no need to transmit it.
Encoding / decoding is done simultaneously with the traversal of the automaton. Input letters are treated one at a time. The current state contains the frequencies used by arithmetic coding to encode the current letter.
The encoding mechanism needs the probability of the letter, and the cumulative probability of all alphabetically preceding letters (any order will work, if it is identical for the encoding and decoding mechanism). To speed up treatment, accumulated sums are stored in edges: an edge contains the accumulated frequencies of the preceding letter in alphabetical order, the total sum is stored in the default edge. For precision purpose we store frequencies and not probabilities.

3.1. Treatment of a symbol
The algorithm works on a pair $(S;1)$ where S is the current state and 1 the current letter.
Initially the state S is set to I the initial state and a letter 1 is read.
- If 1 is the label of an edge of S, it is encoded with the stored frequencies. The edge is used to go to the next state S', a new letter $1'$ is read from the input. We start over with the new pair $(S';1')$.
- If 1 is not the label of an edge of S, the empty word transition is used and encoded. The next state is initial state, no letter is input. We start over with the pair $(I;1)$.
The decoding scheme executes the same moves. As any given file labels a unique path in the automaton, the decoding is correct.

3.2. Adaptivity
For a better description of the specific behavior of the letters, we adapt the frequencies used while the text is being treated.
After the use of an edge the frequency associated to it is incremented. This ensures the adaptivity. The decoding mechanism adapts the model in the same way. After having treated the same part of the file, it uses the same probabilities as the encoding mechanism. This guarantees that the decoding is possible and correct.

3.3. Initialization of the frequencies
We have tested different initialization schemes, but the best one to date is to initialize all edges of every state with an identical probability, a frequency of use set to 1. The reasons are:
1- the adaptive mechanism is more rapidly efficient, because unused edges become rapidly very improbable;
2- the algorithm must be universal, therefore it must not prejudge too strongly about the kind of text it will compress;
3- last and most important there is no need to store standard frequencies.

3.4. Special features for text
We want our algorithm to be very efficient on natural language texts so we add some features to take into account some common text formatting rules:
- words are usually separated by space and/or punctuation marks;
- punctuation marks are followed by a space;
- sentences start with upper case letters.

198

For this:
- an edge (T,space,I) is added to each terminal state **T**;
- a punctuation state **P** is created with the edge (P,space,I) and two edges, (T,'.',P) and (T,',',P), are added to all terminal states;
- for every edge (I,a,Q) the edge (I,A,Q) is added.
An average 2% gain is obtain on natural language text with these features, and little effect if any is done on other kind of files.has those states are never reach.

Use of an order 1 Markov Chain.
An important improvement to our initial algorithm is to use a order 1 Markov chain to store the probabilities used on the initial state. This allows a better description of the behavior of unexpected letters (not prefixes of French words). The compression results are better on French texts and much better on other kinds of texts. With this improvement result are always better than Abrahamson's as our model is at least as efficient as his.

4. The evolutive scheme

We present two version of an evolutive algorithm, they add news words to the lexicon. One that uses the same lexicon as the non evolutive algorithm and one that starts with a reduced lexicon (the set of symbols). In this section only the second one is presented as both methods works the same way (but not with the same compression ratio).

As the text is treated it is parsed in to words and strings of non-alphabetic symbols. A word is a sequence of letters. The new words are added to the lexicon as they are parsed. The lexicon is represented by a lexicographical tree (remark: a trie structure will be tested in the future). The edges of each node of the tree are labeled by a pair made of a letter and the probability of appearance of this letter after the prefix that leads to the node.
Example 2
We consider the input text "we work on words.".
After it has been treated the dynamic lexicon is represented by the following tree :
Figure 2.

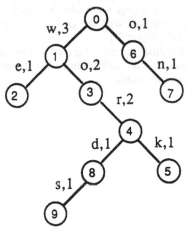

In node 0 (the beginning of a word) the frequency of the letter 'o' is 1 and it is 3 for the letter 'w'. So at the beginning of a word we estimate of the probability of letter 'o' to be 1/4 and that of the letter 'w' to be 3/4.

The probability of a letter depends on its position inside a word. For instance with this tree (Figure 2.) the estimate of the probability of letter 'o' is 1/4 at the beginning of a word and 2/3 in second position of a word starting with a 'w' (in node 1).

◊

4.1. Using the tree and arithmetic coding

In the same way as in section 2 we have to make the tree able to parse any text, from this follow two additions : (i)the initial state starts with all possible edges (the non-alphabetic loop on the root node). (ii) Every new inner node is built with a default edge, labelled by the empty node that sends back to the root node. This is not needed on leaf nodes as the only possible movement is to go back to the root node.

Default edges are used if the current symbol cannot be parsed by the current node.

The transformed tree is an automaton that can recognize any string of ASCII characters and this in a unique way.

4.2. Dynamic Construction of the Lexicon

The algorithm works in the same fashion has with a fixed automaton with the following additions:

On any given node N, different from the root, if the current symbol l_0 is not a label of an edge the default edge is used. The symbol ε is encoded with the probabilities stored on node N. The symbol l_0 is encoded with the probabilities stored on the root and the learning mechanism starts.

If l_0 is a letter, a new node N_0 and new edge (N,l_0,N_0) are created. While the next symbols $l_1...l_k$ are letters, they are encoded with the probabilities memorized on the root and the a new branch $(N_0,l_1,N_1) ... (N_{k-1},l_k,N_k)$ is created.

The frequency associated to new edges is 1.

The decoding mechanism starts the same learning mechanism after having decoded an ε edge. It will add the same edges on the same nodes.

Example 2, continued, learning the word "will"

Suppose the next word to be encoded is "will" followed by a space.

The letter 'w' is encoded (on the root), the corresponding edge is used, it leads to node 1.

The 'i' is not an edge label of node 1. The ε edge is encoded and used.

Symbols 'ill' are letters so they are encoded with the probabilities given by the root and the edges (1,i,10) (10,l,11) (11,l,12) are created (Figure 3.).

The space is encoded with the probabilities given by the root but no edge is created.

The length of the added word is limited, this checks unbounded sequences of letters that are not words of a natural language (DNA sequences for example).

Figure 3.

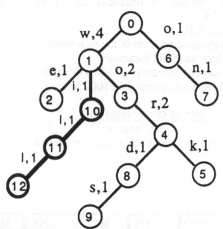

The dynamic lexicon is constructed in the same way by the encoding and decoding mechanisms. No extra information in the encoded message is needed for the decoder to rebuild the lexicon.

◊

4.3. Practical complexity

The present implementation could be optimized for speed, but we can already note that on French text it is faster than Abrahamson's implementation of arithmetic coding because of the small local alphabet found after a given prefix. On a large text the time necessary to load the lexicon is made up. Our algorithm is 10 to 20 times slower than compress on the two machines tested (NeXT, IBM PS2).

The present overall memory usage is 1.6 M bytes but can be reduced to 800K.

5. Results

The following figure shows results obtained on the set of files:
French texts:

> Droit is an article on international law (1989) ;
> Verne is a novel written by French author Jules Verne;
> AFP is a concatenation of news from the Agence France Press.
> A 10% gain against ZL on the average with the French lexicon.

English text:

> DOC is the Documentation given with the PKZIP PC software
> (43% of the bytes are spaces this is good for compress)
> Brown is the concatenation of English newspapers articles (1989).
> Hamlet is Shakespear's play in RTF.

other:

> C is the sources of the unix Moria game;
> Moria is the striped executable of the game;
> those two files contains a lot of English words.

The tested algorithms are:

> Abrahamson: Arithmetic coding with an order 1 Markov chain [1].
> Ziv & Lempel: the Unix command compress [10],[13].

All different version the DZ (Dominique & Zip) algorithms use an order 1 Markov chain and the rules described in § 3.4.

DZ: using the L.E.L.A.F.

EVO: DZ with the dynamic addition of the new words

NODIC: EVO without an initial dictionary

ENG: DZ with an English dictionary of 128666 words.

ITA: DZ with and Italian dictionary of 81000 words.

The percentage is the ratio: $\dfrac{\text{size of compressed file}}{\text{size of source}}$

As expected the best gains are on French texts with the French lexicon and on English text with the English dictionary.

name	Droit	Verne	AFP	Brown	DOC	Hamlet	Moria	C
Size	1705435	363631	744794	790662	140355	234836	410256	709226
compress	660367	147107	333911	344639	40310	90541	210237	244001
	38,72%	40,46%	44,83%	43,59%	28,72%	38,55%	51,25%	34,40%
Abrahamson	753678	163356	355469	359456	49992	108491	212431	309195
	44,19%	44,92%	47,73%	45,46%	35,62%	46,20%	51,78%	43,60%
DZ	455912	112835	259408	306092	43518	101506	209518	280136
	26,73%	31,03%	34,83%	38,71%	31,01%	43,22%	51,07%	39,50%
EVO	530782	123093	267165	295522	42789	91972	210007	234406
	31,12%	33,85%	35,87%	37,38%	30,49%	39,16%	51,19%	33,05%
NODIC	606864	137627	297520	271554	40439	89357	208140	232366
	35,58%	37,85%	39,95%	34,35%	28,81%	38,05%	50,73%	32,76%
ENG	460511	139664	309309	236407	37379	85847	201934	257574
	27,00%	38,41%	41,53%	29,90%	26,63%	36,56%	49,22%	36,32%
ITA	494268	151003	333279	331483	45669	105520	211150	292414
	28,98%	41,53%	44,75%	41,92%	32,54%	44,93%	51,47%	41,23%

BIBLIOGRAPHY

[1] D. ABRAHAMSON, An adaptive dependency source model for data compression, *Commun. ACM* , 32,1 (1989), 77-83.

[2] T.C. BELL, J. G. CLEARY, I. H. WITTEN, *Text Compression* , Prentice Hall advanced reference series, 1990, ISBN 0-13-911991-4.

[3] P.F BROWN, S.A. DELLA PIETRA, V. J. DELLA PIETRA, J.C. LAI, R.L. MERCER, An Estimate of an Upper Bound for the Entropy of English, *Preprint* (1991).

[4] G.V. CORMACK, R.N.S. HORSPOOL, Data compression using dynamic Markov modelling, *Comput. J.*, 30,6 (1987), 541-550.

[5] M. GUAZZO, A general minimum-redundancy source-coding algorithm, *I.E.E.E. Trans. on Inform. Theory*, 26,1 (1980), 15-25,January.

[6] D. A. HUFFMAN, A method for the construction of minimum redundancy codes, *Proc. IRE*, 40 (1952), 1098-1101,September.

[7] D. REVUZ, *Dictionnaires et Lexiques Méthodes et Algorithmes*, These de Doctorat Université Paris 7, (1991).

[8] J. RISSANEN, Generalized Kraft inequality and Arithmetic coding, *IBM J. Res. Dev.* , 20 (1976), 198-203,May.

[9] J. RISSANEN, G.G. LANGDON Jr., Arithmetic coding, *IBM J. Res. Dev.*, 23,2 (1979), 149-162,March.

[10] T.A. WELCH, A technique for high-performance data compression, *IEEE Computer* , 17,6 (1984), 8-19,June.

[11] I.H. WITTEN, R.M. NEAL, J.G. CLEARY, Arithmetic coding for data compression, *Commun. ACM*, 30,6 (1987), 520-540, June.

[12] J. ZIV, A. LEMPEL, A Universal algorithm for sequential data compression, *I.E.E.E. Trans. Inform.Theory*, 23,3 (1977), 337-343, May.

[13] J. ZIV,A. LEMPEL, Compression of individual sequences via variable-rate coding, *I.E.E.E. Trans. Inform.Theory*, 24,5 (1978), 530-536, September.

Appendix. Arithmetic coding

Arithmetic coding is a more general method than a substitution encoding like Huffman's. It consists in representing a text with an interval of real numbers included in the [0,1[interval. The interval is reduced step by step while translating the text, hence the number of bits needed to specify the interval grows. Each new symbol reduces the current interval in accordance with its probability. The more likely symbols reduce the range of the interval by less than the unlikely symbols and hence add fewer bits to the representation of the text.

Example

Encoding

Let the alphabet be {a,b} and suppose we encode a text where the probability of appearance of the letter "a" is 3/4 and the one of the letter "b" is 1/4. Thus the letter "a" will be associated with the three lower quarters of the current interval, and the letter "b" will be associated with the upper quarter.

Suppose we want to encode "abb".

Initially the interval is [0,1[. After reeding "a" it is reduced to [0,3/4[. Reading the first "b" reduces this interval to its upper quarter : [9/16,3/4[. Encoding the second "b" reduces it again to [45/64,3/4[.

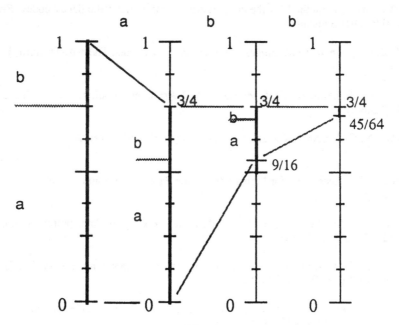

abb is associated with the interval [45/64;3/4[.

Decoding

To decode an interval, the only thing to do is to find which letter is associated with an interval that contains this interval.

Thus, initially, the current interval is [0,1[, "a" is associated with [0,3/4[and "b" with [3/4,1[; the encoding interval is [45/64,3/4[, included in [0,3/4[it means that "a" has been encoded. The current interval is reduced to [0,3/4[and the process continues until the current interval is equal to the encoding interval.

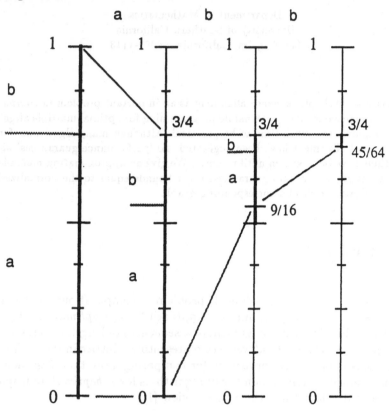

Remark

This description is only a presentation of arithmetic coding. When implementing arithmetic coding, one has to resolve the problem of the precision of the representation of real numbers. The usual solution is to use intervals of long integers and shifts are done when the high bit of both bounds are equal. This choice is an approximation of arithmetic coding.

◊

Multiple Alignment with Guaranteed Error Bounds and Communication Cost

Pavel A. Pevzner

Department of Mathematics
University of Southern California
Los Angeles, California, 90089-1113

Abstract. Multiple sequence alignment is an important problem in computational molecular biology. Dynamic programming for optimal multiple alignment requires too much time to be practical. Although many algorithms for suboptimal alignment have been suggested, no 'performance guarantees' algorithms have been known until recently. We give an approximation multiple alignment algorithm with guaranteed error bounds equal to the normalized communication cost of a corresponding graph.

1 Introduction

Multiple sequence alignment is a difficult problem in computational molecular biology. The classical dynamic programming approach for the *optimal* multiple alignment problem reduces multiple alignment of k sequences of length n to the minimum path problem in the graph with n^k vertices representing a lattice in the k-dimensional space. Since this approach is impractical for comparing more than 3 sequences, the length of an average protein, a number of algorithms for *suboptimal* multiple alignment have been developed (see the recent review [1]).

A common approach in computer science to solve hard optimization problems is to develop fast approximation algorithms whose maximum possible deviation from the optimal solution can be proven to be bounded by a small multiplicative constant c. Gusfield [5] first proposed an approximation algorithm for the multiple alignment problem with $c = 2 - 2/k$. It is known that models currently employed to align sequences are not quite adequate; thus, for practical sequence alignment it is not always necessary to produce an optimal alignment but only one that is plausible. The Gusfield [5] approximation algorithm produces plausible alignments; a computational experiment with an alignment of 19 sequences gave a suboptimal solution only 2% worse than the optimal one .

This paper develops Gusfield's approximation algorithm. An approximation algorithm with $c = 2 - 3/k$ generalizing Gusfield's centered tree approach is suggested. Running time of the algorithm is defined by the running time of all triple alignments among k sequences and equals $O(n^3 k^3 + k^4)$. We also formulate an open problem of devising polynomial approximation algorithms with $c = 2 - l/k$ for arbitrary fixed $l \leq k$. A paper treating these and related topics in more detail will appear in [8].

2 Alignment Graph

Let $\mathcal{A}$ be an alphabet of α letters. Denote $\sum = \mathcal{A} \bigcup \Delta$; Δ is often said to represent an *'indel'* or *insertion/deletion* of a letter. Let $S_1 = s_1(1)s_1(2)\ldots s_1(l_1)$, $S_2 = s_2(1)s_2(2)\ldots s_2(l_2)$, $S_k = s_k(1)s_k(2)\ldots s_k(l_k)$ be a set of sequences over the alphabet $\mathcal{A}$ of length $l_1, l_2, \ldots, l_k$ respectively.

Consider a set of k-dimensional integer vectors $V = \{\mathbf{v} : \ 0 \le \mathbf{v} \le \boldsymbol{\ell}\}$ and define the *alignment digraph* $G(V, E)$ by the rule

$$(\mathbf{v}, \mathbf{v}')\epsilon E \Longleftrightarrow 0 < \mathbf{v}' - \mathbf{v} \le 1$$

(here $\boldsymbol{\ell} = (l_1 l_2 \ldots l_k)$, $\mathbf{0} = (00\ldots 0)$, $\mathbf{1} = (11\ldots 1)$). Denote by μ the set of all paths in G from $\mathbf{0}$ to $\boldsymbol{\ell}$.

Each arc e in G between $\mathbf{v} = (v_1 v_2 \ldots v_k)$ and $\mathbf{v}' = (v_1' v_2' \ldots v_k')$ corresponds to the k-tuple $f(e) = (a_1 a_2 \ldots a_k)$ in the alphabet $\sum$ according to the rule:

$$a_i = \begin{cases} s_i(v_i'), & \text{if } v_i < v_i' \\ \Delta, & \text{otherwise} \end{cases}$$

Each path $(e_1 e_2 \ldots e_n)\epsilon\mu$ can be represented as a $k \times n$ matrix where the i-th column of this matrix corresponds to $f(e_i)$. This representation is known as a *multiple alignment* of $S_1 S_2 \ldots S_k$ and n is referred to as the number of columns of the multiple alignment. The sequences $S_1 S_2 \ldots S_k$ can be used for defining the lengths on μ; in this case the *minimum path problem* in G is referred to as the *optimal (multiple) alignment problem*.

Let $(d(a_i, a_j))$ be a $|\sum| \times |\sum|$ 2-dimensional *'distance'* matrix. If both $a_i, a_j \ne \Delta$, $d(a_i, a_j)$ is the weight of the *substitution* a_j for a_i, if $a_i = \Delta$, $d(\Delta, a_j)$ is the weight of *insertion* a_j, if $a_j = \Delta$, $d(a_i, \Delta)$ is the weight of *deletion* a_i. For the case of two sequences S_1 and S_2 the distance matrix can be used for defining the arc lengths in G

$$d(e) = d(f(e)) = d(a_1, a_2)$$

and the length of the path $(e_1 e_2 \ldots e_n)\epsilon\mu$ can be defined simply as the sum of the lengths of its arcs:

$$d(e_1 e_2 \ldots e_n) = \sum_{i=1}^{n} d(e_i) \tag{1}$$

For example, the k-dimensional matrix defined by the rule

$$d(e) = d(a_1 a_2 \ldots a_k) = \begin{cases} 1, & \exists a\epsilon\mathcal{A}: \quad a_i = a \text{ or } a_i = \Delta \text{ for all } i\epsilon\{1, 2, \ldots, k\} \\ \infty, & \text{otherwise} \end{cases}$$

corresponds to the *Multiple Shortest Common Supersequence* (MSCS) problem. The decision version of MSCS was shown to be NP-complete for alphabet size $\alpha \ge 5$ (Maier [7]). Timkovsky [9] proved the NP-complexity of the MSCS for a few particular cases and suggested an approximation algorithm without guaranteed bounds.

The k-dimensional matrix defined by the rule

$$d(e) = d(a_1 a_2 \ldots a_k) = \begin{cases} -1, & \text{if } a_1 = a_2 = \ldots = a_k \ne \Delta \\ 0, & \text{otherwise} \end{cases}$$

corresponds to the *Multiple Longest Common Subsequence* (MLCS) problem. Maier [7] proved that the decision version of MLCS is NP-complete. Timkovsky [9] raised the open problems of devising the efficient approximation algorithms for MSCS and MLCS with guaranteed error bounds .

To generalize the definition (1) for $k > 2$ sequences, we can define a k-dimensional distance matrix $d(a_1 a_2 \ldots a_k)$ and assume

$$d(e) = d(f(e)) = d(a_1 a_2 \ldots a_k). \tag{2}$$

In the biological applications, $d(a_1 a_2 \ldots a_k)$ is often defined through the 2-dimensional matrix by

$$d(a_1 a_2 \ldots a_k) = \sum_{1 \leq i < j \leq k} d(a_i, a_j). \tag{3}$$

In this case, the length of the path $P \epsilon \mu$ equals the sum of the lengths of *projections* of P onto all pairs of sequences S_i and S_j. This function is called Sum-of-the-Pairs or *SP-score* . A natural generalization of (3) is the *weighted SP-score*

$$d(a_1 a_2 \ldots a_k) = \sum_{1 \leq i < j \leq k} c_{i,j} \cdot d(a_i, a_j), \tag{4}$$

where $c_{i,j}$ is the 'weight' of the pairwise alignment of S_i and S_j.

In this paper, we assume that $d(\Delta, \Delta) = 0$ and d satisfies the *triangle inequality*

$$\forall a_1, a_2, a_3 : \quad d(a_1, a_3) \leq d(a_1, a_2) + d(a_2, a_3) \tag{5}$$

Denote by $[1 : k]$ the set of integers $1 \leq i \leq k$. Given a set of k sequences $S_1 S_2 \ldots S_k$ and $\Omega = \{i_1 i_2 \ldots i_t\} \subset [1 : k]$ we denote $D^{opt}(\Omega)$ $(D^{opt}(S_{i_1}, S_{i_2}, \ldots, S_{i_t}))$ the score of optimal alignment of $S_{i_1}, S_{i_2}, \ldots, S_{i_t}$. In particular $D^{opt}(S_i, S_j)$ denotes the score of the optimal pairwise alignment of S_i and S_j, while $D^{opt}(S_i, S_j, S_k)$ denotes the score of the optimal triple alignment of S_i, S_j and S_k. Given a multiple alignment A, we denote by $D(A)$ the score of alignment A and by $D(A|\Omega)$ the score of the multiple alignment of $S_{i_1}, S_{i_2}, \ldots, S_{i_t}$ *induced* by A. Obviously, for an arbitrary alignment $D(A|\Omega) \geq D^{opt}(\Omega)$. A is called Ω-consistent if $D(A|\Omega) = D^{opt}(\Omega)$.

Comment. It is quite common in computational molecular biology to refer to multiple alignment as a NP-complete problem. It is worth noting that questions about the computational complexity of the optimal alignment with SP-score or evolutionary tree score are still open. (The Maier's [7] reduction of MLCS to a vertex cover problem is not generalized for these scores.)

3   Configurations and l-stars

In this section we generalize the Feng and Doolitle [2] construction for multiple alignment consistent with a tree.

Let $V = [1 : k]$, $G(V, E)$ be an *undirected graph* and $\{\Omega_1, \Omega_2, \ldots, \Omega_t\}$ be the list of all *cliques* of G. Denote $W_1 = [1 : t]$, $W_2 = \{v : v \epsilon \Omega_i \bigcap \Omega_j \text{ for } 1 \leq i < j \leq t\}$. Define a *bipartite* graph $G^c(W_1 \bigcup W_2, E^c)$ with the parts W_1 and W_2 and the edge set E^c:

$$(i, v) \epsilon E^c \iff v \epsilon \Omega_i \bigcap \Omega_k \text{ for a set } \Omega_k.$$

The graph G is called a *configuration* if G^c is a tree (it implies $|\Omega_i \cap \Omega_j| \leq 1$ for $1 \leq i < j \leq t$).

A configuration fulfilling the conditions

(i) $|\Omega_1| = |\Omega_2| \ldots = |\Omega_t| = l$,

and

(ii) $\Omega_i \cap \Omega_j = \{x\}$ for $1 \leq i < j \leq t$

is called *l-star* and x is called the *center* of the *l-star*.

Feng and Doolitle [2] observed that given any tree T where each vertex is labeled with a distinct sequence, there is a multiple alignment A of these sequences which is consistent with the optimal pairwise alignment corresponding to the edges of T. That is , if S_i and S_j are sequences corresponding to any two adjacent vertices of T, then the pairwise alignment of S_i and S_j induced by A has score exactly $D^{opt}(S_i, S_j)$.

The following lemma generalizes the Feng and Doolitle [2] construction.

Lemma 1. *If G is a configuration with the cliques $\Omega_1, \Omega_2, \ldots, \Omega_t$, then there exists a multiple alignment of the sequences $S_1, S_2, \ldots, S_k$ which is consistent with $\Omega_1, \Omega_2, \ldots, \Omega_t$.*

4 Communication Cost

For a connected graph $G(V, E)$, define the *communication cost* $c(G)$ as

$$c(G) = \sum_{i,j \in V} l(i,j),$$

where $l(i,j)$ denotes the number of edges in the shortest path between i and j in G and the sum is taken over all unordered pairs of distinct vertices i and j. This definition generalizes *communication spanning tree cost* (Hu [6]). Fix shortest paths $\gamma(i,j)$ for each $i, j \in V$, and denote by $\Gamma(G|e)$ the number of paths $\gamma(i,j)$ containing the edge e. Notice that

$$c(G) = \sum_{e \in E} \Gamma(G|e). \tag{6}$$

The *complete graph* H_k has minimum communication cost among all k-vertices graphs. We call $b(G) = \frac{c(G)}{c(H_k)} = 2\frac{c(G)}{k(k-1)}$ the *normalized communication cost* of G.

5 Guaranteed Error Bounds for Multiple Alignment

In this section, for an arbitrary configuration G we construct an alignment with guaranteed error bounds equal to the normalized communication cost of G.

Let $G(V, E)$ be a configuration, and $H_k(V, E_k)$ be a complete (undirected) graph with k vertices (we assume $V = [1 : k]$). Let $\mathcal{G}$ be the set of all subgraphs of H_k isomorphic to G. For $G'(V, E') \in \mathcal{G}$ we denote by $\omega' : V \longrightarrow V$ the *isomorphism* transforming G into G'. Correspondingly, the image of the edge $(i,j)\epsilon E$ under isomorphism ω' will be $(\omega'(i),\omega'(j))\epsilon E'$.

For a configuration $G'(V,E')\epsilon\mathcal{G}$ and an alignment A, denote

$$F(A|G') = \sum_{(i,j)\epsilon E'} \Gamma(G'|(i,j)) \cdot D(A|S_i, S_j). \tag{7}$$

(We assume that the family of shortest paths $(\gamma'_{i,j})$ in G' is induced (through ω') by the family of shortest paths $(\gamma_{i,j})$ in G.)

Lemma 2. *For an arbitrary alignment A and configuration G,*

$$F(A|G) \geq D(A).$$

Lemma 3. *For an arbitrary alignment A and an edge $(i,j)\epsilon E$,*

$$\sum_{G'\epsilon\mathcal{G}} D(A|S_{\omega'(i)}, S_{\omega'(j)}) = \frac{|\mathcal{G}|}{|E_k|} D(A).$$

Now we introduce the notion of optimal G-configuration. For a fixed $G'\epsilon\mathcal{G}$, let $A(G')$ be an alignment yielding minimum

$$F(A(G')|G') = \min_A F(A|G') \tag{8}$$

and let $G^*\epsilon\mathcal{G}$ (*optimal G-configuration*) be a configuration yielding minimum

$$F(A(G^*)|G^*) = \min_{G'\epsilon\mathcal{G}} F(A(G')|G') \tag{9}$$

Denote for simplicity $A^* = A(G^*)$.

Let A^{opt} be the optimal alignment of $S_1, S_2, \ldots, S_k$

Theorem 4. *For an arbitrary configuration G the normalized communication cost $b(G)$ is a guaranteed error bound for alignment A^*:*

$$\frac{D(A^*)}{D(A^{opt})} \leq b(G).$$

Proof. Let M be the number of configurations in $\mathcal{G}$ containing the edge e of E_k. Notice that due to symmetry

$$|\mathcal{G}|/M = |E_k|/|E|. \tag{10}$$

Define

$$W = \frac{1}{M} \cdot \sum_{G'\epsilon\mathcal{G}} F(A^{opt}|G').$$

According to equations (8) and (9)

$$W = \frac{1}{M} \cdot \sum_{G'\epsilon\mathcal{G}} F(A^{opt}|G') \geq \frac{1}{M} \cdot \sum_{G'\epsilon\mathcal{G}} F(A(G')|G') \geq \frac{|\mathcal{G}|}{M} \cdot F(A^*|G^*)$$

Therefore by (10) and lemma 2,

$$W \geq \frac{|E_k|}{|E|} D(A^*).$$ (11)

On the other hand,

$$W = \frac{1}{M} \sum_{G'(V,E') \in \mathcal{G}} \left(\sum_{(i,j) \in E'} \Gamma(G'|(i,j)) \cdot D(A^{opt}|S_i, S_j) \right)$$

$$= \frac{1}{M} \sum_{G' \in \mathcal{G}} \left(\sum_{(l,m) \in E} \Gamma(G'|\omega'(l), \omega'(m)) \cdot D(A^{opt}|S_{\omega'(l)}, S_{\omega'(m)}) \right).$$

As the family of shortest paths $(\gamma'_{i,j})$ in G' is induced through ω' by the family of shortest paths $(\gamma_{i,j})$ in G, we have $\Gamma(G'|\omega'(l), \omega'(m)) = \Gamma(G|l, m)$. Therefore according to lemma 3 and equation (6),

$$W = \frac{1}{M} \sum_{(l,m) \in E} \Gamma(G|l, m) \left(\sum_{G' \in \mathcal{G}} D(A^{opt}|S_{\omega'(l)}, S_{\omega'(m)}) \right)$$

$$= \frac{1}{M} \cdot \frac{|\mathcal{G}|}{|E_k|} D(A^{opt}) \cdot \sum_{(l,m) \in E} \Gamma(G|l, m)$$

$$= \frac{1}{M} \cdot \frac{|\mathcal{G}|}{|E_k|} D(A^{opt}) \cdot c(G).$$

According to equation (10)

$$W = \frac{1}{|E|} D(A^{opt}) \cdot c(G),$$

and therefore, according to inequality (11),

$$\frac{D(A^*)}{D(A^{opt})} \leq \frac{c(G)}{|E_k|} = b(G). \quad \square$$

The following corollary from the theorem 4 yields the Gusfield's guaranteed error bound for multiple alignment ([5]).

Corollary 5. *If G is 2-star with k vertices, then*

$$\frac{D(A^*)}{D(A^{opt})} \leq 2 - \frac{2}{k}.$$

Corollary 6. *If G is l-star with k vertices, then*

$$\frac{D(A^*)}{D(A^{opt})} \leq 2 - \frac{l}{k}.$$

Corollary 7. *If G is 3-star with an additional edge having k vertices, then*

$$\frac{D(A^*)}{D(A^{opt})} \leq 2 - \frac{3}{k} + \frac{2}{k(k-1)}.$$

6 Search of Optimal 3-star

Theorem 4 reduce the problem of devising an approximation algorithm with guaranteed error bound to the search of optimal G-configuration. In this section we give a polynomial algorithm for the search of optimal configuration for 3-star (*optimal 3-star*).

Let $H_k(V, E_k, D)$ be the complete weighted graph with the weights $D(i, j) = D^{opt}(S_i, S_j)$. Gusfield [5] defined a *center star* to be a 2-star of minimum weight in $H_k(V, E_k, D)$. For the case when G is a 2-star with $k = t+1$ vertices, $\Gamma(G|(i,j)) = t$ for each edge (i, j). Therefore, a center star is an optimal configuration for 2-star (it yields the minimum in (9) and gives an upper bound for the score of A^* equal to $t \sum_{i \neq x} D(S_x, S_i)$, where x denotes the center of the center star).

The computation of the weight function D needs $O(n^2 \cdot k^2)$ operations. According to corollary 5, a center star method gives a $O(n^2 \cdot k^2)$ algorithm for a multiple alignment problem with guaranteed upper bound $c = 2 - \frac{2}{k}$. Notice that a *minimum communication spanning tree* (Hu [6]) in $H_k(V, E_k, D)$ gives, in general, better alignment than the centered tree, especially in the case when among k sequences there exist triples S_i, S_j, S_l such that S_i is an ancestor of S_j and S_j is an ancestor of S_k (see Hu [6], theorem 3). It can be proved also that the Waterman and Perlwitz [10] 'line geometry' algorithm for constructing multiple alignments gives, in general, a better alignment than the multiple alignment consistent with a tree if the order of pairwise alignments in the Waterman-Perlwitz algorithm corresponds to the tree.

Next, we establish the following result:

Theorem 8. *If G is 3-star with $2t + 1$ vertices, an optimal 3-star G^* and an alignment A^* yielding the minimum in (9) can be found in time $O(n^3 \cdot k^3 + k^4)$.*

Consider the set of graphs $\mathcal{G}_x = \{G' : G' \text{ is 3-star with center } x\}$. Let G_x^* be a graph satisfying

$$F(A(G_x^*)|G_x^*) = \min_{G' \in \mathcal{G}_x} F(A(G')|G'). \tag{12}$$

Notice that $\mathcal{G} = \bigcup_{x \in V} \mathcal{G}_x$ and therefore

$$F(A(G^*)|G^*) = \min_{x \in V} F(A(G_x^*)|G_x^*). \tag{13}$$

To find a configuration G_x^*, consider the weighted complete graph $H_{k-1}(V \setminus \{x\}, E_{k-1}, w)$ with $k - 1$ vertices. The weight function $w(i, j)$ is defined as the score of triple alignment A for sequences S_i, S_j, S_x minimizing :

$$w(i, j) = D(A|S_i, S_j) + (k - 2)(D(A|S_i, S_x) + D(A|S_j, S_x)) \tag{14}$$

The alignment A can be found in $O(k^3)$ time as a triple weighted SP-alignment according to (4).

Theorem 9. *Let $(i_1, j_1), (i_2, j_2), \ldots, (i_t, j_t)$ be a perfect matching of minimum weight in H_{k-1}. Let $G_x^*(V, E_x^*)$ be the 3-star defined by the edge set:*

$$E_x^* = \{(i_1, j_1), (i_2, j_2), \ldots, (i_t, j_t), (i_1, x), (i_2, x), \ldots, (i_t, x), (j_1, x), (j_2, x), \ldots, (j_t, x)\}$$

Then $G_x^*(V, E_x^*)$ yield minimum in (12):

$$F(A(G_x^*)|G_x^*) = \min_{G' \in \mathcal{G}_x} F(A(G')|G').$$

Proof. Let $G(V, E)$ be a 3-star with center x. Notice that for each edge $(x, i) \epsilon E$, $\Gamma(G|(x, i)) = k - 2$ and for each edge $(i, j) \epsilon E$ with $i, j \neq x$, $\Gamma(G|(i, j)) = 1$. Therefore

$$F(A|G) = \sum_{(i,j)\epsilon E, \ i,j \neq x} D(A|S_i, S_j) + (k - 2)(D(A|S_i, S_x) + D(A|S_j, S_x)) =$$

$$\sum_{(i,j)\epsilon E, \ i,j \neq x} w(i, j)$$

The last equation implies that the value $F(A|G)$ equals the score of perfect matching in H_{k-1} defined by edges $\{(i, j) \epsilon E : \ i, j \neq x\}$. □

Applying an algorithm for the weighted matching problem (Gabow [3], Galil [4]), this theorem implies an $O(n^3 \cdot k^3 + k^4)$ approximation algorithm for multiple alignment with guaranteed bound $c = 2 - \frac{3}{k}$ (for odd k). The approximation algorithm for even k with $c = 2 - \frac{3}{k} + \frac{2}{k(k-1)}$ can be also implemented with $O(n^3 \cdot k^3 + k^4)$ running time.

Although corollary 6 raises the possibility of devising an approximation algorithm based on l-sequence alignments with $c = 2 - \frac{l}{k}$, we do not know of a polynomial algorithm for optimal l-star search for $l > 3$.

Conjecture. For an arbitrary fixed l, there exists a polynomial approximation algorithm for multiple alignment of k sequences with guaranteed upper bound $c \leq 2 - \frac{l}{k}$.

7 Acknowledgements

The research was supported in part by the National Science Foundation (DMS 90-05833) and the National Institute of Health (GM36230). I am grateful to Michael Waterman for helpful discussions.

References

1. Chan S.C., Wong A.K.C., Chiu D.K.Y.: A survey of multiple sequence comparison methods. Bull. Math. Biol. (1992) (in press)
2. Feng D., Doolittle R.: Progressive sequence alignment as a prerequisite to correct phylogenetic trees. Journal of Molec. Evol. **25** (1987) 351-360
3. Gabow H.H.: An efficient implementation of Edmonds' algorithm for maximum matching on graphs. J.ACM. **23** (1976) 221-234
4. Galil Z.: Sequential and parallel algorithms for finding maximal matching in graphs. Ann. Rev. of Comp. Sci. **1** (1986) 197-224
5. Gusfield D.: Efficient method for multiple sequence alignment with guaranteed error bounds. Technical Report, Computer Science Division, University of California, Davis, CSE-91-4(1991) (to appear in Bull. of Math. Biol.)
6. Hu T.C.: Optimum communication spanning trees. SIAM J. Comput. **3** (1974) 188-195
7. Maier D.: The complexity of some problems on subsequences and supersequences. J. ACM **25** (1978) 322-336

8. P.A. Pevzner.: Multiple alignment, communication cost, and graph matching. SIAM J. Appl. Math. (1992) (in press)
9. Timkovsky V.G.: The complexity of subsequence, supersequences and related problems. Kibernetika. **5** (1989) 1-13 (in Russian)
10. Waterman M.S., Perlwitz M.D.: Line geometries for sequence comparison. Bull. of Math. Biol. **46** (1984) 567-577

Two Algorithms for the Longest Common Subsequence of Three (or More) Strings

Robert W. Irving and Campbell B. Fraser*

Computing Science Department,
University of Glasgow,
Glasgow,
Scotland.

Abstract. Various algorithms have been proposed, over the years, for the longest common subsequence problem on 2 strings (2-LCS), many of these improving, at least for some cases, on the classical dynamic programming approach. However, relatively little attention has been paid in the literature to the k-LCS problem for $k > 2$, a problem that has interesting applications in areas such as the multiple alignment of sequences in molecular biology.

In this paper, we describe and analyse two algorithms with particular reference to the 3-LCS problem, though each algorithm can be extended to solve the k-LCS problem for general k. The first algorithm, which can be viewed as a "lazy" version of dynamic programming, has time and space complexity that is $O(n(n-l)^2)$ for 3 strings, and $O(kn(n-l)^{k-1})$ for k strings, where n is the common length of the strings and l is the length of an LCS. The second algorithm, which involves evaluating entries in a "threshold" table in diagonal order, has time and space complexity that is $O(l(n-l)^2 + sn)$ for 3 strings, and $O(kl(n-l)^{k-1} + ksn)$ for k strings, where s is the alphabet size. For simplicity, the algorithms are presented for equal-length strings, though extension to unequal-length strings is straightforward.

Empirical evidence is presented to show that both algorithms show significant improvement on the basic dynamic programming approach, and on an earlier algorithm proposed by Hsu and Du, particularly, as would be expected, in the case where l is relatively large, with the balance of evidence being heavily in favour of the threshold approach.

Key words: string algorithms, longest common subsequence.

1 Introduction

The longest common subsequence problem for 2 strings (2-LCS), and the related, more general, minimum edit-distance problem, have been extensively studied, and various algorithms for these problems have been proposed — see, for example, [14], [4], [5], [7], [9], [11], [13], [10], [1], [3], [15]. However relatively little attention has been paid to the general k-LCS problem for k (> 2) strings, although there are

---

* Supported by a postgraduate research studentship from the Science and Engineering Research Council

some important areas of application, particularly in multiple sequence alignment in molecular biology.

The well-known basic dynamic programming scheme for the 2-LCS problem ([4], [12], [14]) was explicitly extended to the k strings case by Itoga [8]. The time and space requirements of the resulting algorithm are $O(k|S_1|.|S_2|\ldots|S_k|)$, where $|S_i|$ is the length of the ith string, or $O(kn^k)$ if all strings have length n. (In what follows, we generally suppose, for simplicity, that all strings have length n, though all of the results can be extended in obvious ways to the case of unequal-length strings.)

In the case of the 2-LCS problem, many variations of and alternatives to the basic dynamic programming algorithm have been proposed — see the citations above. Masek and Paterson's "Four Russians" approach [9] achieves a worst-case complexity that is $O\left(n^2\frac{\log\log n}{\log n}\right)$, but is of theoretical rather than practical interest. None of the other proposed algorithms gives an improvement on the $O(n^2)$ performance of dynamic programming in the *worst* case, but generally their complexities can be expressed in terms of other parameters, such as the length of an LCS, and this leads to significant improvements in practice in many circumstances.

Hsu and Du [6] addressed explicitly the k-LCS problem, and proposed a general algorithm that involves systematic enumeration in a so-called *CS-tree* — a tree containing an explicit representation of all the common subsequences of the set of strings. By suitably pruning the tree to avoid repeating earlier computations, Hsu and Du's algorithm requires $O(ksn+ksr)$ time and $O(ksn+r)$ space, where s is the alphabet size and r is the number of tuples $(i_1, i_2, \ldots, i_k)$ such that $S_1[i_1] = S_2[i_2] = \cdots = S_k[i_k]$. This algorithm will clearly be at its most effective when r is relatively small. However, when the alphabet size is small, as will be the case in at least some areas of application, we can expect r to be relatively large, with a detrimental effect on the efficiency of the algorithm. Hsu and Du presented no empirical evidence in support of their algorithm.

In this note, we describe and analyse, with particular reference to the 3-LCS problem, two new algorithms, which can be seen as building on the approaches to the 2-LCS problem of Ukkonen [13], Myers [10] and Wu et al [15] in the one case, and of Hunt and Szymanski [7] and Nakatsu et al [11] in the other case, in the context of the 2-LCS problem. Each of these algorithms can be extended in fairly obvious ways to deal with k strings for any fixed $k > 3$. The first algorithm, which we refer to as the "lazy" algorithm, uses $O(n(n-l)^2)$ time and space in the case $k = 3$, and $O(kn(n-l)^{k-1})$ time and space for general k. The second algorithm, which we refer to as the "diagonal threshold" approach, uses $O(l(n-l)^2 + sn)$ time and space in the case $k = 3$, and $O(kl(n-l)^{k-1} + ksn)$ time and space for general k. Our empirical evidence for the case $k = 3$ suggests that, as would be expected, the new algorithms improve substantially on basic dynamic programming and on the Hsu and Du algorithm when l is relatively large, and the dynamic threshold method is also very effective when l is very small. However, this empirical work also indicates strongly that, in most circumstances when $k \geq 3$, the major constraint determining which LCS instances can and which cannot be solved in practice, at least using current methods, is likely to be space rather than time.

The remainder of the paper is structured as follows. Section 2 contains a description and analysis of the lazy algorithm, section 3 a description and analysis of the diagonal threshold algorithm, and section 4 reports the empirical findings and conclusions.

2 The "Lazy" Approach to Dynamic Programming

Now consider the problem of finding an LCS of 3 strings A, B and C, each of length n. We shall assume equal length strings throughout, for simplicity, but all of our results can be extended in a straightforward way to cases where the strings are of different lengths. Denote by $L(A, B, C)$ the length of an LCS of A, B and C, and by $L(i, j, k)$ the length of an LCS of the ith prefix $A^i = A[1 \ldots i]$ of A, the jth prefix $B^j = B[1 \ldots j]$ of B and the kth prefix $C^k = C[1 \ldots k]$ of C. As in the classical case of two strings [14], a basic dynamic programming scheme, using $\Theta(n^3)$ time and space in this case, can be set up, based on the recurrence

$$L(i, j, k) = \begin{cases} L(i-1, j-1, k-1) + 1 & \text{if } A[i] = B[j] = C[k] \\ \max(L(i-1, j, k), L(i, j-1, k), L(i, j, k-1)) & \text{otherwise} \end{cases} \quad (1)$$

and the initial conditions

$$L(i, j, 0) = L(i, 0, k) = L(0, j, k) = 0 \quad \text{for all } i, j, k.$$

We now investigate how this dynamic programming approach can be speeded up by suppressing unnecessary evaluations. Our approach is similar to that employed in [10], [15] and [13] for the case of two strings, and we refer to our algorithm as the "lazy" approach to dynamic programming.

We first look at the problem from a slightly different point of view. Define the *difference factor* $D = D(A, B, C)$ of the three strings to be the smallest total number of elements that need be deleted from the strings in order to leave three identical strings. Clearly D is a multiple of 3 if the strings are of equal lengths, since the same number of elements must be deleted from each string.

Lemma 1. $D(A, B, C) = 3(n - L(A, B, C))$.

Proof. The identical strings that remain when D elements are deleted from A, B, C form an LCS of A, B and C. Hence the deleted symbols, together with the 3 copies of the LCS constitute the $3n$ symbols of the original strings.

It follows that evaluation of $D(A, B, C)$ gives the value of $L(A, B, C)$, and that identification of a minimum set of deletions reveals an LCS. This is the approach that we shall now follow.

Let $D(i, j, k)$ be the difference factor of the prefixes A^i of A, B^j of B and C^k of C. Then the recurrence corresponding to 1 above is

$$D(i, j, k) = \begin{cases} D(i-1, j-1, k-1) & \text{if } A[i] = B[j] = C[k] \\ 1 + \min(D(i-1, j, k), D(i, j-1, k), D(i, j, k-1)) & \text{otherwise} \end{cases} \quad (2)$$

subject to

$$D(i, j, 0) = i + j, \quad D(i, 0, k) = i + k, \quad D(0, j, k) = j + k \quad \text{for all } i, j, k.$$

Clearly, in the light of this equation, cells (i, j, k) such that $A[i] = B[j] = C[k]$ have a special significance — we refer to such a cell as *a match position*.

We now consider the task of evaluating $D(n, n, n)$ while calculating as few as possible of the other D values. We refer to the 3-dimensional table $D(i, j, k)$ $(1 \leq i, j, k \leq n)$ as the *D-table* and we say that the cell (i, j, k) has *D-value* $D(i, j, k)$. Note that we will include coordinate values of -1 and 0 when implementing the table to facilitate appropriate initialisation in our algorithm.

By a *diagonal* of the D-table we mean an ordered sequence of cells with coordinates $(i + l, j + l, k + l)$ where $\min(i, j, k) = -1$ and $l = 0, \ldots, n - \max(i, j, k)$. The diagonal containing cell (i, j, k) will be denoted by $\langle i, j, k \rangle$. We say that a given cell (i, j, k) occupies *position* $i + j + km$ on its diagonal. Note that the positions on diagonal $\langle i, j, k \rangle$ increase by 3 from cell to cell, and are all $\equiv i + j + k \pmod{3}$.

We also define

$$\ll i, j, k \gg = \{\langle i, j, k \rangle, \langle i, k, j \rangle, \langle j, i, k \rangle, \langle j, k, i \rangle, \langle k, i, j \rangle, \langle k, j, i \rangle\} \ ,$$

noting that this set of diagonals is of size 6, 3 or 1 depending whether the number of distinct values in the set $\{i, j, k\}$ is 3, 2 or 1. Any such set of diagonals has a *canonical* representation as $\ll i, j, 0 \gg$ with $i \geq j \geq 0$.

The following lemma is immediate:

Lemma 2. *If* $\langle x, y, z \rangle \in \ll i, j, 0 \gg$, *with* $x, y, z \geq 0$ *and* $i \geq j \geq 0$, *then* $D(x, y, z) \geq i + j$.

The *neighbours* of diagonal $\langle i, j, k \rangle$ are the 3 diagonals $\langle i - 1, j, k \rangle$, $\langle i, j - 1, k \rangle$ and $\langle i, j, k - 1 \rangle$ and the *neighbourhood* of the set $\ll i, j, k \gg$ consists of all the diagonals that neighbour at least one diagonal in that set. (Note that the neighbour relation is not a symmetric one — indeed it is antisymmetric, in that if diagonal $\langle i', j', k' \rangle$ is a neighbour of diagonal $\langle i, j, k \rangle$ then $\langle i, j, k \rangle$ is not a neighbour of $\langle i', j', k' \rangle$. However, the following lemma may easily be verified.

Lemma 3. *If diagonal* Δ' *is a neighbour of diagonal* Δ, *then there is a diagonal* Δ'' *such that* Δ'' *is a neighbour of* Δ' *and* Δ *is a neighbour of* Δ''.

The *neighbours* of a cell (i, j, k), likewise, are defined to be the cells $(i - 1, j, k)$, $(i, j - 1, k)$ and $(i, j, k - 1)$. Obviously, the cell in position p in a given diagonal has, as its neighbours, the cells in position $p - 1$ in each of the neighbouring diagonals.

For the cell in position p in a given diagonal Δ, let us denote the D-value of that cell by $D(p, \Delta)$. The following technical lemmas may be established easily from the definitions of difference factor, neighbour and position.

Lemma 4. *Let diagonal* Δ' *be a neighbour of diagonal* Δ. *Then*

$$D(p, \Delta) \leq 1 + D(p - 1, \Delta') \ .$$

Lemma 5. *The values in any diagonal of the D-table form a non-decreasing sequence, with the difference between successive elements in that sequence being 0 or 3.*

```
max := 0; {length of LCS found so far}
d := -1 ; { d is the current diagonal }
repeat
        d := d + 1 ;
        i := d ; { i is the row number in the τ table }
        τ_{i,0} := {(0,0)} ;
        m := 0 ; { m is the column number in the τ table }
        repeat
                i := i + 1 ;
                m := m + 1 ;
                if m > max then
                        τ_{i-1,m} := ∅ ; { further initialisation }
                σ := τ_{i-1,m} ∪ ⋃_{(j,k)∈τ_{i-1,m-1}} (b_{i,j}, c_{i,k}) ;
                τ_{i,m} := σ̂ ;
        until (τ_{i,m} = ∅) or (i = n) ;
        if m - 1 > max then
                max := m - 1
until i = n ;
if τ_{n,m} = ∅ then
        l := m - 1
else
        l := m
```

Fig. 5. The threshold algorithm for the length of the LCS of 3 strings

The crux of the main part of the algorithm is the evaluation of $\tau_{i,m}$ from $\tau_{i-1,m-1}$ and $\tau_{i-1,m}$. This can be achieved in $O(|\tau_{i-1,m}| + |\tau_{i-1,m-1}|)$ time by maintaining each set as a linked list with (j, k) preceding (j', k') in the list if and only if $j < j'$. For then the lists may easily be merged in that time bound to form a list representing $\sigma_{i,m}$, and as observed earlier, the list representing $\tau_{i,m} = \hat{\sigma}_{i,m}$ may be generated by a further single traversal of that list, deleting zero or more elements in the process.

It follows that the total number of operations involved in the forward pass of the algorithm is bounded by a constant times the number of pairs summed over all the sets $\tau_{i,m}$ evaluated. The backward pass involves tracing a single path through the τ table, examining a subset of the entries generated during the forward pass, and hence the overall complexity is dominated by the forward pass.

A trivial bound on the number of pairs in $\tau_{i,m}$ is $n - m + 1$, since the first component of each pair must be unique and in the range $m, \ldots, n$. Hence, since the number of sets $\tau_{i,m}$ evaluated is bounded by $l(n - l + 1)$ — at most l sets in each of $n - l + 1$ diagonals — this leads to a trivial worst-case complexity bound of $O(l(n - l)(n - \frac{l}{2}))$. In fact, it seems quite unlikely that all, or even many, of the sets $\tau_{i,m}$ can be large simultaneously, and we might hope for a worst-case bound of, say, $O(l(n - l)^2)$. But it seems to be quite difficult to establish such a bound for the algorithm as it stands. To realise this bound, we adapt the algorithm to a rather

2.1 Description of the "Lazy" Algorithm

The objective of the pth iteration of our so-called "lazy" algorithm is the determination of the last (i.e., highest numbered) position in the main diagonal (diagonal $\langle 0, 0, 0 \rangle$) of the D-table occupied by the value $3p$. But this will be preceded by the determination of the last position in each neighbouring diagonal occupied by $3p-1$, which in turn will be preceded by the determination of the last position in each neighbouring diagonal of these occupied by $3p-2$, and so on. As soon as we discover that the last position in the main diagonal occupied by $3p$ is position $3n$ (corresponding to cell (n, n, n)), we have established that $D(A, B, C) = 3p$, and therefore, by Lemma 1, that $L(A, B, C) = n - D(A, B, C)/3 = n - p$.

In general, when we come to find the last position in diagonal $\langle i, j, k \rangle$ occupied by the value r, we will already know the last position in each neighbouring diagonal occupied by $r-1$. If we denote by $t(\Delta, r)$ the last position in diagonal Δ occupied by the value r, then we can state the fundamental lemma that underpins the lazy algorithm.

Lemma 6. *If $x = \max t(\Delta', r-1)$, where the maximum is taken over the neighbours Δ' of diagonal Δ, then*

$$t(\Delta, r) = x + 1 + 3s \ ,$$

where s is the largest integer such that positions $x+4, x+7, \ldots, x+3s+1$ on diagonal Δ are match positions. (These positions constitute a "snake" in the terminology of Myers [10])

Proof. It follows from Lemma 4 that $D(x+1, \Delta) \leq r$, and clearly if its value is $< r$ then it is $\leq r - 3$. But in this latter case, by Lemmas 3 and 4, $D(x+3, \Delta') \leq r-1$ for Δ' a neighbour of Δ, contradicting the maximality of x. Hence $D(x+1, \Delta) = r$. Further, if $D(y, \Delta) = r$ for any $y > x + 1$ then, since no neighbouring diagonal has a value less than r beyond position x, according to the basic dynamic programming formula (2) for D, it must follow that y is a match position in that diagonal.

We define the *level* of a diagonal Δ to be the smallest m such that there is a sequence $\Delta_0 = \langle 0, 0, 0 \rangle, \Delta_1, \ldots, \Delta_m = \Delta$ with Δ_s a neighbour of Δ_{s-1} for each s, $s = 1, \ldots, m$. It may easily be verified that the level of $\Delta = \langle i, j, k \rangle$ is $f_{ijk} = 3 \max(i, j, k) - (i + j + k)$, and therefore that the level of diagonal $\Delta = \langle i, j, 0 \rangle$ $(i \geq j)$ is $2i - j$. This can be expressed differently, as in the following lemma.

Lemma 7. *The diagonals at level x are those in the sets $\ll x - y, x - 2y, 0 \gg$ for $y = 0, 1, \ldots, \lfloor \frac{x}{2} \rfloor$.*

The next lemma is an easy consequence of Lemmas 2 and 7.

Lemma 8. *The smallest D-value in any cell of a diagonal at level x is $\frac{x}{2}$ if x is even, and $\frac{x+3}{2}$ if x is odd.*

During the pth iteration of our algorithm, we will determine, in decreasing order of x, and for each relevant diagonal Δ at level x, the value of $t(\Delta, 3p - x)$. Since, by

Lemma 8, no diagonal at level $> 2p$ can contain a D-value $\leq p$, we need consider, during this pth iteration, only diagonals at levels $\leq 2p$. Furthermore, by Lemma 2, the only diagonals at level x that can include a D-value as small as $3p - x$ are those diagonals in the family $\ll x - y, x - 2y, 0 \gg$ with $(x - y) + (x - 2y) \leq 3p - x$, i.e., with $y \geq x - p$.

Hence, to summarise, the pth iteration involves the evaluation, for $x = 2p, 2p - 1, \ldots, 0$, of $t(\Delta, 3p - x)$ for $\Delta \in \ll x - y, x - 2y, 0 \gg$ with $\max(0, x - p) \leq y \leq \lfloor \frac{x}{2} \rfloor$.

Lemma 9. *If the t values are calculated in the order specified above, then, for arbitrary Δ, r, the values of $t(\Delta', r - 1)$, for each neighbour Δ' of Δ, will be available when we come to evaluate $t(\Delta, r)$.*

Proof. Consider, without loss of generality, $\Delta = \langle i, j, k \rangle$ with $i \geq j \geq k$. If the t values are calculated according to the scheme described, then $t(\langle i, j, k, \rangle, r)$ is evaluated during iteration $p = \frac{r + 2i - j - k}{3}$. Furthermore, $t(\langle i, j - 1, k \rangle, r)$ and $t(\langle i, j, k - 1 \rangle, r)$ are also evaluated during iteration p, and $t(\langle i - 1, j, k \rangle, r)$ is evaluated during iteration $p - 1$ or p according as $i > j$ or $i = j$. Since, during iteration p, any $t(., r - 1)$ is evaluated before any $t(., r)$, it follows that $t(\Delta', r - 1)$ is evaluated before $t(\langle i, j, k \rangle, r)$ for every neighbour Δ' of $\langle i, j, k \rangle$.

We can now express the algorithm as in Fig. 1.

```
p := -1 ;
repeat
    p := p + 1 ;
    Initialise t values for pth iteration ;
    for x := 2p downto 0 do
        for y := max(0, x − p) to ⌊ x/2 ⌋ do
            for Δ ∈ ≪ x − y, x − 2y, 0 ≫ do
                evaluate t(Δ, 3p − x)
    until t(⟨0, 0, 0⟩, 3p) = n
```

Fig. 1. The lazy algorithm for the length of the LCS of 3 strings

The evaluation of $t(\Delta, 3p - x)$ is carried out using Lemma 6, following a snake, as in Fig. 2 for diagonal $\langle i, j, k \rangle$.

The relevant initialisation for the pth iteration involves those diagonals not considered in the $(p - 1)$th iteration, and these turn out to be the diagonals in the set $\ll p, r, 0 \gg$ for $p \geq r \geq 0$. We initialise t for such a diagonal so that the (imaginary) last occurrence of the value $p + r - 3$ is in position $p + r - 3$ — see Fig. 3.

Recovering an LCS Recovering an LCS involves a (conceptual) trace-back through the n-cube from cell (n, n, n) to cell $(0, 0, 0)$ under the control of the t values. At

$u := 1 + \max t(\Delta', 3p - x - 1)$ over neighbours Δ' of $\langle i, j, k \rangle$;
$v := (u - i - j - k)$ div 3 ;
while $(A[i + v + 1] = B[j + v + 1] = C[k + v + 1])$ do {Match position}
$\quad v := v + 1$; {Assuming sentinels}
$t(\langle i, j, k \rangle, 3p - x) := i + j + k + 3v$

Fig. 2. Evaluation of $t(\Delta, 3p - x)$ for $\Delta = \langle i, j, k \rangle$

for $r := 0$ to p do
$\quad$ for $\langle i, j, k \rangle \in \ll p, r, 0 \gg$ do
$\qquad t(\langle i, j, k \rangle, p + r - 3) := p + r - 3$

Fig. 3. Initialisation for the pth iteration of the lazy algorithm

each step, we are in a cell whose D-value became known during the execution of the algorithm. We step back one position in the current diagonal if that cell is in a snake, and otherwise step into a neighbouring cell whose D-value (which can be discovered from the t array) is exactly one smaller. In the former case, we will have found one more character in the LCS, and as this backward trace unfolds, the LCS will be revealed in reverse order.

The precise algorithm is shown in Fig. 4.

Time Complexity of the Algorithm We now consider the time and space requirements of the algorithm. As far as time is concerned, it is not hard to see that the worst-case complexity is dependent on the total number of elements of the t table that are evaluated. In iteration p, the diagonals for which a t value is calculated for the first time are precisely the diagonals in the sets $\ll p, r, 0 \gg$ for $r = 0, \ldots, p$. There are $6p$ such diagonals in total, since $| \ll p, r, 0 \gg | = 3$ for $r = 0$ or $r = p$, and $| \ll p, r, 0 \gg | = 6$ for $1 \leq r \leq p - 1$. The number of iterations is the value of p for which $D(A, B, C) = 3p$, and by Lemma 1, this is exactly equal to $n - l$, where $l = L(A, B, C)$.

So, the total number of diagonals involved during the execution of the algorithm is

$$\sum_{p=0}^{n-l} 6p = 3(n - l)(n - l + 1) .$$

Furthermore, the $6p$ diagonals that are started at the pth iteration each contains at most $n - p$ entries in the n-cube, so the total number of t elements evaluated cannot

```
{ Throughout, Δ represents diagonal ⟨i, j, k⟩ }
i := n ; j := n ; k := n ;
d := 3p ;
m := n − p ;
repeat
    if A[i] = B[j] = C[k] then
        begin
        LCS[m] := A[i] ; m := m − 1 ;
        i := i − 1 ; j := j − 1 ; k := k − 1 ;
        end
    else
        begin
        find neighbour Δ' of Δ such that t(Δ', d − 1) = i + j + k − 1 ;
        d := d − 1 ;
        decrement i, j or k according as Δ' differs from Δ
            in its 1st, 2nd or 3rd coordinate
        end
    until i = 0
```

Fig. 4. Recovering an LCS from the t table

exceed

$$\sum_{p=0}^{n-l} 6p(n - p) = (n - l)(n - l + 1)(n + 2l - 1). \tag{3}$$

As a consequence, the worst-case complexity of the algorithm is $O(n(n-l)^2)$, showing that we can expect it to be much faster than the naive dynamic programming algorithm in cases where the LCS has length close to n.

Space Complexity Diagonal $\langle i, j, k \rangle$ may be uniquely identified by the ordered pair $(i - j, j - k)$ (say), and such a representation leads in an obvious way to an array based implementation of the algorithm using $O(n^3)$ space. But appropriate use of dynamic linked structures enables just one node to be created and used for each t value calculated, and so by Equation 3, this yields an implementation that uses both $O(n(n-l)^2)$ time and space in the worst case. A little more care is required, in that case, in recovering the LCS from the linked lists of t values.

As with standard algorithms for the LCS of two strings, the space requirement can be reduced to $O((n-l)^2)$ if only the length of the LCS is required — in that case, only the most recent t value in each diagonal need be retained.

Extension to ≥ 4 Strings The lazy algorithm described in the previous section may be extended in a natural way to find the LCS of a set of k strings for any fixed $k \geq 3$. Such an extended version can be implemented to use $O(kn(n-l)^{k-1})$ time and space in the worst case, the factor of k arising from the need to find the smallest of k values in the innermost loop.

3   A Threshold Based Algorithm

Hunt and Szymanski [7] and Nakatsu et al [11] introduced algorithms for the LCS of two strings based on the so-called "threshold" approach. In the case of 3 strings A, B and C of length n, we define the *threshold set* $\tau_{i,m}$ to be the set of ordered pairs (j, k) such that A^i, B^j and C^k have a common subsequence of length m, but neither A^i, B^j and C^{k-1} nor A^i, B^{j-1} and C^k have such a common subsequence.

For example, for the sequences

$$A = abacbcabbcac \quad B = bbcabcbaabcb \quad C = cabcacbbcaba$$

we have $\tau_{5,2} = \{(2,7),(3,4),(5,3)\}$, corresponding to the subsequences bb, bc, and ab or cb respectively.

If (x, y), (x', y') are distinct ordered pairs of non-negative integers, we say that (x, y) *dominates* (x', y') if $x \le x'$ and $y \le y'$. It is immediate from the definition of the set $\tau_{i,m}$ that it contains no two pairs one of which dominates the other — we say that it is *domination-free*.

It is also clear from the definition of $\tau_{i,m}$ that the length of a longest common subsequence of A, B and C is the largest value of m for which $\tau_{n,m}$ is non-empty. Hence, evaluation of the threshold sets $\tau_{i,m}$ in some appropriate order will enable us to determine the length of the LCS, and by suitable back tracing, to find an actual LCS of the 3 strings.

We shall now describe an algorithm for the LCS of 3 strings based on the evaluation of the threshold sets $\tau_{i,m}$ in diagonal order — i.e., in general we evaluate $\tau_{i,m}$ immediately after $\tau_{i-1,m-1}$. To explain the algorithm we need some additional terminology and notation.

Let S be a set of ordered pairs of non-negative integers. The *dominating reduction* of S, denoted $\hat{S}$ is the minimal subset of S such that

$$(x', y') \in S \Rightarrow \exists (x, y) \in \hat{S} \text{ such that } (x, y) \text{ dominates } (x', y').$$

Clearly $\hat{S}$ is domination-free for any set S, and can be found from S by successively removing from S any pair that is dominated by another pair in S. If the pairs in S are arranged in a list so that (x, y) precedes (x', y') in the list whenever $x < x'$, then a similarly ordered list representing $\hat{S}$ can be obtained by a single scan of the original list, comparing successive neighbours on the list and deleting any pair found to be dominated by its neighbour. So deriving $\hat{S}$ from S can be done in time linear in the size of S.

For any character α in the string alphabet, and any position i ($0 \le i \le n$), we define the *next-occurrence* table N_B for string B by

$$N_B[\alpha, i] = \begin{cases} \min j : j > i \text{ and } B[j] = \alpha \text{ if such a } j \text{ exists} \\ \infty \qquad\qquad\qquad\qquad\qquad \text{otherwise} \end{cases}$$

So, as a special case, $N_B[\alpha, 0]$ is the position of the first occurrence in string B of character α. The next-occurrence table N_C for string C is defined analogously.

For any positions i and j in strings A and B respectively, we define $b_{i,j}$ to be the first position after j in B that is occupied by character $A[i]$, i.e.,

$$b_{i,j} = N_B[A[i], j] .$$

Similarly, we define

$$c_{i,j} = N_C[A[i], j] .$$

The basis of our algorithm is the following Lemma.

Lemma 10. *If* $\sigma_{i,m} = \tau_{i-1,m} \cup \bigcup_{(j,k) \in \tau_{i-1,m-1}} (b_{i,j}, c_{i,k})$ *then* $\tau_{i,m} = \hat{\sigma}_{i,m}$.

Proof. Let $(j, k) \in \tau_{i,m}$, so that A^i, B^j and C^k have a common subsequence of length m but neither A^i, B^{j-1}, C^k nor A^i, B^j, C^{k-1} have such a common subsequence. If A^{i-1}, B^j, C^k have a common subsequence of length m, then $(j, k) \in \tau_{i,m-1}$. Otherwise any common subsequence of A^i, B^j, C^k of length m contains $A[i] = B[j] = C[k]$ as its last element. So there is some $(j', k') \in \tau_{i-1,m-1}$ such that $b_{i,j'} = j$, $c_{i,k'} = k$. As a consequence, any element of $\tau_{i,m}$ is in the set $\sigma_{i,m}$ defined in the lemma, and since $\tau_{i,m}$ is domination-free, it is included in $\hat{\sigma}_{i,m}$.

On the other hand, any member of $\tau_{i-1,m}$ and any pair $(b_{i,j'}, c_{i,k'})$ for $(j', k') \in \tau_{i-1,m}$ is bound to be dominated by a member of $\tau_{i,m}$, so that there are no elements in $\hat{\sigma}_{i,m}$ that are not in $\tau_{i,m}$.

Our algorithm begins the evaluation of the sets $\tau_{i,m}$ on the main diagonal — i.e., it uses the above lemma to evaluate $\tau_{1,1}$, $\tau_{2,2}$, ... until $\tau_{i,i} = \emptyset$, this last condition indicating that the ith prefix of A is not a common subsequence of A, B and C. Next comes the evaluation of $\tau_{2,1}$, $\tau_{3,2}$, ..., again continuing until the first empty set is reached.

The process continues until $\tau_{n,m}$ is evaluated for some m. At that point the algorithm terminates, for no $\tau_{n,m'}$ can be non-empty for any $m' > m$. If $\tau_{n,m} \neq \emptyset$ then m is the length of the LCS of A, B and C, otherwise its length is $m - 1$, since $\tau_{n-1,m-1}$ must have been non-empty. To enable the recurrence scheme to work correctly, at the beginning of the ith iteration $\tau_{i,0}$ is initialised to $(0, 0)$, and whenever an unevaluated set $\tau_{i,m}$ is required (in the diagonal above the current one) it is initialised to the empty set.

The algorithm to find the length l of the LCS by this method is summarised in Fig. 5.

Recovering an LCS If the algorithm terminates with $\tau_{n,m}$ non-empty then we start the trace-back at an arbitrary pair (j, k) in $\tau_{n,m}$, otherwise at an arbitrary pair (j, k) in $\tau_{n-1,m-1}$. In either case, $B[j] = C[k]$ is the last character of our LCS. At each stage, when we have reached $\tau_{i,m}$ say, we repeatedly decrement i until the current pair (j, k) is not a member of $\tau_{i-1,m}$. This indicates that, when constructing the τ table, we had $A[i] = B[j] = C[k]$, and (j, k) must have arisen from an element (j', k') in $\tau_{i-1,m-1}$ that dominates it — so we locate such a pair, and record $A[i]$ as an element in the LCS. We then decrement i and m, and the pair (j', k') becomes the new (j, k). This part of the algorithm is summarised in Fig. 6.

Time Complexity of the Algorithm As described, the algorithm requires the pre-computation of the N_B and N_C tables, each requiring $\Theta(sn)$ steps, where s is the alphabet size.

```
if τ_{n,m} = ∅ then
    begin
        choose (j, k) ∈ τ_{n-1,m-1} ;
        i := n - 1 ;   r := m - 1
    end
else
        begin
            choose (j, k) ∈ τ_{n,m} ;
            i := n ;   r := m
        end;
    while m > 0 do
        begin
            while (j, k) ∈ τ_{i-1,r} do
                i := i - 1 ;
            {A[i] = B[j] = C[k] in LCS}
            i := i - 1 ;   r := r - 1 ;
            choose (j', k') ∈ τ_{i,r} dominating (j, k) ;
            (j, k) := (j', k') ;
        end
```

Fig. 6. Recovering an LCS from the threshold table

different form, with the aid of a trick also used by Apostolico et al [2] in a slightly different context.

The crucial observation is that, if a pair (j, k) in set $\tau_{i,m}$ is to be part of an LCS of length l, then we must have $0 \le j - m \le n - l$ and $0 \le k - m \le n - l$, since there cannot be more than $n - l$ elements in either string B or string C that are not part of the LCS. So, if we knew the value of l in advance, we could immediately discard from each set $\tau_{i,m}$ any pair (j, k) for which $j > n - l + m$ or $k > n - l + m$. Hence each set would have effective size $\le n - l + 1$, and since the number of sets is $O(l(n - l))$, this would lead to an algorithm with $O(l(n - l)^2)$ worst-case complexity.

Of course, the problem with this scheme is that we do not know the value of l in advance — it is precisely this value that our algorithm is seeking to determine. However, suppose we parameterise the forward pass of our algorithm with a bound p, meaning that we are going to test the hypothesis that the length of the LCS is $\ge n - p$. In that case, we need evaluate only $p + 1$ diagonals (at most), each of length $\le l$, and in each set $\tau_{i,m}$ we need retain at most $p + 1$ pairs. So this algorithm requires $O(lp^2)$ time in the worst case.

Suppose that we now apply the parameterised algorithm successively with $p = 0, 1, 2, 4, \ldots$ until it returns a "yes" answer. At that point we will know the length of the LCS, and we will have enough of the threshold table to reconstruct a particular LCS. As far as time complexity is concerned, suppose that $2^{t-1} < n - l \le 2^t$. Then the algorithm will be invoked $t + 1$ times, with final parameter 2^t. So the overall worst-case complexity of the resulting LCS algorithm is $O(l \sum_{i=0}^{t} (2^i)^2) = O(l2^{2t+2}) = O(l(n - l)^2)$, as claimed.

As it turns out, the above worst-case analysis of the threshold algorithm does not depend on the fact that the sets $\tau_{i,m}$ are domination-free, merely on the fact that if $(j,k),(j',k') \in \tau_{i,m}$ then $j \neq j'$ and $k \neq k'$ (though the maintenance of the sets as domination-free, by the method described earlier, will undoubtedly speed up the algorithm in practice.)

As an aside, we observe that, when restricted to the case of two strings, this threshold approach differs from, say, the approach of Nakatsu et al [11], in the order of evaluation of the elements of the threshold table. The resulting algorithm has $(O(l(n-l) + sn)$ complexity, a bound that does not appear to have been achieved previously.

Space Complexity As described above, the space complexity of the algorithm is also $O(l(n-l)^2 + sn)$, the first term arising from the bound on the number of elements summed over all the $\tau_{i,m}$ sets evaluated, and the second term from the requirements of the next-occurrence tables. The first term can be reduced somewhat by changing the implementation so that, for each pair (j,k) that belongs to some $\tau_{i,m}$, a node is created with a pointer to its "predecessor". In each column of the τ table, a linked list is maintained of the pairs in the most recent position in that column. By this means, each pair generated uses only a single unit of space, so the overall space complexity is big oh of the number of (different) pairs generated in each column, summed over the columns. An LCS can be re-constructed by following the predecessor pointers.

A Time-Space Tradeoff As described by Apostolico and Guerra [3], the next-occurrence tables can be reduced to size $1 \times n$ rather than of size $s \times n$, for an alphabet of size s, at the cost of an extra $\log s$ factor in the time complexity — each lookup in the one-dimensional next-occurrence table takes $\log s$ time rather than constant time. In fact, for most problem instances over small or medium sized alphabets, the size of the next-occurrence tables is likely to be less significant than the space needed by the threshold table.

Extension to ≥ 4 Strings The idea of the threshold algorithm can be extended to find an LCS of k strings for any fixed $k \geq 4$, using sets of $(k-1)$-tuples rather than sets of pairs. In the general case, it is less clear how the sets $\tau_{i,m}$ can be efficiently maintained as domination-free, but as observed above for the case of three strings, the worst-case complexity argument does not require this property. In general, the extended version of the algorithm has worst-case time and space complexity that is $O(kl(n-l)^{k-1} + ksn)$, with the same time-space trade-off option available as before.

4 Empirical Evidence and Conclusions

To obtain empirical evidence as to the relative merits of the various algorithms, we implemented, initially for 3 strings, the basic dynamic programming algorithm (DP), the lazy algorithm (Lazy), the diagonal threshold algorithm (Thresh), and the algorithm of Hsu and Du (HD) [6]. In the case of the lazy algorithm, we used a

Table 1. Cpu times when LCS is 90% of string length

		String lengths						
		100	200	400	800	1600	3200	6400
Algorithms	DP	2.5	19.2	-	-	-	-	-
	Lazy	0.05	0.18	1.5	11.9	-	-	-
	Thresh $s=4$	0.03	0.1	0.4	2.3	21.7	136.2	-
	Thresh $s=8$	0.03	0.1	0.3	1.7	8.8	54.8	-
	Thresh $s=16$	0.02	0.1	0.3	1.3	7.1	33.6	212.5
	HD $s=4$	0.9	10.5	-	-	-	-	-
	HD $s=8$	0.4	5.0	73.4	-	-	-	-
	HD $s=16$	0.2	2.8	36.7	-	-	-	-

Table 2. Cpu times when LCS is 50% of string length

		String lengths					
		100	200	400	800	1600	3200
Algorithms	DP	2.3	18.9	-	-	-	-
	Lazy	3.0	23.1	-	-	-	-
	Thresh $s=4$	0.2	1.5	11.7	107.8	-	-
	Thresh $s=8$	0.1	0.7	5.5	48.3	-	-
	Thresh $s=16$	0.05	0.4	2.5	18.8	184.0	-
	HD $s=4$	0.8	9.7	-	-	-	-
	HD $s=8$	0.3	4.5	59.6	-	-	-
	HD $s=16$	0.2	2.3	31.8	-	-	-

dynamic linked structure, as discussed earlier, to reduce the space requirement. We implemented both the straightforward threshold algorithm of Figs. 5 and 6, and the version with the guaranteed $O(l(n-l)^2 + sn)$ complexity, and found that the former was consistently between 1.5 and 2 times faster in practice — so we have included the figures for the faster version. The Hsu-Du algorithm was implemented using both a 3-dimensional array and a dynamic structure to store match nodes. There was little difference between the two in terms of time, but the latter version allowed larger problem instances to be solved, so we quote the results for that version.

Table 3. Cpu times when LCS is 10% of string length

		String lengths					
		100	200	400	800	1600	3200
Algorithms	DP	2.4	18.0	-	-	-	-
	Lazy	17.0	-	-	-	-	-
	Thresh $s=4$	0.02	0.4	2.6	17.1	142.2	-
	Thresh $s=8$	0.02	0.3	2.3	15.3	128.5	-
	Thresh $s=16$	0.02	0.2	1.5	10.6	117.2	-
	HD $s=4$	0.06	0.5	-	-	-	-
	HD $s=8$	0.06	0.4	9.7	-	-	-
	HD $s=16$	0.06	0.3	5.0	-	-	-

The algorithms were coded in Pascal, compiled and run under the optimised Sun Pascal compiler on a Sun 4/25 with 8 megabytes of memory. We used alphabet sizes of 4, 8 and 16, string lengths of 100, 200, 400, ... (for simplicity, we took all 3 strings to be of the same length), and we generated sets of strings with an LCS of respectively 90%, 50% and 10% of the string length in each case. In all cases, times were averaged over 3 sets of strings.

The results of the experiments are shown in the tables, where the cpu times are given in seconds. Times for DP and Lazy were essentially independent of alphabet size, so only one set of figures is included for each of these algorithms. In every case, we ran the algorithms for strings of length 100, and repeatedly doubled the string length until the program failed for lack of memory, as indicated by "-" in the table.

Conclusions. As far as comparisons between the various algorithms are concerned, the diagonal threshold method appears to be the best across the whole range of problem instances generated. The lazy method is competitive when the LCS is close to the string length, and the Hsu-Du algorithm only when the LCS is very short.

However, the clearest conclusion that can be drawn from the empirical results is that, at least using the known algorithms, the size of 3-LCS problem instances that can be solved in practice is constrained by space requirements rather than by time requirements. Preliminary experiments with versions of the threshold and Hsu/Du algorithms for more than 3 strings confirm the marked superiority of the former, and show that the dominance of memory constraint is likely to be even more pronounced for larger numbers of strings. So there is clearly a need for algorithms that use less space. One obvious approach worthy of further investigation is the application of space-saving divide-and-conquer techniques [4] [2], and other possible time-space trade-offs, particulary, in view of the evidence, to the threshold algorithm.

References

1. A. Apostolico. Improving the worst-case performance of the Hunt-Szymanski strategy for the longest common subsequence of two strings. *Information Processing Letters*, 23:63–69, 1986.
2. A. Apostolico, S. Browne, and C. Guerra. Fast linear-space computations of longest common subsequences. *Theoretical Computer Science*, 92:3–17, 1992.
3. A. Apostolico and C. Guerra. The longest common subsequence problem revisited. *Algorithmica*, 2:315–336, 1987.
4. D.S. Hirschberg. A linear space algorithm for computing maximal common subsequences. *Communications of the A.C.M.*, 18:341–343, 1975.
5. D.S. Hirschberg. Algorithms for the longest common subsequence problem. *Journal of the A.C.M.*, 24:664–675, 1977.
6. W.J. Hsu and M.W. Du. Computing a longest common subsequence for a set of strings. *BIT*, 24:45–59, 1984.
7. J.W. Hunt and T.G. Szymanski. A fast algorithm for computing longest common subsequences. *Communications of the A.C.M.*, 20:350–353, 1977.
8. S.Y. Itoga. The string merging problem. *BIT*, 21:20–30, 1981.
9. W.J. Masek and M.S. Paterson. A faster algorithm for computing string editing distances. *J. Comput. System Sci.*, 20:18–31, 1980.

10. E.W. Myers. An O(ND) difference algorithm and its variations. *Algorithmica*, 1:251–266, 1986.
11. N. Nakatsu, Y. Kambayashi, and S. Yajima. A longest common subsequence algorithm suitable for similar text strings. *Acta Informatica*, 18:171–179, 1982.
12. D. Sankoff. Matching sequences under deletion insertion constraints. *Proc. Nat. Acad. Sci. U.S.A.*, 69:4–6, 1972.
13. E. Ukkonen. Algorithms for approximate string matching. *Information and Control*, 64:100–118, 1985.
14. R.A. Wagner and M.J. Fischer. The string-to-string correction problem. *Journal of the A.C.M.*, 21:168–173, 1974.
15. S. Wu, U. Manber, G. Myers, and W. Miller. An O(NP) sequence comparison algorithm. *Information Processing Letters*, 35:317–323, 1990.

Color Set Size Problem
with Applications to String Matching [*]

Lucas Chi Kwong Hui

Computer Science
University of California, Davis
Davis, CA 95616
hui@cs.ucdavis.edu

Abstract. The *Color Set Size* problem is: Given a rooted tree of size n with l leaves colored from 1 to m, $m \leq l$, for each vertex u find the number of different leaf colors in the subtree rooted at u. This problem formulation, together with the *Generalized Suffix Tree* data structure has applications to string matching. This paper gives an optimal sequential solution of the color set size problem and string matching applications including a linear time algorithm for the problem of finding the longest substring common to at least k out of m input strings for all k between 1 and m. In addition, parallel solutions to the above problems are given. These solutions may shed light on problems in computational biology, such as the multiple string alignment problem.

1   Introduction

We consider the following problem: Given m strings, for every pattern P that occurs in any one of the m strings, find the number of strings that contain P. We give an $O(n)$ time (implicit) solution to the above problem (denoted as the *all-patterns* problem) where n is sum of the lengths of the m strings. By implicit, we mean that we do not count the time to output the solution. As n is the size of the input, $O(n)$ time is optimal.

We show that the implicit all-patterns problem can be viewed as a *color-set-size* problem on the Generalized Suffix Tree representing the m strings. The color-set-size (abbrev. CSS) problem on trees will be defined precisely in §2. We give a linear time solution to the CSS problem, which implies a linear time solution (linear in the sum of length of the m strings) to the all-patterns problem. The CSS solution has an underlying structure that is concise and clear, and the running time constant is small. This makes it very practical. We believe that the CSS problem will have other applications in string matching and other areas.

1.1   Applications

In §3 we give a list of string matching problems are variations of the all-patterns problem. All use the all-patterns solution. The problem that motivated us is: Given

---

[*] This work was partially supported by NSF Grant CCR 87-22848, and Department of Energy Grants DE-AC03-76SF00098 and DE-FG03-90ER60999.

m strings of total length n, find the longest substring that is common to at least k strings for every k between 1 and m. Our all-patterns solution implies an $O(n + q)$ solution to the above, where q is the total output size.

A special case of the above problem is to find the longest substring that is common to at least k strings for a *fixed* k between 1 and m. We refer this as the *k-out-of-m* problem. Historically, this problem was claimed to be solved by [17] but no details are given. Also the repeat finding algorithm of [14] (which sorts the suffixes in lexicographic ordering) can solve this problem in $O(n \log n)$ expected time. We present an optimal solution to the k-out-of-m problem, which is a special simplified case of the solution to the all-patterns problem. This solution is the first optimal one that appears in literature and is practical as the components (building a Generalized Suffix Tree and solving the CSS problem) are easy to implement.

Multiple string alignment

One possible motivation of the all-patterns problem and its variations is the multiple string (sequence) alignment problem, which is a difficult problem of great value in computational biology. The CSS solution does not solve the multiple string alignment directly, but it may shed light. A simplified description of the multiple string alignment problem follows. Details can be found in [4, 6, 8].

An alignment $\mathcal{A}$ of two strings X and Y is obtained by adding spaces into X and Y and then placing the two resulting strings (with length $= l$) one above the other. The value of $\mathcal{A}$, $V(\mathcal{A})$, is defined as $\sum_{i=1}^{l} s(X(i), Y(i))$, where $s(X(i), Y(i))$ is the value contributed by the two oppositing characters in position i. A simple but common scheme is $s(X(i), Y(i)) = 0$ if $X(i)$ and $Y(i)$ are the same character, and $s(X(i), Y(i)) = 1$ otherwise. An optimal alignment of X and Y is an alignment $\mathcal{A}$ which minimizes $V(\mathcal{A})$.

A multiple alignment of $m > 2$ strings $\mathcal{X} = \{X_1, X_2, ..., X_m\}$ is a natural generalization of the pairwise alignment defined above. Chosen spaces are inserted into each string so the resulting strings have the same length l, and then the strings are arrayed in m rows of l columns each so that each character and space of each string is in a unique column. However, the value of a multiple alignment is not so easily generalized. Several approaches are used, for example the value of a multiple string alignment can be defined as the sum of values of the induced pairwise alignments [4, 6, 8]. Recently, Gusfield [8] introduced two computational efficient multiple alignment methods (for two multiple alignment value functions) whose deviation from the optimal multiple alignment value is guaranteed to be less than a factor of two.

By solving the all-patterns problem of $\mathcal{X}$ in linear time, we get the following information: for every sub-string occurred in $\mathcal{X}$, we know the number of input strings containing this sub-string. This information is certainly helpful to the construction of an optimal multiple alignment, but research effort is needed to figure out the details. One possible approach is to use a 'greedy' approach to construct the multiple alignment. We start working with the sub-strings which occur in most strings in $\mathcal{X}$. Those sub-strings are used as *anchor* points to align the input strings. This may give us an efficient algorithm which is optimal or nearly optimal.

Computer virus detection

Another area of possible application is in computer virus detection. We consider a computer program as a string of machine instructions. The alphabet is the set of different machine instructions. If a computer system is infected by a computer virus, many computer programs will have the same piece of virus code (again a string of machine instructions) attached to it. Therefore a string of machine instructions which is common to a lot of existing programs is likely to be a virus. We can build a Generalized Suffix Tree for the computer program strings, assign a different color for each program string, and solve the CSS problem. The CSS solution will contain the following information: For every pattern (or piece of code) that occurs in any one of the computer program strings, we know the number of computer programs that contain it [11].

This is certainly useful for the virus detector, and this can be computed in linear time. However, one possible problem is that the alphabet size (size of set of instructions) may be too large for an efficient construction of the Generalized Suffix Tree. Research effort is needed to solve this problem.

This paper is organized as follows: §2 describes the CSS problem and its solution. §3 describes the *Generalized Suffix Tree*, the all-patterns problem, the k-out-of-m problem and other string matching applications. §4 and §5 give parallel algorithms for the CSS and the string matching problems.

2 The Color Set Size problem

Let $C = \{1, \ldots, m\}$ be a set of colors and T be a general rooted tree with n leaves where each leaf is associated with a color c, $c \in C$. The *color set size* for a vertex v, denoted as $css(v)$, is the number of different leaf colors in the subtree rooted at v. The Color Set Size (CSS) problem is to find the color set size for every internal vertex v in T. For example in Figure 1 $css(x) = 2$ and $css(y) = 3$.

We start by defining the following. Let *LeafList* be the list of leaves ordered according to a post-order traversal of the input tree T. For v a leaf of T with color c, let $lastleaf(x)$ be the last leaf which precedes x in *LeafList* and has color c. If no leaf with color c precedes x in *LeafList*, then $lastleaf(x) = Nil$. See Figure 1 for an example. For vertices x and y of T, let $lca(x, y)$ be the least common ancestor of x and y. For an internal vertex u of T let $subtree(u)$ be the subtree of T rooted at u, and $leafcount(u)$ be the number of leaves in $subtree(u)$.

An internal vertex u is a *color-pair-lca* (of color c) if there exists a leaf x of color c, such that $u = lca(lastleaf(x), x)$. For example in Figure 1 vertex w is a color-pair-lca of color **3**. Note that a single vertex can be a color-pair-lca for several different colors. For an internal vertex u of T, define $CPLcount(u) = k$ if among all colors, there are k occurrences of leaf pairs for which u is their color-pair-lca ($CPLcount()$ stands for the color-pair-lca counter). Also define $duplicate(u) = \sum_{v \in subtree(u)} CPLcount(v)$ $= CPLcount(u) + \sum_{w \in W} duplicate(w)$ where W is set of children of u.

2.1 Core idea for a linear CSS solution

We want an $O(n)$ solution to the CSS problem. The direct approach is to compute whether a leaf colored c exists in $subtree(u)$ or not, for every internal vertex u and color c. However even if we could answer the above query in constant time it

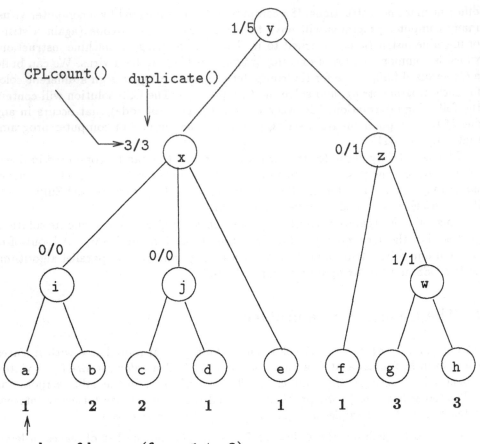

color of leaves (from 1 to 3)

```
lastleaf(a) = Nil        lastleaf(b) = Nil
lastleaf(c) = b          lastleaf(d) = a
lastleaf(e) = d          lastleaf(f) = e
lastleaf(g) = Nil        lastleaf(h) = g
```

The two numbers associated with each internal
vertex are CPLcount() and duplicate() in the
format of : CPLcount()/duplicate().

Fig. 1. CSS Problem Example

still takes $O(mn)$ time to compute all of them. We achieve the $O(n)$ run time by avoiding explicit computation of the above queries. Instead we compute $css()$ values by counting leaves and color-pair-lca's.

Lemma 1 (CSS Counting Lemma). $css(u) = leafcount(u) - duplicate(u)$.

Proof. The color set size of an internal vertex u is determined by the colors of leaves in $subtree(u)$. If each color appears in either zero or one leaf in $subtree(u)$, then $css(u) = leafcount(u)$. However $leafcount(u)$ is higher than $css(u)$ when there are $k > 1$ leaves with the same color in $subtree(u)$. Among those k leaves, we should only count the leaf that appears first in $LeafList$ and treat the other $k - 1$ leaves as duplicate leaves. For example in Figure 1, we should only count leaves a, b, and g to compute $css(y)$.

Our method of counting the leaves that 'appear first in $LeafList$' is first to count all leaves and then subtract from it the number of 'duplicate leaves'. If x is a 'duplicate' leaf in $subtree(u)$, then both x and $lastleaf(x)$ must be in $subtree(u)$. Otherwise x is not a 'duplicate' leaf. Therefore $lca(lastleaf(x), x)$ must be in $subtree(u)$. Conversely if $lca(lastleaf(x), x)$ is in $subtree(u)$ then both $lastleaf(x)$ and x must be in $subtree(u)$. Thus x is a 'duplicate' leaf in $subtree(u)$. In summary, we have,

Fact 1 x is a 'duplicate' leaf in $subtree(u)$ *iff* $lca(lastleaf(x), x)$ is in $subtree(u)$.

Suppose that in $subtree(u)$ for an internal vertex u, there are $i \geq 1$ 'duplicate' leaves among all colors. Each 'duplicate' leaf x is associated with a color-pair-lca, $lca(lastleaf(x), x)$. Observe that an internal vertex v in $subtree(u)$ can be associated with more than one 'duplicate' leaf. (For example in Figure 1, x is $lca(a, d)$, $lca(d, e)$, and $lca(b, c)$.) It is obvious that if $CPLcount(v) = j \geq 1$, then there are j 'duplicate' leaves in $subtree(u)$ associated with v. This implies that the sum of all $CPLcount()$ values in $subtree(u) = i =$ number of 'duplicate' leaves in $subtree(u)$. Hence,

$$leafcount(u) - css(u) = \text{number of 'duplicate' leaves in } subtree(u)$$
$$= \sum_{v \in subtree(u)} CPLcount(v)$$
$$= duplicate(u).$$

$\square$

As a result we can find $css(u)$ by subtracting number of color-pair-lca's in $subtree(u)$ from $leafcount(u)$. For example in Figure 1, there are 5 leaves in $subtree(x)$ and 3 color-pair-lca's (x being $lca(a, d)$, $lca(d, e)$, and $lca(b, c)$), so $css(x) = 5 - 3 = 2$.

2.2 Algorithm

Now we use the CSS Counting Lemma to solve the CSS problem. All we need is to compute the *leafcount*() and *duplicate*() values. To compute *duplicate*() values, we need to compute *CPLcount*() values. To compute *CPLcount*() values, we need to compute *lastleaf*() values. So the algorithm is as follows:

```
Algorithm CSS1;
    Create the LeafList
    Compute the leafcount() values
    Compute the lastleaf() values
    Compute the CPLcount() values
    Compute the duplicate() values
    Compute the css() values
end CSS1
```

Creating *LeafList* and computing *leafcount*() values are straight-forward, using a post-order traversal of T. The *lastleaf*() values can be computed by a traversal of the *LeafList* as follows: For every color c we keep $last[c]$, the index of the last leaf colored c that was seen so far. Initially this index is Nil. Every time a new leaf x colored c is encountered, we set *lastleaf*(x) to $last[c]$, and we update $last[c]$.

Now, we compute *CPLcount*() values as follows: Initialize all *CPLcount*() values to zero. For every leaf x, we compute $u = lca(x, lastleaf(x))$ and increment *CPLcount*(u) by 1.

Finally, we use a post-order traversal to compute all *duplicate*() values by the recursive relationship $duplicate(u) = CPLcount(u) + \sum_{w \in W} duplicate(w)$ where W is set of children of u. Having computed *leafcount*() and *duplicate*() values, computing *css*() values is trivial.

2.3 Time and space analysis

Computing *lastleaf*() involves a single scan of *LeafList*, hence can be done in $O(n)$ time and space and $O(m)$ extra space. [19] gives an algorithm that can compute the lca of an arbitrary pair of leaves in $O(1)$ time, with $O(n)$ time preprocessing and $O(n)$ space. So the *CPLcount*() values can be computed in $O(1)$ time per leaf, and $O(n)$ total time including preprocessing.

Creating *LeafList*, computing *leafcount*(), *duplicate*(), and *css*() all takes $O(n)$ time and space. Hence we have the first main result:

Theorem 2 (CSS Linear Theorem). *Algorithm CSS1 solves the CSS problem in* $O(n)$ *time,* $O(n)$ *space, and* $O(m)$ *extra space.*

Remark. If $m \gg n$ we can use hashing to get rid of the $O(m)$ extra space. In this case we have $O(n)$ expected time and $O(n)$ space.

3   String matching problems

The all-patterns problem described in §1 can be solved by the CSS solution. The following is a non-complete list of some pattern matching problems that use the all-patterns solution and their time bounds:

1. Given m input strings, for all k between 1 and m, find the longest pattern which appears in at least k input strings.
2. Given m input strings and a fixed integer k between 1 to m, find the longest pattern which appears in at least k input strings. This is a special case of problem 1.
3. Given m input strings and integers k and l, find a pattern with length l which appears in exactly k input strings.
4. Given m input strings and integers $l_1 < l_2$, find the pattern with length between l_1 and l_2 which appears in as many input strings as possible.
   All the above problems can be solved in $O(n)$ time + time to output the answer.
5. Given m input strings and $O(n)$ preprocessing time, answer the query "How many input strings contain a given pattern of length p". This is solved in $O(p)$ time on-line.

There are other variations that fall in the above list, including many problems involving a pattern selection criteria about the length of patterns and the number of input strings containing the pattern. As the solutions to the above problems are similar, we only present solution of problem 2, the k-out-of-m problem. This problem may find applications in DNA matching and computer virus detection. Other problem solutions can be found by modifying the k-out-of-m solution.

3.1   Generalized Suffix Tree

A Suffix Tree is a trie-like data structure that compactly represents a string and is used extensively in string matching. Details of suffix tree can be found in [1, 2, 15, 22]. We only give a brief review.

A suffix tree of a string $W = w_1, \ldots, w_n$ is a tree with $O(n)$ edges and n leaves. Each leaf in the suffix tree is associated with an index i $(1 \le i \le n)$. The edges are labeled with characters such that the concatenation of the labeled edges that are on the path from the root to leaf with index i is the ith suffix of W. See Figure 2. A suffix tree for a string of length n can be built in $O(n)$ time and space [15]. We say that "vertex v represents the string V" instead of "the path from root to vertex v represents the string V" for simplicity. Properties of suffix tree include:

1. Every suffix of the string W is represented by a leaf in the suffix tree.
2. Length of an edge label of the suffix tree can be found in $O(1)$ time.
3. If a leaf of the suffix tree represents a string V, then every suffix of V is also represented by another leaf in the suffix tree.
4. If a vertex u is an ancestor of another vertex v then the string that u represents is prefix of the string that v represents.
5. If vertices $v_1, v_2, \ldots, v_k$ represent the strings $V_1, V_2, \ldots, V_k$ respectively, then the least common ancestor of $v_1, \ldots, v_k$ represents the longest common prefix of $V_1, \ldots, V_k$.

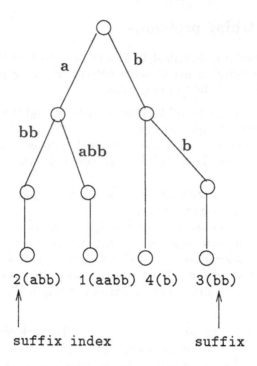

Fig. 2. Suffix Tree Example

The concept of suffix tree can be extended to store more than one input strings. This extension is called the *Generalized Suffix Tree* (GST). Note that sometimes two or more input strings may have the same suffix, and in this case the GST has two leaves corresponding to the same suffix, each corresponds to a different input string. See Figure 3 for an example. All properties of a suffix tree also exist in a GST, except that (**1.**) should be replaced by:

1'. Every suffix of every input string is represented by a leaf in the GST.

A GST can be built in O(n) time and space where n is the total length of all input strings.

238

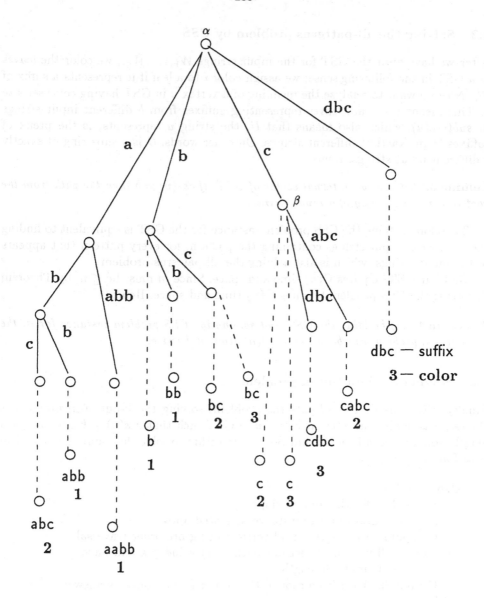

input strings: 1 aabb 2 cabc 3 cdbc
bc and c are common suffixes of 2 and 3
$css(\alpha) = 3$ $css(\beta) = 2$

Fig. 3. Generalized Suffix Tree Example

3.2 Solving the all-patterns problem by CSS

After we have built the GST for the input strings $W_1, \ldots, W_m$, we color the leaves of a GST in the following sense: we assign color i to a leaf if it represents a suffix of W_i. Now we want to analyze the meaning of a vertex u in GST having color set size b. This means there are leaves representing suffixes from b different input strings in $subtree(u)$, which also means that U, the string u represents, is the prefix of suffixes from exactly b different strings. On other words, U is a substring of exactly b different input strings. Hence,

Lemma 3. *Let u be an internal vertex of GST. If $css(u) = b$ then the path from the root to u is a substring of b input strings.*

Therefore, solving the CSS problem instance for the GST is equivalent to finding the number of input strings containing the pattern, for every pattern that appears in the input strings, which is also solving the all-patterns problem.

Building GST requires $O(n)$ time and space. Since m must be $\leq n$, by Theorem 2, solving the CSS problem requires $O(n)$ time and space. Hence,

Theorem 4. *By building the GST and solving the CSS problem instance for it, the all-patterns problem can be solved in $O(n)$ time and space.*

3.3 Solving the k-out-of-m problem

Finally, after solving the all-patterns problem, solving the k-out-of-m problem is the same as finding an internal vertex u in GST such that $css(u) \geq k$ and the path length from root to u is maximum. So the algorithm to solve the k-out-of-m problem is as follows:

    Algorithm KM;
        Build GST for the input strings
        Solve the CSS problem instance as stated above
        Compute path length for all vertices using pre-order traversal
        Among all vertices, select one with css() value ≥ k and have
            maximum path length
        Output the path from root to the vertex found above as answer
    end KM

The first two steps together is the solution for the all-patterns problem.

To compute path length for every vertex in GST, we use two observations: (i) path length of root is zero; and (ii) if u is son of v in the GST, then path length of $u =$ path length of $v +$ edge label length of (u,v). With the above two facts, a pre-order traversal is sufficient to compute path length of all vertices. It takes $O(n)$ time and space. Selecting the vertex as answer takes $O(n)$ time and space. Finally, outputting the answer takes $O(q)$ time and space where q is the output size. Hence,

Theorem 5. *Algorithm KM solves the k-out-of-m problem in $O(n + q)$ time and space, where n is the input size and q is the output size.*

Remark. Other string matching problems listed in this paper can be solved by modifying the k-out-of-m solution. The first three steps: building the GST, solving the CSS problem instance, and computing path lengths are common to all problem solutions. Only the last selection step is different. As designing the selection step is straight-forward, we skip the details.

4 Parallel implementation of CSS1

The CSS1 algorithm is easily parallelizable. There is a PRAM solution of the CSS problem (called the PCSS algorithm) which uses the exact framework of the sequential CSS1. We just parallelize all the steps.

4.1 The Euler Tour technique

The PCSS algorithm uses the *Euler Tour* technique [20, 21] which is extensively used in parallelizing tree algorithms such as computing ear decomposition [16] and the accelerated centroid decomposition technique [7]. A tree T is treated as an undirected graph and a directed graph T' corresponding to T is created in the following sense: (i) The vertices of T and T' are the same. (ii) For each edge $u - v$ in T there are two edges, $u \rightarrow v$ and $v \rightarrow u$, in T'. Since the in-degree and out-degree of each vertex in T' are the same, T' has an Euler path that starts and ends at the root of T'. This is an Euler Tour of T. The Euler Tour for Figure 1 is : y →x →i →a →i →b →i →x →j →c →j →d →j →x →e →x →y →z →f →z →w →g →w →h →w →z →y. For a vertex u, if u has k children then there are $k+1$ appearances of u in the Euler Tour. We call each appearance an *Euler record* of u. The *Euler segment* of vertex u is the segment of the Euler Tour which starts at the first Euler record of u and ends at the Euler record of u. All Euler records of any vertex in $subtree(u)$ must fall in the Euler segment of u.

Suppose we have a numeric field $f(u)$ in every vertex u of T, and we want to compute $\sum_{v \in subtree(u)} f(v)$, for every internal vertex u. Here the Euler Tour technique is applicable. For each vertex we associate its $f()$ value with one of its Euler records in the Euler Tour. After that, for any vertex u, $\sum_{v \in subtree(u)} f(v)$ is equal to the sum of $f()$ values for all Euler records in u's Euler segment. Summation over all sublists (the Euler segments) in a list (the Euler Tour) can be easily reduced to a prefix sum problem, which requires $O(\log n)$ time using $\frac{n}{\log n}$ processors on an EREW PRAM [7]. We apply this technique to compute $leafcount()$ and $duplicate()$. We use parallel sorting [3, 5, 9, 10] and parallel prefix sum to implement the other steps of PCSS.

4.2 Creating *LeafList* and computing the *leafcount()* values

Note that the Euler Tour can be prepared using $\frac{n}{\log n}$ processors, in $O(\log n)$ time, on an EREW PRAM [7].

In the Euler Tour, the order of the leaves is exactly the same as the order of the leaves in *LeafList*. We create the *LeafList* from the Euler Tour by the following method. For the ith node in the Euler Tour, we assign a list L_i to it. If the ith node in the Euler Tour is a leaf x then we set L_i to contain x only; otherwise we set L_i to an empty list. After that, we compute $LeafList = L1||L2||L3||...||L_{\hat{n}}$, where $||$ is the concatenate operator, and $\hat{n}$ is the length of the Euler Tour ($\hat{n}$ is O(n)). Since $||$ is associative, we can use parallel prefix sum [12]. This takes $\frac{n}{\log n}$ processors O($\log n$) time on an EREW PRAM.

By associating a '1' to every leaf and '0' to every internal vertex, computing $leafcount(u)$ means to sum the numbers in $subtree(u)$. Using the Euler Tour technique described in §4.1, we can reduce this to a prefix sum problem on the Euler Tour, which can be solved by $\frac{n}{\log n}$ processors, in O($\log n$) time on an EREW PRAM.

4.3 Computing the *lastleaf*() values

One obvious approach is to group the leaves according to their colors. This can be achieved by sorting. [3, 5] give optimal sorting algorithms which using n processors, takes O($\log n$) time on an EREW PRAM.

In many cases we can assume the number of colors, m, is O(n), so we can use parallel bucket sort to overcome the $\Omega(n \log n)$ work barrier of comparison-based sorting algorithms. A parallel bucket sort can be computed: (i) using $\frac{n}{\log n}$ processors, EO($\log n$) parallel expected time on a priority CRCW PRAM [18] (we will discuss how this is achieved in the next paragraph); (ii) using $\frac{n}{\log n} \log\log n$ processors, O($\log n$) time on a priority CRCW by the algorithm of [9]; (iii) or using $n^{1-\epsilon}$ processors, O(n^ϵ) time on an EREW PRAM for any $\epsilon > 0$ [10];

In the computation of *lastleaf*() values we need stable sorting. The randomized parallel integer sorting algorithm in [18] is unstable. We use the following two steps to achieve the same effect as a stable sort on the colors: (Step 1) We sort the leaves according to their colors. After this step leaves of a particular color are grouped together. (Step 2) For every color c, we sort (in parallel) all leaves of color c according to their position in the *LeafList*. The above two steps together have the same effect as a stable sorting on the colors. The asymptotic running time is not increased.

4.4 Computing the *CPLcount*() values

One difficulty in computing the *CPLcount*() values in parallel is that more than one leaves (of same or different colors) may want to increment the *CPLcount*() value of the same internal vertex at the same time. For example in Figure 1 the leaves c, d, and e want to increment $CPLcount(x)$ simultaneously. Following shows how to solve this problem using integer sorting.

[19] shows that with $\frac{n}{\log n}$ processors we can preprocess a tree in O($\log n$) time such that k lca queries can be answered in constant time using k processors on a CREW PRAM. We use this to parallelize computation of *CPLcount*() values. Each leaf x is assigned a processor. This processor creates a record which contains the vertex id $u = lca(lastleaf(x), x)$. If $lastleaf(x)$ does not exist then the processor creates an empty record. After that all processors form an array, RA, of all the created records. This can be done in constant parallel time using n processors. For an internal vertex u, the number of occurrences of its id in RA is equal to $CPLcount(u)$.

We can use a parallel bucket sort on RA to group all occurrences of the same internal vertex together. Finally a prefix sum can be used to count the number of occurrences of each internal vertex in RA, which is the $CPLcount()$ values of that internal vertex.

As a prefix sum takes $O(\log n)$ time with $\frac{n}{\log n}$ processors, the bottleneck of this step is in the sorting part, which has the same run time as computing $lastleaf()$.

4.5 Computing the *duplicate*() values

By using the relation $duplicate(u) = \sum_{w \in subtree(u)} CPLcount(w)$, we can use the same Euler Tour technique as described before (by associating $CPLcount(v)$ to an Euler record of v in the Euler Tour, for every vertex v) to transform this to a prefix sum problem. The running time is the same as computing $leafcount()$.

4.6 Computing the *css*() values and overall run time

Finally computing the $css()$ values can be easily done by n processors, in $O(1)$ time on an EREW PRAM.

Therefore overall run time of PCSS is (assume m is $O(n)$):

- Using $\frac{n}{\log n}$ processors, $EO(\log n)$ parallel expected time on a priority CRCW PRAM.
- Or using $\frac{n}{\log n} \log \log n$ processors, $O(\log n)$ time on a priority CRCW PRAM.
- Or using $n^{1-\epsilon}$ processors, $O(n^\epsilon)$ time for any $\epsilon > 0$ on a CREW PRAM.

Remark. If m is unbounded, we have to use the parallel sorting of [5], which uses n processors, $O(\log n)$ time on a CREW PRAM. The overall run time of PCSS is using n processors, $O(\log n)$ time on a CREW PRAM.

5 Parallel all-patterns solution

Computing the path length for all vertices can be reduced to a prefix sum problem on the Euler Tour. This is achieved by assigning length of the Euler Tour edges in the following way. Suppose u_1 is parent of u_2 in the GST, and the edge label length of (u_1, u_2) is d, then we assign $+d$ as label length of the edge $u_1 \to u_2$ in the Euler Tour, and assign $-d$ as label length of the edge $u_2 \to u_1$ in the Euler Tour. After this assignment, the path length from root to a vertex v in the GST is the sum of all edge label lengths from the start of the Euler Tour to the first Euler record of v. As stated before, this takes $\frac{n}{\log n}$ processors $O(\log n)$ time on an EREW PRAM.

The critical step here is to construct the generalized suffix tree in parallel. The first parallel algorithm in constructing suffix tree is given in [13]. It runs in $O(\log n)$ time and uses $n^2 / \log n$ processors. The best result is [2]. An arbitrary CRCW PRAM algorithm was given in [2], which runs in $O(\log n)$ time and uses n processors. Its approach can be generalized to construct generalized suffix tree so this is also the current best parallel time bound for solving the all-patterns problems (and other listed strings matching applications).

Acknowledgement. The author wants to thank Chip Martel, Dan Gusfield, and Dalit Naor for their helpful comments about the presentation of this paper, and also Dan for introducing the k-out-of-m problem in his string matching class.

References

1. A. Apostolico. The myriad virtues of subword trees. In A. Apostolico and Z. Galil, editors, *Combinatorial Algorithms on Words, NATO ASI Series, Series F: Computer and System Sciences, Vol. 12*, pages 85–96, Springer-Verlag, Berlin, 1985.

2. A. Apostolico, C. Iliopoulos, G. M. Landau, B. Schieber, and U. Vishkin. Parallel construction of a suffix tree with applications. *Algorithmica*, 3:347–365, 1988.

3. M. Ajtai, J. Komlós, and E. Szemerédi. An $O(n \log n)$ sorting network. In *Proc. of the 15th ACM Symposium on Theory of Computing*, pages 1–9, 1983.

4. S. Altschul and D. Lipman. Trees, stars, and multiple biological sequence alignment. *SIAM Journal on Applied Math*, 49:197–209, 1989.

5. R. Cole. Parallel merge sort. In *Proc. 27nd Annual Symposium on the Foundation of Computer Science*, pages 511–516, 1986.

6. H. Carrillo and D. Lipman. The multiple sequence alignment problem in biology. *SIAM Journal on Applied Math*, 48:1073–1082, 1988.

7. R. Cole and U. Vishkin. The accelerated centroid decomposition technique for optimal parallel tree evaluation in logarithmic time. *Algorithmica*, 3:329–346, 1988.

8. D. Gusfield. *Efficient methods for multiple sequence alignment with guaranteed error bounds*. Technical Report CSE-91-4, Computer Science, U. C. Davis, 1991.

9. T. Hagerup. Towards optimal parallel bucket sorting. *Information and Computation*, 75:39–51, 1987.

10. C. Kruskal, L. Rudolph, and M. Snir. The power of parallel prefix. *IEEE Trans. Comput.* C-34:965-968, 1985.

11. R. Lo. personal communications. 1991.

12. L. Ladner and M. Fischer. Parallel prefix computation. *J.A.C.M.*, 27:831–838, 1980.

13. G. M. Landau and U. Vishkin. Introducing efficient parallelism into approximate string matching and a new serial algorithm. In *Proc. of the 18th ACM Symposium on Theory of Computing*, pages 220–230, 1986.

14. H. M. Martinez. An efficient method for finding repeats in molecular sequences. *Nucleic Acids Research*, 11(13):4629–4634, 1983.

15. E. M. McCreight. A space-economical suffix tree construction algorithm. *J.A.C.M.*, 23(2):262–272, 1976.

16. Y. Maon, B. Schieber, and U. Vishkin. Open ear decomposition and s-t numbering in graphs. *Theoretical Computer Science*, 1987.

17. V. R. Pratt. Improvements and applications for the weiner repetition finder. 1975. unpublished manuscript.

18. S. Rajasekaran and J. H. Reif. Optimal and sublogarithmic time randomized parallel sorting algorithms. *SIAM Journal on Computing*, 18:594–607, 1989.

19. B. Schieber and U. Vishkin. On finding lowest common ancestors: simplification and parallelization. *SIAM Journal on Computing*, 17:1253–1262, 1988.

20. R. E. Tarjan and U. Vishkin. An efficient parallel biconnectivity algorithm. *SIAM Journal on Computing*, 14:862–874, 1985.

21. U. Vishkin. On efficient parallel strong orientation. *I.P.L.*, 20:235–240, 1985.

22. P. Weiner. Linear pattern matching algorithms. In *Proc. 14th IEEE Symp. on Switching and Automata Theory*, pages 1–11, 1973.

Computing Display Conflicts in String and Circular String Visualization *

Dinesh P. Mehta[1,2] and Sartaj Sahni[1]

[1] Dept. of Computer and Information Sciences, University of Florida
Gainesville, FL 32611
[2] Dept. of Computer Science, University of Minnesota
Minneapolis, MN 55455

Abstract. Strings are used to represent a variety of objects such as DNA sequences, text, and numerical sequences. Circular strings are used to represent circular genomes in molecular biology, polygons in computer graphics and computational geometry, and closed curves in computer vision. We have proposed a model for the visualization of strings and circular strings [1], where we introduced the problem of display conflicts. In this paper, we provide efficient algorithms for computing display conflicts in linear strings. These algorithms make use of the scdawg data structure for linear strings [4]. We also extend the scdawg data structure to represent circular strings. The resulting data structure may now be employed to compute display conflicts in circular strings by using the algorithms for computing conflicts in linear strings.

1 Introduction

The string data type is used to represent a number of objects such as text strings, DNA or protein sequences in molecular biology, numerical sequences, etc. The circular string data type is used to represent a number of objects such as circular genomes, polygons, and closed curves. Research in molecular biology, text analysis, and interpretation of numerical data involves the identification of recurring patterns in data and hypothesizing about their causes and/or effects [2, 3]. Research in pattern recognition and computer vision involves detecting similarities within an object or between objects [5]. Detecting patterns visually in long strings is tedious and prone to error. We have proposed [1] a model to alleviate this problem. The model consists of identifying all recurring patterns in a string and highlighting identical patterns in the same color.

We first discuss the notion of maximal patterns. Let abc be a pattern occurring m times in a string S. Let the only occurrences of ab be those which occur in abc. Then pattern ab is not maximal in S as it is always followed by c. The notion of maximality is motivated by the assumption that in most applications, longer patterns are more significant than shorter ones. Maximal patterns that occur at least twice are known as displayable entities. The problem of identifying all displayable entities

* This research was supported in part by the National Science Foundation under grant MIP 91-03379.

and their occurrences in S can be solved from the results in [4]. Once all displayable entities and their occurrences are obtained, we are confronted with the problem of color coding them. In the string, $S = abczdefydefxabc$, abc and def are the only displayable entities. So, S would be displayed by highlighting abc in one color and def in another as shown in Figure 1.

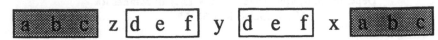

Fig. 1. Highlighting displayable entities

In most strings, we encounter the problem of conflicts: Consider the string $S = abcicdefcdegabchabcde$ and its displayable entities, abc and cde (both are maximal and occur thrice). So, they must be highlighted in different colors. Notice, however, that abc and cde both occur in the substring $abcde$, which occurs as a suffix of S. Clearly, both displayable entities cannot be highlighted in different colors in $abcde$ as required by the model. This is a consequence of the fact that the letter c occurs in both displayable entities. This situation is known as a prefix-suffix conflict (because a prefix of one displayable entity is a suffix of the other). Note, also, that c is a displayable entity in S. Consequently, all occurrences of c must be highlighted in a color different from those used for abc and cde. But this is impossible as c is a subword of both abc and cde. This situation is referred to as a subword conflict. The problem of subword conflicts may be partially alleviated by employing more

Fig. 2. Single-copy subword-overlap model

sophisticated display models as in Figure 2.

Irrespective of the display model used, it is usually not possible to display all occurrences of all displayable entities because of display conflicts. We are therefore forced into having to choose which ones to display. There are three ways of achieving this: Interactive, Automatic, and Semi-Automatic [1]. All three methods require knowledge about conflicts. Automatic methods would require a list of all the conflicts, while interactive methods require information about conflicts local to a particular segment of the string. Since prefix suffix and subword conflicts are handled differently by different display models, separate lists for each are required.

In this paper we identify a family of problems relating to the identification of conflicts at various levels of detail. Problems relating to statistical information about conflicts are also identified. Efficient algorithms for these problems are presented. All algorithms make use of the symmetric compact directed acyclic word graph (scdawg) data structure [4] and may be thought of as operations or traversals of the scdawg. The scdawg, which is used to represent strings and sets of strings evolved from other string data structures such as position trees, suffix trees, and directed acyclic word graphs [6, 7, 8, 9].

One approach for extending our algorithms to circular strings is to arbitrarily break the circular string at some point so that it becomes a linear string. Algorithms for linear strings may then be applied to it. However, this has the disadvantage that some significant patterns in the circular string may be lost because the patterns were broken when linearizing the string. Indeed, this would defeat the purpose of representing objects by circular strings.

Another approach is concatenate two copies of the linear string obtained by breaking the circular string and then applying our algorithms for linear strings to the resulting string. This approach has the disadvantage that some of our algorithms which were optimal for linear strings are not optimal with respect to circular strings.

Tanimoto [5] defined a polygon structure graph, which is an extension of suffix trees to circular strings. However, the suffix tree is not as powerful as the scdawg and cannot be used to solve some of the problems that the scdawg can solve. In particular, it does not appear to be possible to efficiently compute display conflicts using suffix trees. In this paper, we define an scdawg for circular strings. Algorithms which make use of the scdawg for linear strings can now be extended to circular strings with minor modifications. The extended algorithms continue to have the same efficient time and space complexities. Further, the extensions take the form of postprocessing or preprocessing steps which are simple to add on to a system built for linear strings, particularly in an object oriented language.

Section 2 contains preliminaries and Section 3 discusses algorithms for computing conflicts in linear strings. Sections 4, 5, and 6 consider size-restricted, pattern-oriented, and statistical queries, respectively. Finally, Section 7 describes the extension of the scdawg data structure to circular strings.

2 Preliminaries

2.1 Definitions

Let S represent a string of length n, whose characters are chosen from a fixed alphabet, Σ, of constant size. A pattern in S is said to be *maximal* iff its occurrences are not all preceded by the same letter, nor all followed by the same letter. A pattern is said to be a *displayable entity* (or displayable) iff it is maximal and occurs more than once in S (all maximal patterns are displayable entities with the exception of S, which occurs once in itself). A *subword conflict* between two displayable entities, D_1 and D_2, in S exists iff D_1 is a substring of D_2. A *prefix-suffix conflict* between two displayable entities, D_1 and D_2, in S exists iff there exist substrings, S_p, S_m, S_s in S such that $S_p S_m S_s$ occurs in S, $S_p S_m = D_1$, and $S_m S_s = D_2$. The string, S_m is known as the *intersection* of the conflict; the conflict is said to occur between D_1 and D_2 with respect to S_m.

Let s denote a circular string of size n consisting of characters from a fixed alphabet, Σ, of constant size. We shall represent a circular string by a linear string enclosed in angle brackets "<>" (this distinguishes it from a linear string) . The linear string is obtained by traversing the circular string in clockwise order and listing each element as it is traversed. The starting point of the traversal is chosen arbitrarily. Consequently, there are up to n equivalent representations of s. We characterize the relationship between circular strings and linear strings by defining the functions, *linearize* and *circularize*. *linearize* maps circular strings to linear strings. It is a one-many mapping as a circular string can, in general, be mapped to more than one linear string. For example, $linearize(<abcd>) = \{abcd, bcda, cdab, dabc\}$. We will assume, for the purpose of this paper, that *linearize* chooses the linear string obtained by removing the angle brackets "<>". So, $linearize(<abcd>) = abcd$. *circularize* maps linear strings to circular strings. It is a many-one function and represents the inverse of *linearize*. The definitions of maximality, displayable entities, subword and prefix-suffix conflicts for circular strings are identical to those for linear strings.

2.2 Symmetric Compact Directed Acyclic Word Graphs (SCDAWGs)

An scdawg, $SCD(S)$, corresponding to a string S is a directed acyclic graph defined by a set of vertices, $V(S)$, and sets, $R(S)$ and $L(S)$, of labeled directed edges called right extension (re) and left extension (le) edges respectively. Each vertex of $V(S)$ represents a substring of S. Specifically, $V(S)$ consists of a source (which represents the empty word, λ), a sink (which represents S), and a vertex corresponding to each displayable entity of S. Let $de(v)$ denote the string represented by vertex, v ($v \in V(S)$). Define the *implication*, $imp(S, \alpha)$, of a string α in S to be the smallest superword of α in $\{de(v): v \in V(S)\}$, if such a superword exists. Otherwise, $imp(S, \alpha)$ does not exist. Re edges from a vertex, v_1, are obtained as follows: for each letter, x, in Σ, if $imp(S, de(v_1)x)$ exists and is equal to $de(v_2) = \beta de(v_1)x\gamma$, then there exists an re edge from v_1 to v_2 with label $x\gamma$. If β is the empty string, then the edge is known as a *prefix extension edge*. Le edges and *suffix extension edges* are defined analogously. Figure 3 shows $V(S)$ and $R(S)$ corresponding to $S = cdefabcgabcde$.

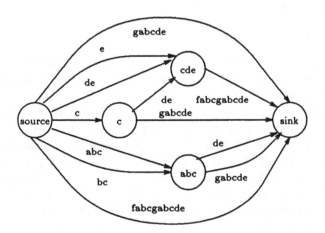

Fig. 3. Scdawg for $S = cdefabcgabcde$, re edges only

abc, cde, and c are the displayable entities of S. There are two outgoing re edges from the vertex representing abc. These edges correspond to $x = d$ and $x = g$. $imp(S, abcd)$ $= imp(S, abcg) = S$. Consequently, both edges are incident on the sink. There are no edges corresponding to the other letters of the alphabet as $imp(S, abcx)$ does not exist for $x \, \epsilon \, \{a, b, c, e, f\}$.

The space required for $SCD(S)$ is $O(n)$ and the time needed to construct it is $O(n)$ [6, 4]. While we have defined the scdawg data structure for a single string, S, it can be extended to represent a *set* of strings. Figure 4 presents an algorithm for

Algorithm A
    Occurrences$(v, 0)$

Procedure *Occurrences*$(v$:vertex,i:integer)
    begin
        if $de(v)$ is a suffix of S
        then output($|S| - i$);
        for each right out edge, e, from v **do**
            begin
                Let w be the vertex on which e is incident;
                Occurrences$(w,|label(e)| + i)$;
            end;
    end;

Fig. 4. Algorithm for obtaining occurrences of displayable entities

computing all the occurrences of $de(v)$ in S. This is based on the outline provided in [4]. The complexity of *Occurrences*$(v, 0)$ is proportional to the number of occurrences of $de(v)$ in S.

2.3 Prefix and Suffix Extension Trees

The *prefix extension tree*, $PET(S, v)$, at vertex v in $V(S)$ is a subgraph of $SCD(S)$ consisting of (i) the root, v, (ii) $PET(S, w)$ defined recursively for each vertex w in $V(S)$ such that there exists a prefix extension edge from v to w, and (iii) the prefix extension edges leaving v. The suffix extension tree, $SET(S, v)$, at v is defined analogously. In Figure 3, $PET(S, v)$, $de(v) = c$, consists of the vertices representing c and cde, and the sink. It also includes the prefix extension edges from c to cde and from cde to the sink.

Lemma 1. $PET(S, v)$ $(SET(S, v))$ *contains a directed path from v to a vertex, w, in $V(S)$ iff $de(v)$ is a prefix (suffix) of $de(w)$.*

3 Computing Conflicts in Linear Strings

3.1 Algorithm to determine whether a string is conflict free

Before describing our algorithm to determine if a string is free of conflicts, we establish some properties of conflict free strings that will be used in this algorithm.

Lemma 2. *If a prefix-suffix conflict occurs in a string S, then a subword conflict must occur in S.*

Proof. If a prefix-suffix conflict occurs between two displayable entities, W_1 and W_2, then there exists $W_p W_m W_s$ such that $W_p W_m = W_1$ and $W_m W_s = W_2$. Since W_1 and W_2 are maximal, W_1 isn't always followed by the same letter and W_2 isn't always preceded by the same letter. I.e., W_m isn't always followed by the same letter and W_m isn't always preceded by the same letter. So, W_m is maximal. But, W_1 occurs at least twice in S (since W_1 is a displayable entity). So W_m occurs at least twice (since W_m is a subword of W_1) and is a displayable entity. But, W_m is a subword of W_1. So a subword conflict occurs between W_m and W_1.

Corollary 3. *If string S is free of subword conflicts, then it is free of conflicts.*

Lemma 4. *$de(w)$ is a subword of $de(v)$ in S iff there is a path comprising right or suffix extension edges from w to v.*

Let V_{source} denote the vertices in $V(S)$ such that an re or suffix extension edge exists between the source vertex of $SCD(S)$ and each element of V_{source}.

Lemma 5. *String S is conflict free iff all right extension or suffix extension edges leaving vertices in V_{source} end at the sink vertex of $SCD(S)$.*

The preceding development leads to algorithm *NoConflicts* (Figure 5).

Algorithm *NoConflicts(S)*
1. Construct $SCD(S)$.
2. Compute V_{source}.
3. Scan all right and suffix extension out edges from each element of V_{source}. If any edge points to a vertex other than the sink, then a conflict exists. Otherwise, S is conflict free.

Fig. 5. Algorithm to determine whether a string is Conflict Free

Theorem 6. *Algorithm NoConflicts is both correct and optimal.*

3.2 Subword Conflicts

Let k_s be the number of subword conflicts in S. We use a compact representation for subword conflicts, which we describe below. Let k_{sc} denote its size.

Consider $S=abcdbcgabcdbchbc$. The displayable entities are $D_1 = abcdbc$ and $D_2 = bc$. The ending positions of D_1 are 6 and 13 while those of D_2 are 3, 6, 10, 13, and 16. A list of the subword conflicts between D_1 and D_2 can be written as: $\{(6,3), (6,6), (13,10), (13,13)\}$. The first element of each ordered pair is the last position of the instance of the superstring (here, D_1) involved in the conflict; the second element of each ordered pair is the last position of the instance of the substring (here, D_2) involved in the conflict.

The cardinality of the set is the number of subword conflicts between D_1 and D_2. This is given by: $frequency(D_1) * number\ of\ occurrences\ of\ D_2\ in\ D_1$. Since each conflict is represented by an ordered pair, the *size* of the output is $2(frequency(D_1) * number\ of\ occurrences\ of\ D_2\ in\ D_1)$.

Observe that the occurrences of D_2 in D_1 are in the same relative positions in all instances of D_1. It is therefore possible to write the list of subword conflicts between D_1 and D_2 as: $(6,13):(0,-3)$. The first list gives all the occurrences in S of the superstring (D_1), and the second gives the relative positions of all the occurrences of the substring (D_2) in the superstring (D_1) from the right end of D_1. The size of the output is now: $frequency(D_1) + number\ of\ occurrences\ of\ D_2\ in\ D_1$. This is more economical than our earlier representation.

In general, a substring, D_i, of S will have conflicts with many instances of a number of displayable entities (say, $D_j, D_k, \ldots, D_z$) of which it (D_i) is the superword. We would then write the conflicts of D_i as:
$(l_i^1, l_i^2, \ldots, l_i^{m_i}) : (l_j^1, l_j^2, \ldots, l_j^{m_j}), (l_k^1, l_k^2, \ldots, l_k^{m_k}), \ldots, (l_z^1, l_z^2, \ldots, l_z^{m_z})$. Here, the l_i's represent all the occurrences of D_i in S; the $l_j's, l_k's, \ldots, l_z's$ represent the relative positions of all the occurrences of $D_j, D_k, \ldots, D_z$ in D_i. One such list will be required for each displayable entity that contains other displayable entities as subwords. The following equalities are easily obtained:

Size of Compact Representation $= \sum_{D_i \epsilon D} (f_i + \sum_{D_j \epsilon D_i^s} (r_{ij}))$.

Size of Original Representation $= 2 \sum_{D_i \epsilon D} (f_i * \sum_{D_j \epsilon D_i^s} (r_{ij}))$.

f_i is the frequency of D_i (only D_i's that have conflicts are considered). r_{ij} is the frequency of D_j in one instance of D_i. D represents the set of all displayable entities of S. D_i^s represents the set of all displayable entities that are subwords of D_i.

$SG(S, v)$, $v \epsilon V(S)$, is defined as the subgraph of $SCD(S)$ which consists of the set of vertices, $SV(S, v) \subset V(S)$ which represents displayable entities that are subwords of $de(v)$ and the set of all re and suffix extension edges that connect any pair of vertices in $SV(S, v)$. Define $SG_R(S, v)$ as $SG(S, v)$ with the directions of all the edges in $SE(S, v)$ reversed.

Lemma 7. $SG(S, v)$ *consists of all vertices, w, such that a path comprising right or suffix extension edges joins w to v in $SCD(S)$.*

Algorithm B of Figure 6 computes the subword conflicts of S. The subword conflicts are computed for precisely those displayable entities which have subword displayable entities. Lines 4 to 6 of *Algorithm B* determine whether $de(v)$ has subword

Algorithm B
```
1       begin
2           for each vertex, v, in SCD(S) do
3               begin
4                   v.subword = false;
5                       for all vertices, u, such that a right or suffix extension edge, < u, v >,
                        is incident on v do
6                           if u ≠ source then v.subword = true;
7                   end
8               for each vertex, v, in SCD(S) such that v ≠ sink and v.subword is true do
9                   GetSubwords(v);
10      end
```

Procedure *GetSubwords(v)*
```
1       begin
2           Occurrences(v,0);
3           output(v.list);
4           v.sublist = {0};
5           SetUp(v);
6           SetSuffixes(v);
7           for each vertex, x (≠ source), in reverse topologicalorder of SG(S, v) do
8               begin
9                   if de(x) is a suffix of de(v) then x.sublist = {0} else x.sublist = {};
10                  for each vertex w in SG(S, v) on which re edge e from x is incident do
11                      begin
12                          for each element, l, in w.sublist do
13                              x.sublist = x.sublist ∪ {l − |label(e)|};
14                      end;
15                  output(x.sublist);
16              end;
17      end
```

Fig. 6. Optimal algorithm to compute all subword conflicts

displayable entities. Each incoming right or suffix extension edge to v is checked to see whether it originates at the source. If any incoming edge originates at a vertex other than *source*, then $v.subword$ is set to *true* (Lemma 4). If all incoming edges originate from *source*, then $v.subword$ is set to *false*. Procedure $GetSubwords(v)$, which computes the subword conflicts of $de(v)$ is invoked if $v.subword$ is *true*.

Procedure $Occurrences(S, v, 0)$ (line 2 of $GetSubwords$) computes the occurrences of $de(v)$ in S and places them in $v.list$. Procedure $SetUp$ in line 5 traverses $SG_R(S, v)$ and initializes fields in each vertex of $SG_R(S, v)$ so that a reverse topological traversal of $SG(S, v)$ may be subsequently performed. Procedure $SetSuffixes$ in line 6 marks vertices whose displayable entities are suffixes of $de(v)$. This is accomplished by following the chain of reverse suffix extension pointers starting at v and marking the vertices encountered as suffixes of v.

A list of relative occurrences, *sublist*, is associated with each vertex, x, in $SG(S, v)$.

x.sublist represents the relative positions of $de(x)$ in an occurrence of $de(v)$. Each relative occurence is denoted by its position relative to the last position of $de(v)$ which is represented by 0. If $de(x)$ is a suffix of $de(v)$ then *x.sublist* is initialized with the element, 0. The remaining elements of *x.sublist* are computed from the *sublist* fields of vertices, w, in $SG(S, v)$ such that a right extension edge goes from x to w. Consequently, *w.sublist* must be computed before *x.sublist*. This is achieved by traversing $SG(S, v)$ in reverse topological order [10].

Lemma 8. *x.sublist for vertex, x, in $SG(S, v)$ contains all relative occurrences of $de(x)$ in $de(v)$ on completion of GetSubwords(v).*

Theorem 9. *Algorithm B correctly computes all subword conflicts and takes $O(n + k_{sc})$ time and space, which is optimal.*

3.3 Prefix Suffix Conflicts

Let w and x, respectively, be vertices in $SET(S, v)$ and $PET(S, v)$. Let $de(v) = W_v$, $de(w) = W_w W_v$, and $de(x) = W_v W_x$. Define $Pshadow(w, v, x)$ to be the vertex representing $imp(S, W_w W_v W_x)$, if such a vertex exists. Otherwise, $Pshadow(w, v, x) = nil$. We define $Pimage(w, v, x) = Pshadow(w, v, x)$ iff $Pshadow(w, v, x) = imp(S, W_w W_v W_x) = W_a W_w W_v W_x$ for some (possibly empty) string, W_a. Otherwise, $Pimage(w, v, x) = nil$. For each vertex, w in $SET(S, v)$, a *shadow prefix dag*, $SPD(w, v)$, rooted at vertex w is comprised of the set of vertices $\{Pshadow(w, v, x)|$ x on $PET(S, v)$, $Pshadow(w, v, x) \neq nil\}$. Figure 7 illustrates these concepts. Bro-

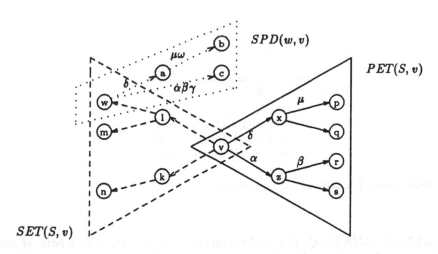

Fig. 7. Illustration of prefix and suffix trees and a shadow prefix dag

ken lines represent suffix extension edges, dotted lines represent right extension edges, and solid lines represent prefix extension edges. $SET(S, v)$, $PET(S, v)$, and $SPD(w, v)$ have been enclosed by dashed, solid, and dotted lines respectively. We have: $Pshadow(w, v, v) = Pimage(w, v, v) = w.$ $Pshadow(w, v, z) = Pshadow(w, v, r)$ $= c.$ However, $Pimage(w, v, z) = Pimage(w, v, r) = nil.$ $Pshadow(w, v, q) =$ $Pimage(w, v, q) = nil.$

Lemma 10. *A prefix suffix conflict occurs between two displayable entities, $W_1 = de(w)$ and $W_2 = de(x)$ with respect to a third displayable entity $W_m = de(v)$ iff (i) w occurs in $SET(S, v)$ and x occurs in $PET(S, v)$, and (ii) $Pshadow(w, x, v) \neq nil$. The number of conflicts between $de(w)$ and $de(x)$ with respect to $de(v)$ is equal to the number of occurrences of $de(Pshadow(w, v, x))$ in S.*

Lemma 11. *If a prefix suffix conflict does not occur between $de(w)$ and $de(x)$ with respect to $de(v)$, where w occurs in $SET(S, v)$ and x occurs in $PET(S, v)$, then there are no prefix suffix conflicts between any displayable entity which represents a descendant of w in $SET(S, v)$ and any displayable entity which represents a descendant of x in $PET(S, v)$ with respect to $de(v)$.*

Lemma 12. *In $SCD(S)$, if (i) $y = Pimage(w, v, x)$, (ii) there is a prefix extension edge, e, from x to z with label $a.\alpha$. (iii) there is a right extension edge, f, from y to u with label $a\beta$, then $Pshadow(w, v, z) = u$.*

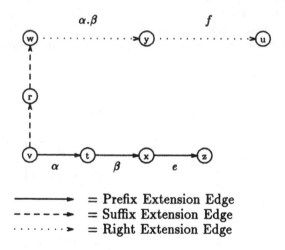

Fig. 8. Illustration of conditions for Lemma 12

Lemma 13. *In $SCD(S)$, if (i) $y = Pimage(w, v, x)$, (ii) there is a path of prefix extension edges from x to x_1 (let the concatenation of their labels be $a\alpha$), (iii) there is a prefix extension edge from x_1 to z with label $b\gamma$, and (iv) there is a right extension edge, f from y to u with label $a\alpha b\beta$, then $u = Pshadow(w, v, z) \neq nil$.*

Lemma 14. *In Lemma 13, if $|label(f)| \leq$ sum of the lengths of the labels of of the edges on the prefix extension edge path P from x to z, then $label(f) =$ concatenation of the labels on P and $u = Pimage(w, v, z)$.*

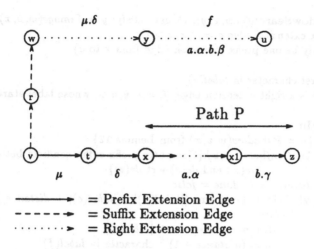

$$\longrightarrow \quad = \text{Prefix Extension Edge}$$
$$\text{-----} \rightarrow \quad = \text{Suffix Extension Edge}$$
$$\cdots\cdots\rightarrow \quad = \text{Right Extension Edge}$$

Fig. 9. Illustration of conditions for Lemmas 13 and 14

Lemma 15. *If $Pshadow(w, v, x) = nil$ then $Pshadow(w, v, y) = nil$ for all descendants, y, of x in $PET(S, v)$.*

Algorithm C
1 Construct $SCD(S)$.
2 **for** each vertex, v, in $SCD(S)$ **do**
3      $NextSuffix(v,v)$; {Compute all prefix suffix conflicts wrt $de(v)$}

Procedure NextSuffix*(current,v)*;
1 **for** each suffix extension edge $< v, w >$ **do**
2 {there can only be one suffix extension edge from v to w}
3   **begin**
4          $exist = false$; {$exist$ is a global variable}
5          $ShadowSearch(v,w,v,w)$; {Traverses $SPD(w,v)$}
6   **if** $exist$ **then** $NextSuffix(w,v)$;
7   **end**;

Fig. 10. Optimal algorithm to compute all prefix suffix conflicts

Theorem 16. *Algorithm C of Figure 10 computes all prefix suffix conflicts of S in $O(n + k_p)$ space and time, which is optimal (k_p denotes the number of prefix suffix conflicts).*

Procedure ShadowSearch(v, w, x, y); {Must satisfy: $y = Pimage(w, v, x)$}
1   **for each** prefix extension edge $e = <x, z>$ **do**
2   {There can only be one prefix extension edge from x to z}
3   **begin**
4            $fc :=$ first character in $label(e)$;
5   **if** there is a right extension edge, $f = <y, u>$, whose label starts with fc
6   **then**
7   **begin**
8   {$u = Pshadow(w, v, x)$ from Lemma 12}
9                    $ListConflicts(u, z, w)$;{Computes prefix suffix conflicts between
                                    $de(w)$ and $de(z)$ wrt $de(v)$}
10                    $distance := 0$; $done = false$
11   **while** (**not** $done$) **and** ($|label(f)| > |label(e)| + distance$)) **do**
12   **begin**
13                            $distance := distance + |label(e)|$;
14                            $nc := (distance + 1)^{th}$ character in $label(f)$.
15   **if** there is a prefix extension edge $<z, r>$ starting with nc
16   **then**
17   **begin**
18                                    $z := r$;
19   {$u = Pshadow(w, v, z)$ from Lemma 13};
20                                    $ListConflicts(u, z, w)$;
21   **end**
22   **else**
23                                $done := true$;
24   **end**
25   **if** (**not** $done$) **then**
26                        $ShadowSearch(v, w, z, u)$; {Correctness follows from Lemma 14}
27                    $exist := true$;
28   **end**
29   **end**

Fig. 11. Algorithm for shadow search

3.4   Alternative Algorithms

In this section, an algorithm for computing *all* conflicts (i.e., both subword and prefix-suffix conflicts) is presented. This solution is relatively simple and has competitive running time. However, it lacks the flexibility required to solve many of the problems listed in Sections 4, 5, and 6 . The algorithm is presented in Figure 12.

Theorem 17. *Algorithm D takes $O(n + k)$ time, where $k = k_p + k_s$.*

Algorithm D can be modified so that the size of the output is $k_p + k_{sc}$. However, its time complexity remains suboptimal.

Algorithm D

Step 1: Obtain a list of *all* occurrences of *all* displayable entities in the string. This list is obtained by first computing the lists of occurrences corresponding to each vertex of the scdawg (except the source and the sink) and then concatenating these lists.

Step 2: Sort the list of occurrences using the start positions of the occurrences as the primary key (increasing order) and the end positions as the secondary key (decreasing order). This is done using radix sort.

Step3:

```
for i:= 1 to (number of occurrences) do
    begin
        j:= i + 1;
        while(lastpos(occ_i) ≥ firstpos(occ_j) do
            begin
                if (lastpos(occ_i) ≥ lastpos(occ_j))
                then occ_i is a superword of occ_j
                else (occ_i, occ_j) have a prefix-suffix conflict;
                j:= j + 1;
            end;
    end;
```

Fig. 12. A simple algorithm for computing conflicts

4 Size Restricted Queries

We consider the following problems:

P1: List all occurrences of displayable entities whose length is greater than k.

P2: Compute all prefix suffix conflicts involving displayable entities of length greater than k.

P3: Compute all subword conflicts involving displayable entities of length greater than k.

The *overlap* of a conflict is defined as the string common to the conflicting displayable entities. The overlap of a subword conflict is the subword displayable entity. The overlap of a prefix-suffix conflict is its intersection. The *size* of a conflict is the length of the overlap. An alternative formulation of the size restricted problem is based on reporting only those conflicts whose *size* is greater than k. This formulation is particularly relevant when the conflicts are of more interest than the displayable entities. It also establishes that all conflicting displayable entities reported have size greater than k. We have the following problems:

P4: Obtain all prefix-suffix conflicts of size greater than some integer k.

P5: Obtain all subword conflicts of size greater than some integer k.

Optimal solutions to **P1, P3, P4,** and **P5** may be obtained by modifying algorithms in the previous section. Note that **P3** and **P5** are actually the same problem.

5 Pattern Oriented Queries

These queries are useful in applications where the fact that two patterns have a conflict is more important than the number and location of conflicts. The following problems arise as a result:

P6: List all pairs of displayable entities which have subword conflicts.

P7: List all triplets of displayable entities (D_1, D_2, D_m) such that there is a prefix suffix conflict between D_1 and D_2 with respect to D_m.

Optimal solutions to **P6** and **P7** may be obtained by modifying algorithms in Section 3.

6 Statistical Queries

These queries are useful when conclusions are to be drawn from the data based on statistical facts. Let $rf(D_1, D_2)$ denote the relative frequency of D_1 in D_2; and $psf(D_1, D_2, D_m)$, the number of prefix suffix conflicts between D_1 and D_2 with respect to D_m. $rf(D_1, D_2)$ may be computed optimally for all D_1, D_2, by modifying procedure $GetSubwords(v)$. $psf(D_1, D_2, D_m)$ is computed optimally, for all D_1, D_2, and D_m, where D_1 has a prefix suffix conflict with D_2 with respect to D_m, by modifying $ListConflicts$ of Figure 11.

7 Extension to Circular Strings

The definition for an scdawg for circular strings is identical to that for linear strings with the following exceptions: the sink represents the periodic infinite string denoted by the circular string s. The labels of edges incident on the sink are themselves infinite and periodic and may be represented by their start positions in s. We also associate with $SCD(s)$ the *periodicity p* of s. This is the value of $|\alpha|$ for the largest value of k such that $S = \alpha^k$. Figure 17 shows $SCD(s)$ for $s = <cabcbab>$.

The scdawg for circular string s is constructed by the algorithm of Figure 13. It is obtained by first constructing the scdawg for the linear string $T = SS$ (where $S = linearize(s)$).

A bit is associated with each re edge in $R(T)$ indicating whether it is a prefix extension edge or not. Similarly, a bit is associated with each le edge in $L(T)$ to identify suffix extension edges. Two pointers, a suffix pointer and a prefix pointer are associated with each vertex, v in $V(T)$. The suffix (prefix) pointer points to a vertex, w, in $V(T)$ such that $de(w)$ is the largest suffix (prefix) of $str(v)$ represented by any vertex in $V(T)$. Suffix (prefix) pointers are the reverse of suffix (prefix) extension edges and are derived from them. Figure 14 shows $CSD(T) = CSD(SS)$ for $S = cabcbab$. The broken edge from vertex c to vertex abc is a suffix extension edge, while the solid edge from vertex ab to vertex abc is a prefix extension edge.

In step 2, we determine the periodicity of s which is equal to the length of the label on any outgoing edge from the vertex representing S in $CSD(T)$.

Next, in step 3, suffix and prefix redundant vertices of $CSD(T)$ are identified. A suffix (prefix) redundant vertex is a vertex v that has exactly one outgoing re (le) edge.

Algorithm CircularScdawg
Step1: Construct $SCD(T)$ for $T = SS$.

Step2:
$v_S :=$ vertex representing S in $CSD(T)$
$e_S :=$ any outgoing edge from v_S
$p := |e_S|$

Step3(a):
{Identify Suffix Redundant Vertices}
$v := sink$;
while $v \neq source$ do
   begin
      $v := v.suffix$;
      if v has exactly one outgoing re edge then mark v suffix redundant;
      else exit Step 3(a);
   end;

Step3(b):
{Identify Prefix Redundant vertices}
{Similar to Step 3(a)}

Step4:
$v := sink$;
while $(v <> source)$ do
   begin
      case v of
         suffix redundant but not prefix redundant: *ProcessSuffixRedundant(v)*;
         prefix redundant but not suffix redundant: *ProcessPrefixRedundant(v)*;
         suffix redundant and prefix redundant : *ProcessBothRedundant(v)*;
         not redundant : {Do nothing };
      endcase;
      $v := NextVertexInReverseTopologicalOrder$;
   end;

Fig. 13. Algorithm for constructing the scdawg for a circular string

A vertex is said to be *redundant* if it is either prefix redundant or suffix redundant or both. In Figure 14, vertex c is prefix redundant only, while vertex ab is suffix redundant only. The vertex representing S is both prefix and suffix redundant since it has one re and one le out edge. The fact that step 3 does, in fact, identify all redundant vertices is established later.

Vertices of $SCD(T)$ are processed in reverse topological order in step 4 and redundant vertices are eliminated. When a vertex is eliminated, the edges incident to/from it are redirected and relabeled as described in Figures 15 and 16. The other cases are similar. The resulting graph is $SCD(s)$.

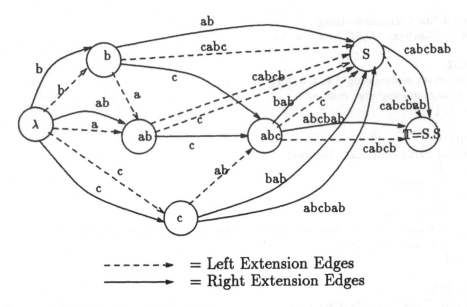

= Left Extension Edges
= Right Extension Edges

Fig. 14. $SCD(T)$ for $T=cabcbabcabcbab$

Procedure ProcessSuffixRedundant(v)

1. Eliminate all left extension edges leaving v (there are at least two of these).
2. There is exactly one right extension edge, e, leaving v. Let the vertex that it leads to be w. Let the label on the right extension edge be $x\gamma$. Delete the edge.
3. All right edges incident on v are updated so that they point to w. If $w \neq sink$, their labels are modified so that they represent the concatenation of their original labels with $x\gamma$. Otherwise, the the edges are represented by their start positions in S.
4. All left edges incident on v are updated so that they point to w. If $w \neq sink$, their labels are not modified. Otherwise, the edges are represented by their start position in s. However, if any of these were suffix extension edges, the bit which indicates this should be reset as these edges are no longer suffix extension edges.
5. Delete v.

Fig. 15. Algorithm for processing a vertex which is suffix redundant

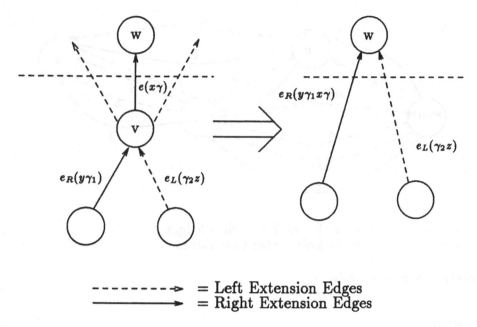

```
- - - - - - - - ->    = Left Extension Edges
————————————→    = Right Extension Edges
```

Fig. 16. v is suffix redundant

Theorem 18. *A vertex v in $V(T)$ is non redundant iff $de(v)$ is a displayable entity of s.*

Lemma 19. *(a) A vertex, v, in $V(T)$ will have exactly one re (le) out edge only if $de(v)$ is a suffix (prefix) of T.*
(b) If a vertex, v, such that $de(v)$ ($|de(v)| < n$) is a suffix (prefix) of T has more than one re (le) out edge, then no vertex, w, such that $de(w)$ is a suffix (prefix) of $de(v)$ can be suffix (prefix) redundant.

We can now show that step 3(a) of **Algorithm CircularScdawg** identifies all suffix redundant vertices in $V(T)$. Since it is sufficient to examine vertices corresponding to suffixes of T (Lemma 19(a)), step 3(a) follows the chain of suffix pointers starting from the sink. If a vertex on this chain representing a displayable entity of length $< n$ has one re out edge, then it is marked suffix redundant. The traversal of the chain terminates either when the source is reached or a vertex with more than one re out edge is encountered (Lemma 19(b)). Similarly, step 3(b) identifies all prefix redundant vertices in $V(T)$.

8 Conclusions

In this paper, we have described efficient algorithms for the computation of display conflicts among patterns in strings. We have extended these algorithms to circular strings by extending the scdawg data structure for linear strings to circular strings. Extending these techniques to the domain of approximate string matching would be useful, but appears to be difficult.

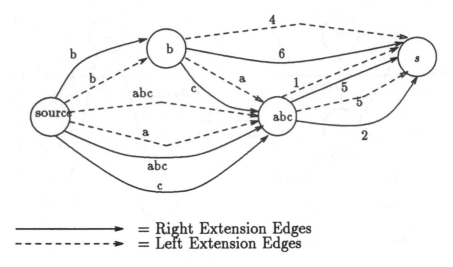

$\longrightarrow$ = Right Extension Edges
$---------\blacktriangleright$ = Left Extension Edges

Fig. 17. Scdawg for $< cabcbab >$

References

1. D. Mehta and S. Sahni, "Models and Techniques for the Visualization of Labeled Discrete Objects," *ACM Symposium on Applied Computing*, pp. 1224–1233, 1991.
2. B. Clift, D. Haussler, T.D. Schneider, and G.D. Stormo , "Sequence Landscapes," *Nucleic Acids Research*, vol. 14, no. 1, pp. 141–158, 1986.
3. G.M. Morris , "The Matching of Protein Sequences using Color Intrasequence Homology Displays," *J. Mol. Graphics*, vol. 6, pp. 135–142, 1988.
4. A. Blumer, J. Blumer, D. Haussler, R. McConnell, and A. Ehrenfeucht, "Complete Inverted Files for Efficient Text Retrieval and Analysis," *J. ACM*, vol. 34, no. 3, pp. 578–595, 1987.
5. S.L. Tanimoto, "A method for detecting structure in polygons," *Pattern Recognition*, vol. 13, no. 6, pp. 389–394, 1981.
6. A. Blumer, J. Blumer, D. Haussler, A. Ehrenfeucht, M.T. Chen, J. Seiferas, "The Smallest Automaton Recognizing the Subwords of a Text," *Theoretical Computer Science*, no. 40, pp. 31–55, 1985.
7. M. E. Majster and A. Reiser, "Efficient on-line construction and correction of position trees," *SIAM Journal on Computing*, vol. 9, pp. 785–807, Nov. 1980.
8. E. McCreight, "A space-economical suffix tree construction algorithm," *Journal of the ACM*, vol. 23, pp. 262–272, Apr. 1976.
9. M. T. Chen and Joel Seiferas, "Efficient and elegant subword tree construction," in *Combinatorial Algorithms on Words* (A. Apostolico and Z. Galil, eds.), NATO ASI Series, Vol. F12, pp. 97–107, Berlin Heidelberg: Springer-Verlag, 1985.
10. E.Horowitz, S. Sahni, *Fundamentals of Data Structures in Pascal, 3'rd Edition.* Computer Science Press, 1990.

Efficient Randomized Dictionary Matching Algorithms

(Extended Abstract)

Amihood Amir[1] Martin Farach[2] Yossi Matias[3]

[1] Georgia Tech[†]
[2] DIMACS[‡]
[3] Univ. of Maryland and Tel Aviv Univ.[§]

Abstract. The standard string matching problem involves finding all occurrences of a single pattern in a single text. While this approach works well in many application areas, there are some domains in which it is more appropriate to deal with *dictionaries* of patterns. A dictionary is a set of patterns; the goal of dictionary matching is to find all dictionary patterns in a given text, simultaneously.

In string matching, randomized algorithms have primarily made use of randomized hashing functions which convert strings into "signatures" or "finger prints". We explore the use of finger prints in conjunction with other randomized and deterministic techniques and data structures. We present several new algorithms for dictionary matching, along with parallel algorithms which are simpler of more efficient than previously known algorithms.

1 Introduction

Traditional *pattern matching* has dealt with the problem of finding all occurrences of a single pattern in a text. A basic instance of this problem is the *exact string matching* problem, the problem of finding all exact occurrences of a pattern string in a text. This problem has been extensively studied. The earliest linear time algorithms include [19] and [7].

While the case of a pattern/text pair is of fundamental importance, the single pattern model is not always appropriate. One would often like to find all occurrences in a given text of patterns from a given set of patterns, called a *dictionary*. This is the *dictionary matching* problem. In addition to its theoretical importance, dictionary matching has many applications. For example, in molecular biology, one is often concerned with determining the sequence of a piece of DNA. Having found the sequence (which is simply a string of symbols) the next step is to compare the string

[†] College of Computing, Georgia Institute of Technology, Atlanta, GA 30332-0280; (404) 853-0083; amir@cc.gatech.edu; partially supported by NSF grant IRI-90-13055.

[‡] DIMACS, Box 1179, Rutgers University, Piscataway, NJ 08855; (908) 932-5928; farach@dimacs.rutgers.edu; supported by DIMACS under NSF contract STC-88-09648.

[§] Institute for Advanced Computer Studies, University of Maryland, College Park, MD 20742; matias@umiacs.umd.edu; partially supported by NSF grants CCR-9111348 and CCR-8906949.

against all known strings to find ones that are related. Clearly one would like an algorithm which is fast, even given some huge dictionary.

Aho and Corasick [1] introduced and solved the exact dictionary matching problem. Given a dictionary D whose characters are taken from the alphabet Σ, they preprocess the dictionary in time $O(|D|\log|\Sigma|)$ and then process text T in time $O(|T|\log|\Sigma|)$. This result is perhaps surprising because the text scanning time is *independent* of the dictionary size (for constant size alphabet). However, if the dictionary changes, that is, if patterns are inserted or deleted, then the work needed to update the dictionary data structure is proportional to the dictionary size, since the entire dictionary needs to be reconstructed. The issue of *dynamic* dictionary matching was introduced in [3]. Related work on dynamic dictionary matching appears in [4, 16].

In this paper we introduce several randomized algorithms that use combinations of useful techniques, such as fingerprints, hash tables, the trie data structure, and tree decomposition. Our algorithms take different approaches than previous algorithms, and they are either simpler or have better performance.

In [18], Karp and Rabin introduced the notion of fingerprints. These are short representative strings which can be used to distinguish, with high probability, between different strings in constant time. Karp and Rabin used the fingerprints to give an elegant algorithm for string matching which runs in linear time, with high probability. Throughout the paper, we follow Rabin and Karp [18] in assuming that the operations allowed in constant time include comparisons, arithmetic operations, and indirect addressing, where the word size is $O(\log|D| + \log|T|)$.

Our results are summarized in Tables 1-2, along with the previously known results. All stated results are asymptotic (the "O" is omitted) and are randomized in the sense that within the stated time the algorithms terminate with one sided error (i.e., the "Monte-Carlo" type). All our algorithms can be easily converted into "Las-Vegas" type algorithms, where there is no error and the randomization is on the running-time only. Let $d = \sum_{i=1}^{k}|P_i|$ be the size of the dictionary, σ be the effective alphabet, that is, the number of distinct characters that occur in D, $t = |T|$ be the text size, $p = |P|$ be the size of the pattern to be inserted or deleted, and m be the size of the largest pattern in D.

	Static		Dynamic	
prep.&updates	$d\log\sigma$	[1]	$p\log d$	[3]
text scan	$t\log\sigma$		$t\log d$	
prep.&updates	d	§2.2	$p\log k$	§5.2
Text Scan	t		$t\log k$	
prep.&updates	d	§4	p	§2.3
text scan	$t\log d$		mt	

Table 1. Serial algorithms

		Static	Dynamic	
prep.&	time		$\log m \log d$ †	
update	work		$p \log d$	[3]
text	time		$\log m \log d$	
scan	work		$t \log m \log d$	
prep.&	time	$\log d \log^* d$	$\log p \log^* p$	
update	work	$d \log d$	$p \log p \log k$	§5
text	time	$\log d$	$\log d$	
scan	work	$t \log d$	$t \log d$	
prep.&	time		$\log p$	
update	work		p	§2.3
text	time		$\log t$	
scan	work		tm	

Table 2. Parallel algorithms. (The preprocessing and update complexities are per pattern.)
† Only semi-dynamic: no deletion.

The rest of the paper is organized as follows. In Section 2, we introduce finger-prints and perfect hashing and show how to combine them into a simple dictionary matching algorithm. In Section 3, we discuss the trie data structure and describe simple trie-based dictionary-matching algorithms. These algorithms serve as a warm-up towards the main algorithm. In Section 4, we present the main algorithm of the paper, which is based on tries and on a tree decomposition technique. This algorithm shares some features with the algorithm presented in [24]; it is however based on a simpler data structure and in fact solves a different problem. Parallel algorithms are presented in Section 5. Our algorithms are more efficient than previous ones, and are the first to support efficiently parallel dynamic dictionary-matching with the delete operation.

2 Fingerprints and Hashing

In this section, we discuss fingerprints, a basic tool for randomized string match-ing. The fingerprints technique will be an essential tool in the algorithms given in the following sections. We show here how they can be used, together with perfect hashing, to get extremely simple algorithms for dictionary matching.

2.1 Fingerprints

Karp and Rabin [18] introduced a simple, yet powerful, technique for randomized string matching. In their algorithm, a given string P is replaced by a fingerprint function $f(P)$: a short string that represents the string. Similarly, each candidate substring S_i in the text T is substituted by a fingerprint $f(S_i)$, where S_i is the substring of length p starting at position i in the text, $i = 1, \ldots t - p + 1$. The idea is to test for a match between P and S_i, by first testing for a match between $f(P)$

and $f(S_i)$. Karp and Rabin showed that one can select a fingerprint function f with the following properties:

1. The function $f(\cdot)$ can be computed for all substrings in T, in $O(t)$ time, and in parallel in $O(\log t)$ time and optimal number of processors.
2. The function f gives a "reliable" fingerprint: if $|f(s)| = O(\log p + \log t)$ then the probability for a false match among t pairs of strings, each of length p, can be bounded by $O(t^{-c})$, for any constant $c > 0$.

The Karp-Rabin algorithm can be described in two steps. First, a fingerprint of length $O(\log t)$ is computed for the pattern P and for all substrings of lengths p. Then, the fingerprint $f(P)$ is checked against each position in the text T; each comparison is assumed to take constant time since the word size is $O(\log t)$. In $O(t)$ sequential time, or in $O(\log t)$ parallel time and optimal work, the algorithm finds all positions i in the text such that $f(P) = f(S_i)$. Each of these positions is a candidate, with high probability, for a match. We can check a candidate match in $O(p)$ steps in the straightforward manner. Since a mismatch of a candidate match only occurs with a polynomially small probability, the number of steps required for finding one occurrence of P in the text or deciding that P does not appear is $O(t)$, with high probability. However, if one wishes to list all occurrences of P without an error, the number of steps would be, with high probability, $O(t + lp)$, where l is the number of times P appears in T. Note that in this case lp is the size of the output.

2.2 Hashing

Given a set S of n keys from a finite universe $U = \{0, \ldots, q - 1\}$, i.e. $S \subset U$ and $|S| = n$, the *hashing* problem is to find a one-to-one function $h : S \mapsto [1, dn]$, for some constant d, such that h is represented in $O(n)$ space and for any $x \in U$, $h(x)$ can be evaluated in constant time. The function h is called a *hash function* and the induced data structure is called a *hash table*. The requirement implies that set-membership queries can be processed in constant time.

The *dynamic hashing* problem is to maintain a data structure which supports the *insert*, *delete*, and *membership query* instructions. That is, the set of keys S is changing dynamically. It is required that at all times, the size of data structure be linear in the number of keys stored in it.

Sequential algorithms In a seminal paper, Fredman, Komlós, and Szemerédi [12] gave a simple algorithm for the hashing problem, whose expected running time is $O(n)$. The general framework introduced in [12] was used in all subsequent papers mentioned here. A dynamic-hashing algorithm with $O(1)$ expected amortized cost per insertion was introduced by [9]. A real-time dynamic-hashing, in which every insertion can be completed in constant time with high probability was introduced in [11]. In a recent paper, [8] show how to modify the FKS hashing algorithm so that its running time would be $O(n)$ with high probability, and so that the number of random bits used by the algorithm is substantially reduced. They also give a simplified real-time dynamic-hashing algorithm that requires substantially fewer random bits than in [11].

266

Parallel algorithms A first optimal[7] parallel dynamic-hashing algorithm was introduced in [10]; its running time is $O(n^\epsilon)$ for any fixed $\epsilon > 0$. An optimal parallel (static) hashing algorithm in $O(\log n)$ expected time was given in [21], followed by an $O(\log\log n)$ expected time algorithm presented in [14]. An $O(\log^* n \log\log^* n)$ expected time[8] algorithm was given in [20]; its time complexity was later shown to actually be $O(\log^* n)$ with high probability [15]. A similar improvement (from $O(\log^* n \log\log^* n)$ to $O(\log^* n)$) was also given by [6]. A real-time parallel dynamic-hashing algorithm was given by [15]; the algorithm supports a batch of n membership-query or delete instructions in constant time, and a batch of n insert instructions in $O(\log^* n)$ time with high probability, using an optimal number of processors.

Applications As noted in [21], hashing can replace sorting-based naming assignment procedures which are used to map potentially large alphabets into a linear size alphabet. In the algorithm given in [1], the $\log\sigma$ factors in the $O(d\log\sigma)$ preprocessing time and in the $O(t\log\sigma)$ text processing time are the result of a binary search through characters which occur in the dictionary patterns. This binary search can be replaced by lookups into a hash table, thus improving the complexities to $O(d)$ for preprocessing and $O(t)$ for text processing.

2.3 Fingerprint based Dictionary Matching

We are ready to present a first dictionary matching algorithm. The algorithm is extremely simple, and its *parallel* complexity is superior to the previously known algorithm.

Assume, for the time being, that all patterns are of the same length p. We will use a fingerprint function of size $O(\log t + \log d)$. This will guarantee that any false match between a substring and a pattern in the dictionary still occurs with sufficiently small probability.

The preprocessing phase of our algorithm consists of the following steps:

1. For each pattern P in the dictionary D compute a fingerprint $f(P)$ of length $|f(P)| = O(\log t + \log d)$.
2. Insert the fingerprints of the dictionary patterns into a perfect hash table $H(D)$.

The text processing consists of the following steps:

3. The fingerprint $f(S)$ is computed for each substring S of size p in the text; this step is done as in the Karp-Rabin algorithm.
4. For each substring S in the text, we perform a membership query into the hash table $H(D)$; for each substring S we find in constant time if it is a candidate for a match. As in the Karp-Rabin algorithm, a candidate has a false match with polynomially small probability.

---

[7] An *optimal* parallel algorithm is an algorithm whose time-processor product is the same, up to a constant factor, as the best known sequential time.
[8] Let $\log^{(i)} x \equiv \log(\log^{(i-1)} x)$ for $i > 1$, and $\log^{(1)} x \equiv \log x$; $\log^* x \equiv \min\{i : \log^{(i)} x \le 2\}$. The function $\log^*(\cdot)$ is extremely slow increasing and for instance $\log^* 2^{65536} = 5$.

In general, the patterns need not be of the same size. Let the length of the longest pattern be m. Then the patterns can be partitioned into at most m equivalence classes based on length. Using the single length algorithm as a subroutine, we can match for each class independently. The hash table can be built using an appropriate hashing algorithm from Section 2.2. We therefore get an algorithm with $O(d)$ time for preprocessing and $O(tm)$ time for text processing.

We note that the fingerprints for a set of patterns can be obtained by concatenating them into a single (long) string and then applying the Karp-Rabin algorithm. This yields an optimal parallel preprocessing algorithm that takes $O(\log d)$ time and $O(d)$ work. In fact, it is easy to obtain an optimal parallel preprocessing algorithm that takes $O(\log m)$ time and $O(d)$ work. The text processing takes $O(\log t)$ time (this can also be improved) and $O(tm)$ work.

For dynamic dictionary matching, we assume that we have a set D' of d' patterns to be inserted or deleted from the dictionary. The update of the dictionary is then similar to steps 1 and 2 above:

1'. For each pattern P in the set D' compute a fingerprint $f(P)$ of length $|f(P)| = O(\log t + \log d)$.

2'. Insert/delete the fingerprints of the new dictionary patterns into/from the perfect hash table $H(D)$.

Step 1' is implemented similarly to Step 1. To implement Step 2', the appropriate dynamic hashing algorithm can be used (see Section 2.2). Therefore, a pattern P can be added into the dictionary or deleted from the dictionary in time $O(p)$.

In parallel, a batch of d' patterns of size at most m can be inserted to or deleted from the dictionary in $O(\log m + \log^* d')$ time and $O(md')$ work.

An input sensitive algorithm In the above algorithm the text scan takes $O(tm)$ time. We present an input-sensitive algorithm which may improve significantly on this bound for certain cases.

Each string P_i in the dictionary D can be decomposed into a "long" prefix $L(P_i)$ and "short" suffix $S(P_i)$, i.e., $P_i = L(P_i)S(P_i)$. $L(P_i)$ is chosen to be the longest prefix whose size is an integer multiple of $\lfloor \sqrt{m} \rfloor$: $L(P_i) = P_i[1], \ldots, P_i[c\sqrt{m}]$, where $c\sqrt{m} \leq p_i < (c+1)\sqrt{m}$, and $S(P_i) = P_i[c\sqrt{m}+1], \ldots, P_i[p_i]$. Note that short strings come in at most $\sqrt{m}$ different sizes, as do long strings. We can therefore use the above algorithm to find all occurrences of long prefixes and of short suffixes of the dictionary patterns in $O(t\sqrt{m})$ time. The remaining issue is how to reconstruct the dictionary strings from the short and long strings that match at each text location.

A naive approach would be to store all pairs $\langle l, s \rangle$ such that long string L_l and short string S_s form a dictionary string $L_l S_s$ in the hash table. Then we could check, for each text location, which of the $O(\sqrt{m})$ long strings and $O(\sqrt{m})$ short strings that matched at that location form a valid pair. Note that at each location we may have match with up to $\sqrt{m}$ long prefixes and up to $\sqrt{m}$ short ones. Therefore, this algorithm takes $O(tm)$ time at the worst case.

However, in many cases only few strings would have the same long prefix. Suppose that no more than q strings in the dictionary have identical $L(\cdot)$ prefixes. Then the algorithm can be made to run in time $O(qt\sqrt{m})$ by simply checking, for each long prefix, if one of its $O(q)$ short suffixes matched. More accurately, let q_i be the number of dictionary patterns sharing the same long prefix as the pattern P_i. Then for a given text the running time is $O(t\sqrt{m}\sum_{i=1}^{d} q_i)$ where the sum is only over the dictionary patterns P_i for which there is a match between the text and their long prefix $L(P_i)$. While the worst case running time is the same as the basic algorithm above, $O(tm)$, it is clear that for many input instances this algorithm will perform much better.

Two dimensional dictionary matching The string matching algorithm of Karp and Rabin extends also for two-dimensional data. Our fingerprints based dictionary matching algorithms can be extended in a similar manner. Details are omitted from this extended abstract.

3 Tries

Consider the following simple algorithm for dictionary matching. For each text location, try to match each pattern by brute force. This algorithm clearly runs in $O(td)$. We will improve on this algorithm in several ways to improve the complexity. First, we use the well known data structure, the trie. A trie is a data structure that allows a set of strings to be stored and updated, and allows membership queries. A straightforward use of a trie for the dictionary problem leads to an algorithm that runs in time $O(tm)$ were m is the length of the longest pattern.

In the following subsections, we present some background needed to manipulate tries. In section 4, we will show how to modify a trie to speed up the search.

3.1 Definition and Construction

Tries are a common data structure for representing sets of strings [2].
Definition: Let $D = \{P_1, \ldots, P_k\}$ where $P_i \in \Sigma^*$ and $\$ \notin \Sigma$. A *trie* T_D of D is a tree such that each edge is labeled with a character from $\Sigma \cup \{\$\}$, and each pattern $P_i \in D$ is represented by a leaf l_i such that the concatenation of the edge labels from the root to l_i is the string $P_i\$$. Furthermore, the only nodes in a trie are exactly those so induced by the k leaves $l_1, \ldots, l_k$.

Constructing and updating a trie is straightforward. Suppose that we want to add string $P[1, \ldots, m]$ to D and to T_D. Then we simply start with the root of the tree as the current node, and at the beginning of P. We check if there is an edge labeled with the next character in P. If such a edge exists then we traverse it, resetting the current node, and advancing one step in the pattern P. When we can no longer traverse the tree, we create a linear subtree for the rest of the pattern and insert it below the current node. Finally, we add a single node at the bottom of the resulting tree with an edge labeled with a $\$$.

To remove a pattern P_i, simply start at leaf l_i and remove it. Proceed to the parent, and, if it now has outdegree zero, remove it. The process continues until a node is reached which still has a child left.

Time: For a bounded alphabet, the outgoing edge labels of each node can be checked in constant time. For unbounded alphabets, the edge labels can be held in a hash table. Then query takes constant time and adding a new edge is implemented by inserting a new element into the hash table (see § 2.2). Therefore the updated and traversal operations can be performed in linear time.

3.2 Compressed Tries

It will sometimes be the case that many nodes in a trie will have out degree one. In such a case, we can use a compressed trie to significantly reduce the space requirements for storing the trie.

In a compressed trie, edges are labeled with substrings of the input patterns rather than simply with single characters. Whenever we have a chain of degree one nodes, we compress the chain into a single edge labeled with the substring obtained by concatenating the deleted edge labels. Clearly the new label will be a substring of some input pattern and so it can be represented by a pair which consists of a pointer into that input pattern and a length. Now every internal node has degree at least two. The size of such a trie on k patterns is therefore $O(k)$ rather than $O(d)$ for the uncompressed trie.

Furthermore, a compressed trie, in conjunction with fingerprints (see § 2.1) can be used to speed up the search through the trie. Let CT_D be the compressed trie for some set D of strings. Let T_D be the corresponding uncompressed trie. If $depth(CT_D) \ll depth(T_D)$ then we can apply the following algorithm.

Proceed as before independently from each text location to search through CT_D. When an edge is labeled with a character, simply compare the character and proceed appropriately. If an edge is labeled with a pattern substring, then compare the fingerprint of that substring with the appropriate fingerprint from the text. Traverse the edge if the two fingerprints are equal.

We still take constant time at each edge to determine if we jump down the tree or if we are done. Therefore, the complexity of this algorithm is $O(t \times depth(CT_D))$.

In general, $depth(CT_D)$ need not differ significantly (or at all) from $depth(T_D)$. In the following section, we present a scheme for producing a shallow search tree based on CT_D, thus speeding of the search algorithm.

4 Trie Based Dictionary Matching Algorithm

Many techniques have been used to balance trees in order to accelerate searching through them. Our problem, as described in Section 3 requires just such a balancing technique. We will consider the separator decomposition of a tree and show how this technique can be used in conjunction with the trie dictionary matching algorithm described above.

4.1 A Separator Decomposition Tree

The notion of the complete separator decomposition of a rooted tree is well known and has been extensively applied (see [17] or [22]). It has been used in the context of string matching by Naor [24] (the problem considered is on-line string matching after the text and the pattern are pre-processed separately).

Definition and construction: In a rooted tree T, $|T| = n$, a node v is called a *separator* if each of the connected components induced by removing v from T is of size at most $\frac{2}{3}n$. It is well known that every tree contains a separator. A *complete separator decomposition tree* $SD(T)$ of T is defined as follows. $SD(T)$ is a rooted tree whose root is a separator v of T. By removing the node v from T we get rooted trees whose roots are the children of v in T as well as the subtree rooted at the root of T (assuming that the root of T is not a separator of T). The recursively defined separators of these rooted subtrees are the children of v in the complete separator decomposition tree $SD(T)$.

It is easy to see that the height of the separator decomposition tree $SD(T)$ is $O(\log n)$. It is known that a complete separator decomposition tree can be constructed in linear time [24].

4.2 Algorithm

We are now ready for the dictionary matching algorithm based on balanced tries.

Suppose that dictionary $D = \{P_1, \ldots, P_k\}$ has compressed trie CT_D and that tree CT_D has separator decomposition $SD(CT_D)$. We will construct a *balanced trie* BT_D as follows. Let $ST(v)$ be the subtree rooted at *node*. Let $ST(v) - ST(v')$ be the subtree rooted at v with the subtree rooted at v' deleted, if v' is a descendant of v, and $ST(v) - ST(v') = ST(v)$ otherwise.

To define BT_D, we take CT_D and perform the following operations. Let v be a separator of CT_D and let $s(v)$ be the concatenation of the labels from the root to v. Let $c_1, c_2, \ldots, c_q$ be the children of the root and let $l(c_i)$ be the label on edge $(root, c_i)$. Then the root of BT_D will have $q + 1$ children $v', c_1', \ldots, c_q'$ where $l(v') = s(v)$ and $l(c_i') = l(c_i)$. Furthermore, the subtree rooted at v' will be the recursively defined balanced trie of $ST(v)$ and the subtrees rooted at each c_i' will be the balanced trie of $ST(c_i) - ST(v)$.

Now the algorithm proceeds as in the case for the compressed trie with one modification. We have a *distinguished* edge which points to the separator u' of the subtree rooted at the current node u. Similarly, we call a child v *distinguished* if the edge between v and $parent(v)$ is the distinguished edge of $parent(v)$. The distinguished edge is labeled by the fingerprint of the string associated with the path between u and u' in the trie. We test this edge first. Whether this edge matches or not, this test eliminates a constant fraction of the remaining trie, thus making the worst case search time logarithmic in k.

To summarize, once a balanced trie has been built for a dictionary, we set $cur \leftarrow root$ and repeat the following:

1. Let v be the distinguished child of cur.
2. If $l(v) =$ next characters in text, that is, their fingerprints match, then
   - set $cur \leftarrow v$
   - advance the pointer in text by $|l(v)|$

   else
   - check which of the remaining edges starts with the next text character.
   - Proceed down that edge if possible.

As with the trie searching procedures described in Section 3, the output of the Balanced Trie search is, for each text position, the longest pattern prefix that matches at that text location. In a straightforward traversal of a trie, this is sufficient to determine which patterns match at that text location, since we traverse all nodes representing prefixes of the longest match. As we descend through the tree, we can check in constant time if there is an outgoing edge labeled with a $. We know that any such edge represents a matching pattern.

However, in a balanced trie, we may jump over such nodes if we traverse the distinguished edge of some node. We can reduce the problem of finding all prefixes of a pattern which are themselves patterns to a tree problem as follows. We call any node in the trie CT_D a *marked* node if it has an outgoing edge labeled with a $. The nodes in BT_D are clearly in a one-to-one correspondence with the nodes of CT_D. Therefore, if s is the longest pattern prefix that matches at text location i, we can find the matching patterns at i by finding all the marked ancestors of the node representing i in CT_D.

Finding Marked Ancestors The algorithm for finding the nearest marked ancestor given an unchanging trie is quite straightforward. By simply traversing the trie in depth first search order, we keep a stack of marked nodes, pushing each marked node as we encounter it and popping it as we exit that node for the last time, i.e. when we finish processing the subtree rooted at that node. For each node, we associate a pointer which we set to be the top node on the marked node stack. Clearly, each node will end up with a pointer to the nearest marked ancestor.

The total time for static dictionary matching is then:

Preprocessing: $O(d)$

Text Scanning: $O(t \log k)$.

4.3 Dynamic Dictionary Matching

The dynamic algorithm is a modification of the static case. Rather than using a balanced trie based on the separator decomposition tree of the dictionary trie, we use a dynamic balanced trie based on the balanced decomposition tree described below. In addition, we can no longer afford to maintain back pointers from each node to their nearest marked ancestor, since new marked nodes may be introduced into the dictionary trie at any time. Instead, we maintain a secondary data structure which allows us to update insertion and deletion information directly, a variation of which appears in [3].

Dynamic Balanced Tries We would like to maintain the balanced trie data structure, when inserting new nodes. To enable efficient computation, we replace this data structure with an alternative one which has more relaxed requirements. A node v in T is called a *pseudo-separator* if each of the subtrees induced by removing v from T is of size at most $\frac{3}{4}n$. We call a balanced trie in which pseudo-separators replace separators a *weakly balanced trie (WBT)*. It is easy to see that the height of a WBT is $O(\log n)$ as well. Such weakly balanced tries can therefore replace balanced tries in our applications. Our objective is to support insertions with an $O(\log n)$ cost per update (since this will meet the complexity bounds of other procedures in our algorithms). Indeed this is done as follows.

We start with a weakly balanced trie that is also a balanced trie. Each node keeps track on the size of its subtree; thus each inserted node updates all nodes along its path from the root in the trie. When a sub-tree becomes too large, let T' be the largest sub-tree that becomes unbalanced. We construct a new complete separator decomposition for T'. While this construction costs $O(|T'|)$ work, we note that T' can become unbalanced only after $O(|T'|)$ insertions (due to the relaxation in the definition of the WBT). Since each node belongs to $O(\log n)$ different sub-trees in the WBT, the update cost of each insertion is $O(\log n)$, as desired.

Finding Marked Ancestor Consider an Euler-tour for the trie CT_D. In such a tour, each edge appears twice, and between those two occurrences are the edges of the subtree rooted at that edge. Similarly, the edges at the subtree rooted at a node v occur between the edges $(parent(v), v)$ and $(v, parent(v))$. Now consider the effect of marking a node v in the tree. As noted above, the subtree rooted at v form a consecutive subsequence in the Euler-tour. It is relatively straightforward to cast the operation of marking nodes, unmarking nodes, and finding the nearest marked ancestor as the sequence operations of **double split**, **union**, and **find**. The analogy is clear since marking a node v means that the descendant edges of v can be **split** away from the rest of the tree. Now finding the nearest marked ancestor involves a **find** operation to find which set the edge currently finds itself in. Finally, unmarking a node reverses the effect of a **double split**, therefore it is implemented by a **join**.

The implementation of such a scheme is straightforward since the induced an ordering on the edges which can be used to maintain a 2-3 tree (see e.g. [2]). 2-3 trees on n nodes support such operations as **split**, **find**, **min**, and **concatenate** in $O(\log n)$ time. The details of the use of **split**, **min**, and **concatenate** to implement the **double split**, **find** and **join** operations on the Euler-tour, are left to the final paper.

Inserting or Deleting P: $O(p \log d)$
Scanning T: $O(t \log d)$

5 Parallel Algorithms

The parallel algorithms described below are based on the algorithms of Section 4. Both the static and dynamic algorithms had the following four components. In the preprocessing we build a trie for the dictionary. We then construct a balanced trie from it. During the text scanning phase, we must traverse the trie and then find the nearest marked ancestor.

5.1 Parallel Static Dictionary Matching

Parallel Trie Construction A suffix tree of a given string is a compressed trie of a dictionary which consists of all the suffixes of the string. Thus, if we take a string to be the concatenation of the dictionary patterns, its suffix tree will contain the trie of the dictionary patterns as a sub-tree, with possibly some paths compressed. Hence, we can use the suffix tree in the same way that we use compressed tries in Sections 3 and 4 for dictionary matching.

Apostolico et al [5] showed how to build a suffix tree for a string of length n in $O(n \log n)$ operations, $O(\log n)$ time, and $n^{1+\epsilon}$ space, for any constant $\epsilon > 0$ (see also [13]). Using a parallel hashing algorithm (Section 2.2), the algorithm can be implemented in linear space; the time then increases by a factor of $O(\log^* n)$ with high probability, but there is no change in the number of operations.

Building the balanced trie In this paper, we will be satisfied with construction which takes $O(\log n)$ time, using n processors for an n node tree T, since we will have other procedure with this complexity. For this performance a simple implementation can be described. We use the Euler-tour technique of Tarjan and Vishkin [26] to compute for each node the size of the subtree of T rooted at this node; this step takes $O(\log t)$ time and $O(t)$ operations. The idea in this technique is to hold a linked list along an Euler circuit that "surrounds" the tree T, starting at the root. The circuit is defined over a graph defined by replacing each edge in T with two anti-symmetric edges. Then, by computing the rank of each edge in the list, a node v can easily find out the size of its subtree by computing the difference in rank between the edge $p(v) \to v$ and $v \to p(v)$, where $p(v)$ is the parent of v in T. Now a separator v can be recognized in constant time and t processors as follows: each of its neighboring edges notify it that the subtree rooted at each of its children is of size at most $\frac{2}{3}n$ and that the subtree rooted at v is of size at least $\frac{1}{3}n$ (equivalently, each connected component induced by removing v has at most $\frac{2}{3}n$ nodes). To construct a balanced trie, we employ the procedure above $O(\log n)$ times. The main observation is that after a separator is removed from a tree, the list of each new sub-tree can be updated in constant time.

Traversing the trie Since the sequential time is $O(\log d)$, we simply assign one processor to each text location and traverse the trie sequentially for each location.

Finding the nearest marked ancestor The final part, that of assigning a pointer from each node to its nearest marked neighbor, can be accomplished in $O(\log d)$ time with d processors using pointer doubling. The overall bounds for preprocessing are $O(\log d \log^* d)$ time and $O(d \log d)$ work.

Scanning the text The text scanning phase has two parts. Scanning through the trie, for each text location, takes $O(\log d)$ sequential time. We can therefore accomplish this phase in $O(\log d)$ with n processors. The second part, that of finding the marked ancestor, takes constant sequential time. The overall bounds are therefore $O(\log d)$ time with n processors.

Overall time bounds For preprocessing, we have time bounds of $O(\log d \log^* d)$ time, $O(d \log d)$ work. To scan the text, it takes $O(\log d)$ time and $O(n \log d)$ work.

5.2 Dynamic Matching

Parallel Trie Update Unlike the sequential suffix tree construction, the parallel construction is not amenable to insertions or deletions of patterns. Therefore, we must resort to indirect methods. In [23], Mehlhorn described a scheme for introducing some dynamization into static data structures, at the expense of amortization and a logarithmic slowdown. Full dynamization is not possible by simply using Mehlhorn's scheme. Only pattern insertion is allows. We discuss pattern deletions separately below. Details of Mehlhorn's scheme are too involved for inclusion in this abstract. We simply note that we can achieve the following complexities by using the above static algorithm, that is, both the suffix tree construction and the trie balancing, as a "black box."

Inserting P: $O(\log p \log^* p)$ time, $O(p \log p \log k)$ work.

Text Scan: $O(\log d + \log t)$ time, $O(t \log d \log k)$ work.

Allowing for pattern deletions To make the algorithm fully dynamic, that is, to allow dictionary patterns to be deleted, we must take into account pattern deletions. As noted above, Mehlhorn's dynamization scheme does not allow for efficient deletions. The problem of pattern deletions can be neatly solved by an additional data structure as follows.

Let m_j be the marked node associated with pattern P_j. Notice that the suffix tree induces a forest partial order amongst the marked nodes as follows: $m_j \leq m_k$ if m_j is an ancestor of m_k. Let $\mathcal{P}(\mathbb{1}_j) = \mathbb{1}_j$ if m_j is the parent of m_i under this forest partial order. If m_i has no parent, the we set $\mathcal{P}(\mathbb{1}_j) = \Lambda$, the root. If we start with each m_j disconnected from the other m_ks and we **join** m_j to $P(m_j)$ whenever m_j is deleted, then finding the nearest marked ancestor for a node v in the suffix tree becomes a two part process. First, the static pointer is followed to get v's nearest marked ancestor m_i. However, since P_i may have been deleted, we check the supplemental data structure for the tree in which m_i currently finds itself. The root of this tree will be the appropriate marked ancestor for n.

Implementation The above additional data structure can be viewed as implementing a UNION/FIND on the set of m_j in the suffix tree. As noted in section 4, many elegant and efficient amortized time algorithms exist for UNION/FIND (see e.g. [25]). However, since it takes $O(\log d)$ to traverse the suffix tree, we can use an $O(\log d)$ worst case time algorithm for UNION/FIND (see [2]).

Comment The general scheme of using the general dynamization paradigm of Mehlhorn for insertions and the union/find operations for deletions, can be used to adapt the Aho-Corasick (static) algorithm into a dynamic one; the preprocessing and updates then become $O(p \log k)$, and the text scan takes $O(t \log k)$.

Overall time bounds

Inserting P: $O(\log p \log^* p)$ time, $O(p \log p \log k)$ work.

Removing P: $O(\log d)$ sequential time initially for the **union** operation followed by an amortized $O(\log d \log^* d)$ time, $O(p \log d)$ work to "clean up" the remnants of the patterns which were not actually deleted from the Mehlhorn dynamic data structure.

Text Scan: $O(\log d + \log t)$ time, $O(t \log d \log k)$ work.

References

1. A. V. Aho and M. J. Corasick. Efficient string matching. *Commun. ACM*, 18(6):333–340, 1975.
2. A. V. Aho, J. E. Hopcroft, and J. D. Ullman. *Data Structures and Algorithms.* Addison-Wesley Publishing Company, 1983.
3. A. Amir and M. Farach. Adaptive dictionary matching. *FOCS '91*, 1991.
4. A. Amir, M. Farach, R. Giancarlo, Z. Galil, and K. Park. Dynamic dictionary matching. Manuscript, 1991.

5. A. Apostolico, C. Iliopoulos, G. M. Landau, B. Schieber, and U. Vishkin. Parallel construction of a suffix tree. *Algorithmica*, 3:347–365, 1988.
6. H. Bast and T. Hagerup. Fast and reliable parallel hashing. In *SPAA '91*, pages 50–61, July 1991.
7. R. Boyer and J. Moore. A fast string searching algorithm. *Commun. ACM*, 20:762–772, 1977.
8. M. Dietzfelbinger, J. Gil, Y. Matias, and N. Pippenger. Polynomial hash functions are reliable. In *ICALP '92*, July 1992.
9. M. Dietzfelbinger, A. Karlin, K. Mehlhorn, F. Meyer auf der Heide, H. Rohnert, and R. E. Tarjan. Dynamic perfect hashing: Upper and lower bounds. In *FOCS '88*, pages 524–531, Oct. 1988. Also, Revised Version: Tech. Report, University of Paderborn, FB 17 Mathematik/Informatik, 1991.
10. M. Dietzfelbinger and F. Meyer auf der Heide. An optimal parallel dictionary. In *SPAA '89*, pages 360–368, 1989.
11. M. Dietzfelbinger and F. Meyer auf der Heide. A new universal class of hash functions and dynamic hahshing in real time. In *ICALP '90*, pages 6–19, 1990.
12. M. L. Fredman, J. Komlós, and E. Szemerédi. Storing a sparse table with $O(1)$ worst case access time. *J. ACM*, 31(3):538–544, July 1984.
13. Z. Galil and R. Giancarlo. Data structures and algorithms for approximate string matching. *J. of Complexity*, 4:33–72, 1988.
14. J. Gil and Y. Matias. Fast hashing on a PRAM—designing by expectation. In *SODA '91*, pages 271–280, Jan. 1991.
15. J. Gil, Y. Matias, and U. Vishkin. Towards a theory of nearly constant time parallel algorithms. In *FOCS '91*, pages 698–710, Oct. 1991.
16. R. Indury and A. Schaeffer. Dynamic dictionary matching with failure functions. In *Combinatorial Pattern Matching*, 1992.
17. C. Jordan. Sur le assemblages des lignes. *J. Reine und Ang. Math.*, 70:185–190, 1869.
18. R. M. Karp and M. O. Rabin. Efficient randomized pattern-matching algorithms. *IBM J. of Research and Development*, 31:249–260, 1987.
19. D. Knuth, J. Morris, and V. Pratt. Fast pattern matching in strings. *SIAM J. Comput.*, 6:323–350, 1977.
20. Y. Matias and U. Vishkin. Converting high probability into nearly-constant time—with applications to parallel hashing. In *STOC '91*, pages 307–316, 1991. Also in UMIACS-TR-91-65, Inst. for Advanced Computer Studies, Univ. of Maryland, April 1991.
21. Y. Matias and U. Vishkin. On parallel hashing and integer sorting. *J. of Alg.*, 12(4):573–606, 1991.
22. N. Megiddo. Applying parallel computation algorithms in the design of serial algorithms. *J. ACM*, 28:852–865, 1983.
23. K. Mehlhorn. *Multi-dimensional Searching and Computational Geometry*. Springer-Verlag, Berlin Heidelberg, 1984.
24. M. Naor. String matching with preprocessing of text and pattern. In *ICALP '91*, pages 739–750, 1991.
25. R. E. Tarjan. Efficiency of a good but not linear set union algorithm. *J. ACM*, 22:215–225, 1975.
26. R. E. Tarjan and U. Vishkin. Finding biconnected components and computing tree functions in logarithmic parallel time. *SIAM J. Comput.*, 14:862–874, 1985.

Dynamic Dictionary Matching
with Failure Functions
(Extended Abstract)

Ramana M. Idury* and Alejandro A. Schäffer**

Department of Computer Science
Rice University
P. O. Box 1892
Houston, Texas 77251 U.S.A.

Abstract. In the dynamic dictionary matching problem, a dictionary D contains a set of patterns that can change over time by insertion and deletion of individual patterns. The user also presents text strings and asks for all occurrences of any patterns in the text.

Amir, Farach, Galil, Giancarlo, and Park [2, 4] used an automaton based on suffix trees to solve the dynamic problem. We show how to match their time bounds for update and search using a failure function framework, similar to that used by Aho and Corasick to solve the static dictionary matching problem. We then show that our approach allows us to achieve faster search times at the expense of the update times. Finally, we show how to speed up the initial dictionary construction.

1 Introduction

Amir, Farach, Galil, Giancarlo, and Park (AFGGP for short) initiated the study of the *dynamic dictionary matching problem* [2, 4]. We are given a collection of patterns $D = \{P_1, P_2, \ldots, P_s\}$, called the *dictionary*, that can change over time. The basic matching operation is to *search* a text $T[1, t]$ and report all occurrences of dictionary patterns in the text. The dictionary can be changed by *inserting* or *deleting* individual patterns.

The *static* or *semi-adaptive dictionary matching problem*, in which inserts and deletes are not supported, was addressed by Aho and Corasick [1]. Algorithms for both the semi-adaptive and dynamic (or adaptive) problems have applications to database searches and molecular biology [1, 4].

The Aho-Corasick algorithm (AC for short) for semi-adaptive dictionary matching can be summarized as follows. We use $|D|$ to denote the total size of all the patterns in the dictionary; we use Σ to denote the alphabet and σ to denote the number of characters that appear in some word of the dictionary. The AC algorithm builds the dictionary D in time $O(|D| \log \sigma)$ and searches a text $T[1, t]$ in $O((t + tocc) \log \sigma)$ time, where $tocc$ is the total number of occurrences reported. The time complexity of this algorithm is considered linear as the $\log \sigma$ factor is usually

---
* Research partially supported by the W. M. Keck Foundation.
** Research partially supported by NSF grant CCR-9010534.

implicit while stating the time bounds. The AC algorithm uses an automaton where states correspond to prefixes of dictionary patterns and transitions are determined by failure functions as in the Knuth-Morris-Pratt [9] string searching algorithm.

The AFGGP algorithm for dynamic dictionary matching uses a suffix tree [12, 10, 7] as an automaton. Each state corresponds to a substring of some pattern. The AFGGP algorithm inserts or deletes a pattern $P[1, p]$ in time $O(p \log |D|)$, and performs a search in time $O((t+tocc) \log |D|)$. If inserts and deletes are frequent enough, this algorithm is better than the simple alternative of using the AC algorithm for searches and rebuilding the dictionary in $O(|D| \log \sigma)$ time at each update. In stating the time bounds, we implicitly use the fact that $\sigma \leq |D|$ because the terms of the form $\log |D|$ really stand for $\log |D| + \log \sigma$. Amir, Farach, and Matias have found efficient randomized algorithms for dynamic dictionary matching [6].

In this paper, we present another algorithm for dynamic dictionary matching that addresses three questions raised by the AFGGP algorithm and time bounds.

First, is the idea of failure functions and the AC automaton of any use at all for dynamic dictionary matching? Our algorithm is based on the AC automaton. With a suitable choice of underlying data structures, our method achieves a search time of $O((t + tocc) \log |D|)$ and an update time of $O(p \log |D|)$ matching the AFGGP bounds.

Second, are other tradeoffs between search and update times possible? If updates are relatively infrequent or the text strings are much longer than the patterns, we would prefer a search time better than $O((t + tocc) \log |D|)$, while tolerating an update time worse than $O(p \log |D|)$. It is interesting to ask: how good can we make the update time if we insist that the search time match the AC bound of $O((t + tocc) \log \sigma)$ for the static problem?

We show that our algorithm can also use a different data structure such that for any constant $k \geq 2$, it can achieve search time $O(t(k + \log \sigma) + (k \cdot tocc))$ and update time $O(p(k|D|^{1/k} + \log \sigma))$. We thereby match the static search time of [1] and have a sublinear update time if the patterns are not very long relative to $|D|$.

Third, can one match the $O(|D| \log \sigma)$ preprocessing time of the AC and CW algorithms? The AFGGP algorithm builds the initial dictionary by repeated insertion of patterns in time $O(|D| \log |D|)$. Regardless of the choice of the data structure, we can build our initial dictionary in $O(|D| \log \sigma)$ time.

After initiating this research, we found in collaboration with A. Amir, M. Farach, and J. A. La Poutré that our dynamic dictionary matching framework has another feature. It can be used to solve a natural two-dimensional version of dynamic dictionary matching [3], while the AFGGP algorithm does not seem to generalize [5].

The rest of this paper is organized as follows. In Sect. 2, we present our basic algorithm for dynamic dictionary matching. In Sect. 3, we show how to modify the underlying data structures, to improve the search time. In Sect. 4, we describe how to construct the initial dictionary in $O(|D| \log \sigma)$ time.

2 Dictionary Automaton and Searching Algorithm

Let $D = \{P_1, \ldots, P_s\}$ be a dictionary of patterns, where each P_i, $1 \leq i \leq s$, is a string over alphabet Σ. We assume that the empty string ϵ is always a pattern in the dictionary. We append to each pattern a special symbol $ that does not occur elsewhere in any pattern or text. We assume that $ $\in \Sigma$ and that $ is the largest symbol in the lexicographic order. We generally use w, x, y, z to denote a pattern prefix and a, b to denote a character of Σ. We use the following dictionary as an example to explain various definitions and concepts.

Example 1. Suppose $\Sigma = \{a, b, \$\}$. Let $\hat{D} = \{\$, b\$, aab\$\}$ be a sample dictionary where every pattern is appended with the special symbol $.

We make the same assumptions on the character set Σ as in the papers on suffix trees [12, 10, 7, 2, 4]:

Assumption 1. Each character is represented by a constant number of bytes.

Assumption 2. The relative order of any two characters can be determined in constant time.

In [1], each state in the automaton corresponds to a prefix of some pattern in D. From now on, we use a prefix to denote its state. If we are in a state x after reading the first j characters of a text, x is the longest prefix of any pattern ending at the jth position of the text. Aho and Corasick defined two important (partial) functions, $goto$ and $fail$, that describe the transitions in their automaton. $goto(x, a) = xa$ if xa is a prefix of some pattern in the dictionary, else it is undefined, which we denote by $\perp$. Therefore, $goto$ is a partial function in that each state may not have a transition on every symbol. For any prefix x, $fail(x) = w$ if w is the longest prefix of some pattern such that w is a proper suffix of x. Whenever a transition cannot be made because $goto$ is undefined, we repeatedly replace the state with $fail(state)$ until a transition can be found. The basic search loop is:

$$
\begin{aligned}
&\textbf{while } goto(state, symbol) = \perp \textbf{ do}\\
&\quad state \leftarrow fail(state)\\
&state \leftarrow goto(state, symbol)\\
&symbol \leftarrow nextsymbol
\end{aligned}
$$

For any given choice of *state* and *symbol* we may have to take the *fail* transition repeatedly, but this shortens the length of the prefix corresponding to the new *state*. The total time needed to scan a text T using this algorithm is $O(|T|(g+f))$, where $g + f$ is the time needed to make one evaluation of $goto$ and $fail$ [1].

Like AC, we store the $goto$ function as a directed rooted tree, where the nodes correspond to automaton states (pattern prefixes). The arcs of the tree are directed away from the root and every arc is labeled with a character from Σ. We organize the outgoing arcs of an internal node into a binary search tree with the arc labels as the keys for search, as suggested in [1]. The *degree*, or the maximum number of children of an internal node, of a $goto$ tree is bounded by σ. With this modification, computing each $goto$ takes $O(\log \sigma)$, or $O(\log |D|)$ time, since $\sigma < |D|$. We also store with each node a pointer to its parent. For each pattern prefix x, we keep a count of how many patterns have x as a prefix.

We call a string a *normal prefix* if it is a prefix of some pattern in D. For each proper prefix x we also define an *extended prefix*, $x\$$, by appending the character $;

the extended prefixes help in detecting patterns. The corresponding states are called normal or extended. In Example 1, the normal prefixes are $\epsilon, a, b, \$, aa, b\$, aab, aab\$$ and the extended prefixes are $\$, a\$, b\$, aa\$, aab\$$. Notice that some prefixes are both normal and extended. We extend the definition of $fail$ to accommodate the extended prefixes as follows: Let w be a prefix (normal or extended). $fail(w) = x$ such that $|x| < |w|$ and x is the longest suffix of w such that x is a normal prefix. In Example 1, $fail(aab\$) = b\$$, but $fail(aa\$) \neq a\$$ since $a\$$ is not a normal prefix.

We recognize patterns as follows. When we reach a position of a text, we pretend that the next symbol is a $\$$. If we can make a transition to some normal prefix ending in $\$$, we know that a pattern has been matched at that position, since any normal prefix ending with a $\$$ is a pattern in the dictionary. By applyng $fail$ repeatedly, we can report all the matching patterns from the longest pattern to the shortest.

Suppose we are searching the text $ababba$ for the occurrences of the patterns in the dictionary $\hat{D}$ of Example 1. After reading the prefix $abab$ we will be in the normal state ab. When we pretend to read $\$$ as the next symbol, we will *temporarily* enter a state $ab\$$ and since this is a normal state in the dictionary we report that a pattern ab is recognized at the current location of the text. If we take $fail(ab\$) = b\$$, we match another (smaller) pattern b. If we take $fail(b\$) = \$$ we realize that no more patterns can be matched as this corresponds to the empty pattern ϵ.

In [1], $fail$ is stored as a directed, rooted tree with the arcs pointing towards the root. If $fail(s) = t$, then there will be an arc $s \to t$. The (in)degree of the $fail$ tree is unbounded. We use a new method to store $fail$ that enables us to insert new patterns, updating the $fail$ function. Before describing our representation of $fail$, we explain some auxiliary prefixes and data structures.

Let $* \notin \Sigma$ be a new symbol such that $* > a$ for any $a \in \Sigma$ (including $\$$) in lexicographic comparison. For every prefix $w \in \Sigma^*$ we define $*w$ as the *complement* of w. We call w a *regular prefix*, and $*w$ a *complementary prefix*. For the purpose of representation, we extend the definition of $fail$ as follows: For a regular prefix x, if $fail(x) = z$ then $fail(*x) = z$. We define a total ordering on the set of prefixes and their complements and call it the *inverted order* denoted by $<_{\text{inv}}$. For two distinct strings w and x, $w <_{\text{inv}} x$ if w^R comes before x^R in the lexicographic ordering, where x^R is the reverse of the string x.

Example 2. Consider the list of prefixes of $\hat{D}$ in Example 1 in the inverted order. It can be represented as:

$$\hat{S} = \epsilon, a, aa, *aa, *a, b, aab, *aab, *b, \$, a\$, aa\$, *aa\$, *a\$, b\$, aab\$, *aab\$, *b\$, *\$, *.$$

The number of extended prefixes is at most equal to the number of normal prefixes. The number of complementary prefixes is equal to the number of regular prefixes. Therefore, the total number of prefixes in our dictionary structure is $O(|D|)$, independent of Σ.

Let S denote the set of all normal and extended, regular and complementary prefixes of patterns in D. An important property of S is that for any string x, if S contains x then S contains every prefix of x. All string comparisons are made w.r.t. $<_{\text{inv}}$ ordering unless stated otherwise. For a nonempty string $x \in S$, $pred(x)$ is the largest string in S smaller than x. We need to compute $pred(x)$ when inserting a new string to know where the new string lies in the $<_{inv}$ order. In Example 2, $pred(aab) = b$ and $pred(b) = *a$. The following lemmas state some important properties of the prefixes in S that follow directly from the definition of $<_{inv}$ and the character $*$.

Lemma 3. *Let $w, x \in S$ be arbitrary regular prefixes. Let $y \in S$ be any prefix.*
1. $w <_{inv} *w$; w is smaller than its complement.
2. $w <_{inv} x <_{inv} *w$ if and only if $w <_{inv} *x <_{inv} *w$; if we replace a regular prefix with a '(' and its complement with a ')' then the prefixes of S in the $<_{inv}$ order yield a list of well balanced parentheses.
3. If $w <_{inv} y <_{inv} *w$ then $y = y'w$ for some y'; w and $*w$ are respectively the smallest and largest prefixes in S with the suffix w.

Lemma 4. *Let xa be the prefix we are inserting into S. Suppose $yb \neq xa$ is a prefix already in S. Then the following relations hold:*
1. $\epsilon <_{inv} xa$.
2. If $b <_{inv} a$ then $yb <_{inv} xa$. Similarly if $a <_{inv} b$ then $xa <_{inv} yb$.
3. If $b = a$ then $yb <_{inv} xa$ if and only if $y <_{inv} x$. Since x and y must already be in S we can determine whether $y <_{inv} x$ from S alone.

Lemma 4 suggests a way to obtain the relative order of two prefixes without making a complete lexicographic comparison. We build an auxiliary search tree, called ST, on the top of all the prefixes of S. The elements of ST are basically pointers to the states of the *goto* tree sorted by the inverted order of the corresponding prefixes. We store ST as an a-b tree [11] with all the prefixes as leaves. For any state corresponding to a prefix yb we can determine the character b and the state with the prefix y in constant time as y is the parent of yb in the *goto* tree. To utilize Lemma 4 we need to know the relative order of all the existing prefixes in S. Since we implement ST as an a-b tree, we can determine the relative order of two prefixes in $O(\log|D|)$ time.

We now describe one way to compute $pred(xa)$ for a nonempty prefix xa not yet in S. We start at the root of the ST and proceed towards the leaves. Suppose yb and zc are two consecutive keys belonging to an internal node. At the next level, we take the child pointer between yb and zc if $yb <_{inv} xa <_{inv} zc$. We need $O(\log|D|)$ tree comparisons to find $pred(xa)$. From now on we assume that we have routine FINDPRED that computes the $pred$ of a new prefix. We can conclude that:

Lemma 5. *Let xa be a prefix to be inserted into S. We can compute $pred(xa)$ using FINDPRED(xa) in $O(\log^2|D|)$ time.*

Proof. We can determine $pred(xa)$ with $O(\log|D|)$ tree comparisons. Since ST is organized as an a-b tree, the relative order of two existing prefixes can be determined in $O(\log|D|)$ time. Therefore, by Lemma 4 each tree comparison takes $O(\log|D|)$ time. From this it follows that we can determine $pred(xa)$ in $O(\log^2|D|)$ time.

To reduce the time to compute $pred$, we use a data structure invented by Dietz and Sleator [8] for solving the *order maintenance problem*. In the order maintenance problem, we define the following operations on a linear list L, initially containing one element.

$L\_Insert(x, y)$: Insert a new element y after the element x in L.

$L\_Delete(x)$: Delete the element x from L.

$L\_Order(x, y)$: Return true if x is before y in L, false otherwise.

Dietz and Sleator [8] provide a solution to the order maintenance problem in which all the three operations take worst-case $O(1)$ time. We store the prefixes of S in an auxiliary list data structure that we call the Dietz-Sleator List, or DSL.

Lemma 6. *Let x and y be two prefixes in S. By building DSL for all the prefixes of S, we can determine the relative order of x and y in constant time. Hence, we can compute pred in $O(\log |D|)$ time.*

When we insert a new prefix xa, we need to find the value of $fail(xa)$. The following two lemmas help in computing $fail$.

Lemma 7. *Suppose w is a normal regular prefix, and $w <_{inv} x <_{inv} *w$. Then $fail(x) = w$ if and only if there is no normal regular y such that $w <_{inv} y <_{inv} x <_{inv} *y <_{inv} *w$.*

In Example 2, $fail(aab) = b$ since there is no normal y such that $b <_{inv} y <_{inv} aab <_{inv} *y <_{inv} *b$, but $fail(aab) \neq \epsilon$ since b can play the role of y in that case.

Lemma 8. *Let x be a regular prefix. Suppose $pred(x) = w$. If w is normal and regular then $fail(x) = w$ else $fail(x) = fail(w)$.*

In Example 2, $fail(aab) = b$ since $b = pred(aab)$ is normal and regular. Similarly $fail(b) = \epsilon = fail(*a)$ and one may observe that $pred(b) = *a$.

We store $fail$ as a forest of a-b trees. Amir and Farach [2] use 2-3 trees in their suffix-tree automaton for a similar purpose. Each tree is called a *fail tree*, and the forest is called a *fail forest*. There is a one-to-one correspondence between the set of a-b trees and the set of normal regular prefixes of S. If w is a normal regular prefix, then T_w denotes the a-b tree associated with w and contains as leaves all strings x such that $fail(x) = w$. In Example 2, T_a contains $\{aa, *aa\}$ as leaves, whereas T_{aab} is an empty tree. The root of the tree T_x contains a pointer to x. Each prefix $x \in S$ is a leaf in exactly one a-b tree, T_y, where $y = fail(x)$. The leaves of any a-b tree are sorted in $<_{inv}$ order. For any x, $fail(x)$ can be computed by starting at the leaf x, finding the root of the tree T_y, and taking the pointer to y.

When we insert a new pattern P into D, we insert its prefixes and extended prefixes into S, in increasing order of length. Thus when we insert a string x into S, we have already inserted all the prefixes of x. Using Lemmas 5 and 6 we can find $w = pred(x)$. From Lemma 8 we can obtain $y = fail(x)$. If $w = y$ insert x into T_y as the leftmost leaf. Otherwise, insert x into T_y right after w. We can similarly find $z = pred(*x)$ (z could be x) and insert $*x$ into T_y right after z. We keep a separate bidirectional link between x and $*x$. Similarly when we delete a pattern P from D, we delete its prefixes in the order of decreasing lengths.

After inserting x and $*x$ into S we must create T_x. For this we must first identify those prefixes whose $fail$ value changes to x. By Lemma 7, these are the prefixes with a suffix x and whose current $fail$ value is y, which are exactly the leaves of T_y properly enclosed between x and $*x$. We change $fail$ for these nodes by a special *split* of T_y into T_y and T_x. We can similarly handle the case when a normal prefix x

with $fail(x) = y$ becomes extended as a result of deletion of some pattern. In this case we fuse T_y and T_x into T_y with a special *concatenate*. These special operations were used similarly in [2].

As an example, consider inserting a new pattern $ab\$$ into the dictionary $\hat{D}$ of Example 1. For this we need to insert the prefixes a, $a\$$, ab, and $ab\$$ into the set $\hat{S}$ of Example 2. Since a and $a\$$ are already present in $\hat{S}$ we simply increment the reference count for them. We insert ab after $pred(ab) = b$, and $*ab$ after $pred(*ab) = *aab$. After this we have $\hat{S} = \{\ldots, b, ab, aab, *aab, *ab, *b, \ldots\}$. T_b contained $\{aab, *aab\}$ as leaves before the insertion of ab and $*ab$. We create T_{ab} by splitting T_b as described above. After this step, T_{ab} contains $\{aab, *aab\}$ as leaves, and T_b contains $\{ab, *ab\}$ as leaves. We similarly insert $ab\$$ and $*ab\$$.

We conclude by giving pseudocode to do search, insert, and delete. We use prefixes instead of states for clarity.

```
SEARCH(T = t₁...tₙ)
    state ← ε
    for i ← 1 to n do
        While goto(state, tᵢ) =⊥
            state ← fail(state)
        state ← goto(state, tᵢ)
        temp ← goto(state, $) /* Pretend a $ is read to look for patterns */
        If temp is not normal then temp ← fail(temp)
        While temp ≠ $ do /* Report all non-empty patterns */
            Print the pattern associated with temp /* temp ends in $ */
            temp ← fail(temp) /* See if any smaller patterns match */
INSERT(P = p₁...pₘ) /* pₘ = $ */
    Suppose p₁...pⱼ is the longest prefix of P shared by some other pattern.
    Increment the reference count for the prefixes of p₁...pⱼ.
    For i ← j + 1 to m do
    1    Let x = p₁...pᵢ₋₁. Let a = pᵢ. /* xa is being inserted. x ∈ S */
    2    Compute y = pred(xa) using FINDPRED.
    3    Compute w = fail(xa) using Lemma 8.
    4    If w ≠ y Then
             insert xa into T_w right after y.
             Else insert xa into T_w as the leftmost leaf.
    5    insert xa into ST after y.
    6    L_Insert xa into DSL after y.
    7    Compute z = pred(*xa) using FINDPRED /* w = fail(*xa). */
    8    insert *xa into T_w right after z.
    9    insert *xa into ST after z.
    10   L_Insert *xa into DSL after z.
    11   Repeat steps 1–10 to insert xa$ and *xa$. /* Extended prefixes */
    12   Form T_{xa} by a split of T_w into T_w and T_{xa}.
DELETE(P = p₁...pₘ) /* pₘ = $ */
    Suppose p₁...pⱼ is the longest prefix of P shared by some other pattern.
    Decrement the reference count for the prefixes of p₁...pⱼ.
```

For $i \leftarrow m$ downto $j + 1$ do
    Let $x = p_1 \ldots p_i$ /* x is a normal prefix */
    If $x\$$ is still in S Then
        delete $x\$$ and $*x\$$ from their fail tree.
        delete $x\$$ and $*x\$$ from ST.
        L_Delete $x\$$ and $*x\$$ from DSL.
    Let $y = fail(x)$. Fuse T_x and T_y into T_y by a *concatenate*.
    delete x and $*x$ from T_y.
    delete x and $*x$ from ST.
    L_Delete x and $*x$ from DSL.

The correctness and running time bounds for the above procedures follow from the previous lemmas.

Theorem 9. *Let D be a dictionary of patterns over an alphabet Σ. We can search a text T for occurrences of patterns of D in time $O(|T|(\log \sigma + \log |D|) + (tocc \cdot \log |D|))$, where tocc is the total number of patterns reported. We can insert or delete a pattern P in time $O(|P| \cdot (\log \sigma + \log |D|))$. Moreover, we require only $O(|D|)$ space to store the automaton.*

3 Linear Time Searching

We show how to improve the search time in this section. From Theorem 9, a text T can be searched in $O(|T|(\log \sigma + \log |D|) + (tocc \cdot \log |D|))$ time. The $\log \sigma$ factor comes from the computation of each *goto* and the $\log |D|$ factors from the computation of each *fail*. We show that it is possible to speed up the computation of *fail* and achieve a faster searching algorithm, and still get an update time sublinear in $|D|$.

In Sect. 2, we used a-b trees to store every fail tree. In an a-b tree, the number of children or the *degree* of a nonroot internal node v, denoted by $\delta(v)$, must be in the range $[a, b]$, and must be in the range $[2, b]$ if v is a root. In this section, we use a variant of a-b trees, which we call *hybrid a-b trees*, to store the fail trees. In a hybrid a-b tree different nodes may have different ranges for the number of children permitted. The ranges depend on the number of leaves in the fail forest.

Our hybrid a-b trees depend on an parameter, which is independent of $|D|$. For any $k \geq 2$, the hybrid trees will allow us to perform a *findroot* in $O(k)$ time, and any update operation in $O(k \cdot n^{\frac{1}{k}})$ time, where n is the total number of leaves in the forest of fail trees. Recall that n is twice the number of states in the automaton.

Let $\alpha \geq 16$ be the smallest power of 2 such that $\alpha^k \leq n \leq (2\alpha)^k$. Each internal node is designated as *small* or *big*, but may change its designation during the algorithm. A small nonroot v has $\delta(v) \in [\alpha, 2\alpha]$, a small root v has $\delta(v) \in [2, 2\alpha]$, a big nonroot v has $\delta(v) \in [2\alpha, 4\alpha]$, and a big root v has $\delta(v) \in [2, 4\alpha]$. We maintain lists of small and big nodes using a separate link. Let #*small* and #*big* denote the number of such nodes. Our ranges imply that any nonroot has $\Theta(n^{1/k})$ children and there will be $O(k)$ levels in any hybrid tree.

The operations *insert, delete, concatenate,* and *split* can be implemented in a similar fashion to that used for regular a-b trees. Each operation visits or modifies at most b nodes at each level of the tree and may cause at most a constant number of nodes per level to violate the constraints on the number of children allowed.

In the rest of the section we show how to handle overflows and underflows in the number of children of a node of a hybrid tree. Define $excess = n - \alpha^k$, and $m = \frac{\alpha^k + (2\alpha)^k}{2}$. We maintain two invariants:

$$\#small + excess \leq (2\alpha)^k - \alpha^k \tag{1}$$

$$\#big \leq excess \tag{2}$$

Note that all nodes will be small when $n = \alpha^k$, and big when $n = (2\alpha)^k$. If n goes above $(2\alpha)^k$, we redefine all the big nodes as small nodes and start operating in the interval $(2\alpha)^k \leq n \leq (4\alpha)^k$ with small nonroot nodes having the range $[2\alpha, 4\alpha]$ and big nonroot nodes having the range $[4\alpha, 8\alpha]$. We redefine $excess$ and m accordingly. We can do a similar thing when n falls below α^k. The following two lemmas simplify invariant maintenance.

Lemma 10. *If there are n leaves in the fail forest, then the number of internal nodes of the fail trees is at most $3n/5$.*

Lemma 11. *If $n \leq m$ then invariant 1 is satisfied. Similarly if $n \geq m$ then invariant 2 is satisfied.*

Lemma 11 implies that when $n \leq m$ we only have to control $\#big$, as invariant 1 is always satisfied in this case, and the value of $\#small$ has no effect on invariant 2. Similarly when $n \geq m$, we only have to control $\#small$.

If $n \leq m$, we do any update operation in such way that $\#big$ never increases. Any update operation can increment or decrement n by at most one. If invariant 2 is satisfied before a *delete* but violated afterwards, we have to decrement $\#big$ by only one to maintain invariant 2. Similarly, when $n \geq m$, we update in such a way that $\#small$ never increases. We may have to decrement $\#small$ by at most one after an *insert*. The operations *concatenate and split* do not change the value of n.

Two primitive operations needed for implementing hybrid trees are those that handle overflow and underflow of an internal node. Those are the cases when the degree of an internal node goes one above or below its declared range. We show how overflow and underflow can be eliminated without violating the invariants. We annotate each line of pseudocode by an ordered pair (i, j) implying that $\#small$ changes by i and $\#big$ by j after executing the line.

overflow(v):
> Suppose we are controlling $\#small$.
>> If v is small redesignate it as a big node. $\{(-1, 1)\}$
>> If v is big break it into two big nodes. $\{(0, 1)\}$
> Suppose we are controlling $\#big$.
> (*)   If v is small break it into two small nodes. $\{(1, 0)\}$
> (*)   If v is big break it into one big and one small node. $\{(1, 0)\}$

underflow(v):
> Suppose v is a root node.

If $degree(v) = 1$ remove v. Make v's child the new root. $\{(-1,0)$ or $(0,-1)\}$

Suppose v is a nonroot node with an immediate sibling w.
Let $[a,b]$ be the range of w.
If $\delta(w) \geq a+1$ transfer one child of w to v. $\{(0,0)\}$
Otherwise
If v and w are both big(small) then
Fuse them into one big(small) node. $\{(0,-1)$ or $(-1,0)\}$
If $v(w)$ is big and $w(v)$ is small fuse them into one big node. $\{(-1,0)\}$

One can check that each change in $\#small$ or $\#big$ does not violate the invariants. Also, $underflow(v)$ never increases $\#small$ or $\#big$. If an invariant is violated by 1 as a result of an $insert$ or $delete$, applying $overflow$ or $underflow$ does not exacerbate the violation, and we restore the violated invariant as explained below.

A single overflow or underflow can be handled in $O(\alpha)$ time. Correcting the overflow(underflow) of a node may cause an overflow(underflow) of its parent. Since there are $O(k)$ levels in any tree, an update operation takes $O(k \cdot \alpha)$ time.

Finally we show how to decrement $\#small$ or $\#big$ by at least one in $O(k \cdot \alpha)$ time. These operations are necessary to restore the invariants that may be violated after an $insert$ or $delete$ operation.

$decreasesmall()$: Pick some small node v. If v is a root, redesignate it as a big node. If v is not a root, it must have an immediate sibling w. If w is small, then fuse v and w into a big node. If w is big, fuse v and w into one or two big nodes depending on $\delta(v)+\delta(w)$. This may propagate underflows to the ancestors of v which can be handled without increasing $\#small$, as we noted earlier that $\#small$ does not increase during an $underflow$ operation.

$decreasebig()$: Pick some big node v. Split v into one or two small nodes depending on $\delta(v)$. Any propagated overflows can be handled without increasing $\#big$ as can be observed from the lines marked (*) of $overflow$.

Theorem 12. *For a fixed integer $k \geq 2$, we can search a text T in $O(|T|(k+\log\sigma)+ (k \cdot tocc))$ time. Furthermore we can insert or delete a pattern P into a dictionary D in $O(((k \cdot |D|^{\frac{1}{k}}) + \log\sigma) \cdot |P|)$ time.*

Proof. For any state v, $fail(v)$ can be obtained with a $findroot$ operation. Since this takes only $O(k)$ time on a hybrid a-b tree we can search T in $O(|T|(k+\log\sigma)+(k \cdot tocc))$ time. Any update operation will be accompanied by at most one $decreasesmall$ or $decreasebig$ operation. Each of these operations take $O(k \cdot \alpha)$ time. Since α is $O(|D|^{\frac{1}{k}})$, it follows that P can be inserted or deleted in the specified time.

4 Building D in Linear Time

In this section, we show how to build the initial dictionary in $O(|D|\log\sigma)$ time using either regular a-b trees or hybrid a-b trees to store the $fail$ function. This is better than doing repeated insertion of patterns which requires $O(|D| \cdot \log|D|)$ (as in [2, 4]) or $O(|D| \cdot |D|^{1/k})$ time depending on the data structure used.

Let $D = \{P_1, \ldots, P_s\}$ be the initial set of patterns. Our first major goal is to sort all the (normal and extended, regular and complementary) prefixes of S in $<_{\text{inv}}$ order. We partition S into two disjoint sets S_1 and S_2. S_1 contains those prefixes of S ending in a \$ and S_2 contains the rest of S.

We start by building a sorted list of the prefixes of S_2. These prefixes are exactly the prefixes of $P_1, \ldots, P_s$ and $*P_1, \ldots, *P_s$ with the \$ stripped from their right ends. We build a suffix tree for the *reverses* of these $2s$ strings using the suffix tree construction of [10] as modified by [4]. There will be a one-to-one correspondence between the leaves of the suffix tree and the prefixes of S_2 as proved in Lemma 2 of Section 2 in [4]. If we sort the children of every internal node of the suffix tree by the labels of the edges connecting the children and the parent, then the left-to-right order of the leaves of the suffix tree is the $<_{\text{inv}}$ order of the prefixes of S_2. We can build the suffix tree in $O(|D| \log \sigma)$ time. Rearranging the children and scanning the leaves in the left-to-right order takes only $O(|D| \log \sigma)$ time.

The prefixes of in S_1 (and S_2) occur consecutively in S. Furthermore the prefixes of S_1 are in one-to-one correspondence with the prefixes of S_2, and occur in exactly the same $<_{\text{inv}}$ order as the corresponding prefixes of S_2. Thus we can build a sorted list of the prefixes of S_1 by scanning the list of the prefixes of S_2, and for each prefix x of S_2 inserting (in the same relative order) the prefix $x\$$ into S_1. Finally we concatenate both S_2 and S_1 to obtain S in the $<_{\text{inv}}$ order. This takes $O(|D|)$ time.

Our second major goal is to compute the *fail* function. Specifically, to each prefix w, we want to associate those strings x, such that $fail(x) = w$. We temporarily keep these strings associated with w in a sorted list in $<_{\text{inv}}$ order, which we call w's *fail list*. The strings on the fail list will become the leaves of the fail tree T_w. By Lemma 7, these are precisely the strings x such that $w <_{\text{inv}} x <_{\text{inv}} *w$, and there is no y such that $w <_{\text{inv}} y <_{\text{inv}} x <_{\text{inv}} *y <_{\text{inv}} *w$. To do this we scan the list S using a stack STK to keep track of the normal prefixes whose complements have not appeared. We use the following scanning rules in the order below:

1   If $x \neq \epsilon$ is regular, then $fail(x) = top(STK)$, so append x to the fail list for $top(STK)$.

2   If x is normal and regular, then $push(x)$ onto STK.

3   If x is normal and complementary, $pop(STK)$.

4   If $x \neq \$$ is complementary, then $fail(x) = top(STK)$, so append x to the fail list for $top(STK)$.

Scanning the list takes constant time per prefix, and $O(|D|)$ time overall.

Finally we organize each fail list into a fail tree. To build T_w we build a search tree with the elements of the fail list as leaves. We can build any standard a-b tree given the sorted list of leaves in linear time [11]. To build hybrid a-b trees, let n be the total number of leaves in all trees. Given $k \geq 2$, choose α as described in Section 3 and compute m. Build the trees with all small nodes if $n \leq m$, and with all big nodes if $n > m$. The correctness of this construction follows from Lemma 11. This also takes only $O(|D|)$ time.

We can summarize our linear time construction algorithm as follows:

BUILD$(D = \{P_1, \ldots, P_s\})$

1	Build suffix tree on the *reverses* of $P_1, \ldots, P_s, *P_1, \ldots, *P_s$ to order S_2.
2	Scan S_2 and build S_1.
3	Concatenate S_2 and S_1 to obtain S.
4	Build ST and DSL for the sorted list of prefixes.
5	Scan the sorted list and compute the fail list of each prefix.
6	Convert each fail list into a (regular or hybrid) a-b tree.

Theorem 13. *The initial dictionary D can be constructed in $O(|D|\log\sigma)$ time.*

Acknowledgments We thank Amihood Amir for pointing out the importance of the large/unbounded alphabet case and for answering our questions about [2, 4].

References

1. Aho, A. V., Corasick, M. J.: Efficient String Matching: An Aid to Bibliographic Search, Comm. ACM **18** (1975), 333–340.
2. Amir, A., Farach, M.: Adaptive Dictionary Matching, Proc. 32nd IEEE Symp. Found. Comp. Sci., (1991), 760–766.
3. Amir, A., Farach, M.:, Two Dimensional Dictionary Matching, submitted for publication, 1991.
4. Amir, A., Farach, M., Galil, Z., Giancarlo, R., Park, K.: Dynamic Dictionary Matching, manuscript, 1991.
5. Amir, A., Farach, M., Idury, R. M., La Poutré, J. A., Schäffer, A. A.: Improved Dynamic Dictionary Matching, manuscript, 1992.
6. Amir, A., Farach, M., Matias, Y., Efficient Randomized Dictionary Algorithms, Proc. 3rd Symp. Combinatorial Pattern Matching, (1992).
7. Chen, M. T., Seiferas, J.: Efficient and Elegant Subword-Tree Construction, in: Apostolico A., Galil, Z. eds.: *Combinatorial Algorithms on Words*, NATO ASI Series, Vol. F12, (Springer-Verlag, Heidelberg, 1985), 97–107.
   Vol. 71 (Springer, Berlin, 1979) 118–132.
8. Dietz, P., Sleator, D. D.: Two Algorithms for Maintaining Order in a List, Proc. 19th ACM Symp. Theor. Comp. Sci., (1987), 365–372.
9. Knuth, D. E., Morris, J. H., Pratt, V. B.: Fast Pattern Matching in Strings, SIAM J. Comp. 6(1977), 323–350.
10. McCreight, E. M., A Space Economical Suffix Tree Construction Algorithm, J. ACM, **23**(1976), 262–272.
11. Mehlhorn, K.: *Data Structures and Algorithms 1: Sorting and Searching*, (Springer-Verlag, Berlin), 1984.
12. Weiner, P.: Linear Pattern Matching Algorithms, Proc. 14th IEEE Symp. on Switching and Automata Theory, (1973), 1–11.

Lecture Notes in Computer Science

For information about Vols. 1–559
please contact your bookseller or Springer-Verlag

Vol. 560: S. Biswas, K. V. Nori (Eds.), Foundations of Software Technology and Theoretical Computer Science. Proceedings, 1991. X, 420 pages. 1991.

Vol. 561: C. Ding, G. Xiao, W. Shan, The Stability Theory of Stream Ciphers. IX, 187 pages. 1991.

Vol. 562: R. Breu, Algebraic Specification Techniques in Object Oriented Programming Environments. XI, 228 pages. 1991.

Vol. 563: A. Karshmer, J. Nehmer (Eds.), Operating Systems of the 90s and Beyond. Proceedings, 1991. X, 285 pages. 1991.

Vol. 564: I. Herman, The Use of Projective Geometry in Computer Graphics. VIII, 146 pages. 1992.

Vol. 565: J. D. Becker, I. Eisele, F. W. Mündemann (Eds.), Parallelism, Learning, Evolution. Proceedings, 1989. VIII, 525 pages. 1991. (Subseries LNAI).

Vol. 566: C. Delobel, M. Kifer, Y. Masunaga (Eds.), Deductive and Object-Oriented Databases. Proceedings, 1991. XV, 581 pages. 1991.

Vol. 567: H. Boley, M. M. Richter (Eds.), Processing Declarative Kowledge. Proceedings, 1991. XII, 427 pages. 1991. (Subseries LNAI).

Vol. 568: H.-J. Bürckert, A Resolution Principle for a Logic with Restricted Quantifiers. X, 116 pages. 1991. (Subseries LNAI).

Vol. 569: A. Beaumont, G. Gupta (Eds.), Parallel Execution of Logic Programs. Proceedings, 1991. VII, 195 pages. 1991.

Vol. 570: R. Berghammer, G. Schmidt (Eds.), Graph-Theoretic Concepts in Computer Science. Proceedings, 1991. VIII, 253 pages. 1992.

Vol. 571: J. Vytopil (Ed.), Formal Techniques in Real-Time and Fault-Tolerant Systems. Proceedings, 1992. IX, 620 pages. 1991.

Vol. 572: K. U. Schulz (Ed.), Word Equations and Related Topics. Proceedings, 1990. VII, 256 pages. 1992.

Vol. 573: G. Cohen, S. N. Litsyn, A. Lobstein, G. Zémor (Eds.), Algebraic Coding. Proceedings, 1991. X, 158 pages. 1992.

Vol. 574: J. P. Banâtre, D. Le Métayer (Eds.), Research Directions in High-Level Parallel Programming Languages. Proceedings, 1991. VIII, 387 pages. 1992.

Vol. 575: K. G. Larsen, A. Skou (Eds.), Computer Aided Verification. Proceedings, 1991. X, 487 pages. 1992.

Vol. 576: J. Feigenbaum (Ed.), Advances in Cryptology - CRYPTO '91. Proceedings. X, 485 pages. 1992.

Vol. 577: A. Finkel, M. Jantzen (Eds.), STACS 92. Proceedings, 1992. XIV, 621 pages. 1992.

Vol. 578: Th. Beth, M. Frisch, G. J. Simmons (Eds.), Public-Key Cryptography: State of the Art and Future Directions. XI, 97 pages. 1992.

Vol. 579: S. Toueg, P. G. Spirakis, L. Kirousis (Eds.), Distributed Algorithms. Proceedings, 1991. X, 319 pages. 1992.

Vol. 580: A. Pirotte, C. Delobel, G. Gottlob (Eds.), Advances in Database Technology - EDBT '92. Proceedings. XII, 551 pages. 1992.

Vol. 581: J.-C. Raoult (Ed.), CAAP '92. Proceedings. VIII, 361 pages. 1992.

Vol. 582: B. Krieg-Brückner (Ed.), ESOP '92. Proceedings. VIII, 491 pages. 1992.

Vol. 583: I. Simon (Ed.), LATIN '92. Proceedings. IX, 545 pages. 1992.

Vol. 584: R. E. Zippel (Ed.), Computer Algebra and Parallelism. Proceedings, 1990. IX, 114 pages. 1992.

Vol. 585: F. Pichler, R. Moreno Díaz (Eds.), Computer Aided System Theory - EUROCAST '91. Proceedings. X, 761 pages. 1992.

Vol. 586: A. Cheese, Parallel Execution of Parlog. IX, 184 pages. 1992.

Vol. 587: R. Dale, E. Hovy, D. Rösner, O. Stock (Eds.), Aspects of Automated Natural Language Generation. Proceedings, 1992. VIII, 311 pages. 1992. (Subseries LNAI).

Vol. 588: G. Sandini (Ed.), Computer Vision - ECCV '92. Proceedings. XV, 909 pages. 1992.

Vol. 589: U. Banerjee, D. Gelernter, A. Nicolau, D. Padua (Eds.), Languages and Compilers for Parallel Computing. Proceedings, 1991. IX, 419 pages. 1992.

Vol. 590: B. Fronhöfer, G. Wrightson (Eds.), Parallelization in Inference Systems. Proceedings, 1990. VIII, 372 pages. 1992. (Subseries LNAI).

Vol. 591: H. P. Zima (Ed.), Parallel Computation. Proceedings, 1991. IX, 451 pages. 1992.

Vol. 592: A. Voronkov (Ed.), Logic Programming. Proceedings, 1991. IX, 514 pages. 1992. (Subseries LNAI).

Vol. 593: P. Loucopoulos (Ed.), Advanced Information Systems Engineering. Proceedings. XI, 650 pages. 1992.

Vol. 594: B. Monien, Th. Ottmann (Eds.), Data Structures and Efficient Algorithms. VIII, 389 pages. 1992.

Vol. 595: M. Levene, The Nested Universal Relation Database Model. X, 177 pages. 1992.

Vol. 596: L.-H. Eriksson, L. Hallnäs, P. Schroeder-Heister (Eds.), Extensions of Logic Programming. Proceedings, 1991. VII, 369 pages. 1992. (Subseries LNAI).

Vol. 597: H. W. Guesgen, J. Hertzberg, A Perspective of Constraint-Based Reasoning. VIII, 123 pages. 1992. (Subseries LNAI).

Vol. 598: S. Brookes, M. Main, A. Melton, M. Mislove, D. Schmidt (Eds.), Mathematical Foundations of Programming Semantics. Proceedings, 1991. VIII, 506 pages. 1992.

Vol. 599: Th. Wetter, K.-D. Althoff, J. Boose, B. R. Gaines, M. Linster, F. Schmalhofer (Eds.), Current Developments in Knowledge Acquisition - EKAW '92. Proceedings. XIII, 444 pages. 1992. (Subseries LNAI).

Vol. 600: J. W. de Bakker, C. Huizing, W. P. de Roever, G. Rozenberg (Eds.), Real-Time: Theory in Practice. Proceedings, 1991. VIII, 723 pages. 1992.

Vol. 601: D. Dolev, Z. Galil, M. Rodeh (Eds.), Theory of Computing and Systems. Proceedings, 1992. VIII, 220 pages. 1992.